Emerging Memory and Computing Devices in the Era of Intelligent Machines

Emerging Memory and Computing Devices in the Era of Intelligent Machines

Special Issue Editor

Pedram Khalili Amiri

MDPI • Basel • Beijing • Wuhan • Barcelona • Belgrade • Manchester • Tokyo • Cluj • Tianjin

Special Issue Editor
Pedram Khalili Amiri
Northwestern University
USA

Editorial Office
MDPI
St. Alban-Anlage 66
4052 Basel, Switzerland

This is a reprint of articles from the Special Issue published online in the open access journal *Micromachines* (ISSN 2072-666X) (available at: https://www.mdpi.com/journal/micromachines/special_issues/Emerging_Memory_Computing_Devices_Era_Intelligent_Machines).

For citation purposes, cite each article independently as indicated on the article page online and as indicated below:

LastName, A.A.; LastName, B.B.; LastName, C.C. Article Title. *Journal Name* **Year**, *Article Number*, Page Range.

ISBN 978-3-03928-502-0 (Pbk)
ISBN 978-3-03928-503-7 (PDF)

Contents

About the Special Issue Editor

Pedram Khalili Amiri is Associate Professor of Electrical and Computer Engineering at Northwestern University, where he is also a faculty member of the Applied Physics graduate program, and Director of the Physical Electronics Research Laboratory. Prior to joining Northwestern, he was an Adjunct Assistant Professor in the department of electrical and computer engineering at the University of California, Los Angeles (UCLA). Since 2009 he has led multiple research programs on voltage-controlled MRAM, spin-transfer-torque MRAM, and nonvolatile spintronic logic working with several major industry and research partners, which resulted in some of the fastest and most energy-efficient magnetic memories to date. Pedram has published over 100 papers in peer-reviewed academic journals, and is an inventor on 15 issued patents. He received the B.Sc. degree from Sharif University of Technology in 2004, and the Ph.D. degree (cum laude) from Delft University of Technology (TU Delft), The Netherlands, in 2008, both in electrical engineering. He serves on the Editorial Board of Journal of Physics: Photonics (IOP), and has served as a Guest Editor for the journals Spin and Micromachines. He has served on the technical program committees and organizing committees of several conferences, including the Joint MMM/Intermag Conference, and is a member of the Flash Memory Summit conference advisory board. He is a Senior Member of the IEEE.

Editorial

Editorial for the Special Issue on Emerging Memory and Computing Devices in the Era of Intelligent Machines

Pedram Khalili Amiri

Department of Electrical and Computer Engineering, Northwestern University, Evanston, IL 60208, USA; pedram@northwestern.edu

Received: 18 December 2019; Accepted: 2 January 2020; Published: 9 January 2020

Computing systems are undergoing a transformation from logic-centric toward memory-centric architectures, where overall performance and energy efficiency at the system level are determined by the density, bandwidth, latency, and energy efficiency of the memory, rather than the logic sub-system. This is driven by the requirements of data-intensive applications in artificial intelligence, autonomous systems, and edge computing. We are at an exciting time in the semiconductor industry where several innovative device and technology concepts are being developed to respond to these demands, and capture shares of the fast growing market for AI-related hardware. The collection of articles in this special issue on "Emerging Memory and Computing Devices in the Era of Intelligent Machines" is devoted to highlighting some of the latest advancements in this area, drawing on work on emerging memory devices including magnetic, resistive, and phase change memories, their related circuit and material-level issues, and emerging architectures based on logic-in memory and in-memory computing concepts. A few articles also highlight some of the recent advances in engineering conventional memories—notably Flash and DRAM—to extend and push their performance limits.

The existing memory hierarchy in electronic systems is characterized by a tradeoff between cost per bit (or, more or less equivalently, bit density per unit area on a chip) and performance (read/write speed). On the slow (highest-density) end is NAND Flash, while the other extreme is (fast but low-density) static random access memory (SRAM), with dynamic RAM (i.e. DRAM) falling in between. Considerable gaps in price per bit and performance exist between NAND and DRAM, and also between DRAM and SRAM.

Much of the wide-ranging ongoing work on emerging memory devices and architectures can be classified into three categories: (i) Memories that fall between DRAM and SRAM in terms of both bit density and speed, i.e., those that are denser than SRAM but not quite as fast, faster than DRAM but more expensive. Magnetic random access memory is the leading contender in this realm, where both discrete and embedded solutions are of interest. Existing spin-transfer torque magnetic RAM (STT-MRAM) is the state-of-the-art magnetic memory that has received much traction within the industry as an embedded nonvolatile memory (eNVM), with the potential to also replace some embedded SRAM (e.g., L3 or L2 Cache) driving much of the ongoing work to further improve its characteristics. (ii) Memories that are targeted to fill in the large performance and cost gap of DRAM and NAND Flash, also referred to as storage-class memories (SCM). These memories are most often geared toward discrete parts (though specialized embedded applications exist), where a cost penalty compared to NAND Flash is acceptable provided a faster read/write performance is achieved. Examples of these memories are many resistive random access memories (RRAM) and phase-change memories reported to date, among others. (iii) Work that draws on the advances in any of the above memory technologies, but explores unconventional computing approaches, examples being logic-in memory, in-memory computing, neuromorphic computing, and probabilistic computing concepts, among others.

This special issue covers examples of work in all three of these areas: (i) One of the key areas of MRAM research is the exploration of alternative write mechanisms with respect to STT, which is based on driving currents through the memory bit. The goal of these efforts is to achieve better tradeoffs between write speed, bit density, and endurance, while reduction of the write energy is also a possible advantage. An important example is MRAM based on voltage control of magnetic anisotropy (VCMA) [1], which completely departs from the current-controlled mechanism of STT and instead uses electric fields to write information. In "Recent Progress in the Voltage-Controlled Magnetic Anisotropy Effect and the Challenges Faced in Developing Voltage-Torque MRAM", T. Nozaki et al. [2] present some of their latest results in the development of this type of voltage-controlled MRAM (i.e., VCM). In "Fine-Grained Power Gating Using an MRAM-CMOS Non-Volatile Flip-Flop", J. Park and Y. Yim [3] explore some of the advantages of MRAM in terms of power management, by taking advantage of its nonvolatility to enable a flip-flop that retains its information without applied voltage. (ii) There are also several examples of RRAM and phase-change memories discussed throughout the selected articles. These range from material- and cell-level studies (X. Lian et al. [4]; Z. Shen et al. [5]; C. Xie et al. [6]; and K. Drake et al. [7]), to the applications of RRAM in processing of biosignals (Y. K. Lee et al. [8]), neural networks (S. Jo et al. [9]; and S. N. Truong [10]), and nonvolatile processors (X. Xue et al. [11]). (iii) Several of the selected articles discuss new computing paradigms that may take advantage of emerging memory devices (G. Santoro et al. [12] and S. Nam et al. [13]), as well as extensions, modifications, or innovations in existing volatile and nonvolatile memory technologies (at both the device and circuit levels), which may add new functionalities or improve their performance for computing applications (H. H. Shin et al. [14]; S. Yang et al. [15]; A. Subbiah and T. Ogunfunmi [16]; H. E. Yantir et al. [17]; and L. Gan et al. [18]). Finally, in "Development of Bioelectronic Devices Using Bionanohybrid Materials for Biocomputation System," J. Yoon et al. [19] review their recent progress in the development of biocompatible memory and computing devices.

The selected papers cover a broad range of research and development activities related to emerging memory devices and computing paradigms. It is hoped that this selection of articles will serve as a resource for researchers in academia and industry, practicing engineers, and students, both as a window into some of the recent advances in emerging memory technologies, as well as to stimulate interest in potential new directions for research.

Conflicts of Interest: The authors declare no conflict of interest.

References

1. Khalili Amiri, P.; Wang, K. Voltage-Controlled Magnetic Anisotropy in Spintronic Devices. *Spin* **2012**, *2*, 1240002. [CrossRef]
2. Nozaki, T.; Yamamoto, T.; Miwa, S.; Tsujikawa, M.; Shirai, M.; Yuasa, S.; Suzuki, Y. Recent Progress in the Voltage-Controlled Magnetic Anisotropy Effect and the Challenges Faced in Developing Voltage-Torque MRAM. *Micromachines* **2019**, *10*, 327. [CrossRef] [PubMed]
3. Park, J.; Yim, Y. Fine-Grained Power Gating Using an MRAM-CMOS Non-Volatile Flip-Flop. *Micromachines* **2019**, *10*, 411. [CrossRef] [PubMed]
4. Lian, X.; Shen, X.; Lu, L.; He, N.; Wan, X.; Samanta, S.; Tong, Y. Resistance Switching Statistics and Mechanisms of Pt Dispersed Silicon Oxide-Based Memristors. *Micromachines* **2019**, *10*, 369. [CrossRef] [PubMed]
5. Shen, Z.; Qi, Y.; Mitrovic, I.Z.; Zhao, C.; Hall, S.; Yang, L.; Luo, T.; Huang, Y.; Zhao, C. Effect of Annealing Temperature for Ni/AlO$_x$/Pt RRAM Devices Fabricated with Solution-Based Dielectric. *Micromachines* **2019**, *10*, 446. [CrossRef] [PubMed]
6. Xie, C.; Li, X.; Chen, H.; Li, Y.; Liu, Y.; Wang, Q.; Ren, K.; Song, Z. Speeding Up the Write Operation for Multi-Level Cell Phase Change Memory with Programmable Ramp-Down Current Pulses. *Micromachines* **2019**, *10*, 461. [CrossRef] [PubMed]
7. Drake, K.; Lu, T.; Majumdar, M.K.H.; Campbell, K.A. Comparison of the Electrical Response of Cu and Ag Ion-Conducting SDC Memristors Over the Temperature Range 6 K to 300 K. *Micromachines* **2019**, *10*, 663. [CrossRef] [PubMed]

8. Lee, Y.K.; Jeon, J.W.; Park, E.-S.; Yoo, C.; Kim, W.; Ha, M.; Hwang, C.S. Matrix Mapping on Crossbar Memory Arrays with Resistive Interconnects and Its Use in In-Memory Compression of Biosignals. *Micromachines* **2019**, *10*, 306. [CrossRef] [PubMed]

9. Jo, S.; Sun, W.; Kim, B.; Kim, S.; Park, J.; Shin, H. Memristor Neural Network Training with Clock Synchronous Neuromorphic System. *Micromachines* **2019**, *10*, 384. [CrossRef] [PubMed]

10. Truong, S.N. Compensating Circuit to Reduce the Impact of Wire Resistance in a Memristor Crossbar-Based Perceptron Neural Network. *Micromachines* **2019**, *10*, 671. [CrossRef] [PubMed]

11. Xue, X.; Wang, C.; Liu, W.; Lv, H.; Wang, M.; Zeng, X. A RISC-V Processor with Area-Efficient Memristor-Based In-Memory Computing for Hash Algorithm in Blockchain Applications. *Micromachines* **2019**, *10*, 541. [CrossRef] [PubMed]

12. Santoro, G.; Turvani, G.; Graziano, M. New Logic-In-Memory Paradigms: An Architectural and Technological Perspective. *Micromachines* **2019**, *10*, 368. [CrossRef] [PubMed]

13. Nam, S.A.; Cho, K.; Bahn, H. Tight Evaluation of Real-Time Task Schedulability for Processor's DVS and Nonvolatile Memory Allocation. *Micromachines* **2019**, *10*, 371. [CrossRef] [PubMed]

14. Shin, H.H.; Chung, E.-Y. In-DRAM Cache Management for Low Latency and Low Power 3D-Stacked DRAMs. *Micromachines* **2019**, *10*, 124. [CrossRef] [PubMed]

15. Yang, S.-D.; Jung, J.-K.; Lim, J.-G.; Park, S.-G.; Lee, H.-D.; Lee, G.-W. Investigation of Intra-Nitride Charge Migration Suppression in SONOS Flash Memory. *Micromachines* **2019**, *10*, 356. [CrossRef] [PubMed]

16. Subbiah, A.; Ogunfunmi, T. A Flexible Hybrid BCH Decoder for Modern NAND Flash Memories Using General Purpose Graphical Processing Units (GPGPUs). *Micromachines* **2019**, *10*, 365. [CrossRef] [PubMed]

17. Yantir, H.E.; Guo, W.; Eltawil, A.M.; Kurdahi, F.J.; Salama, K.N. An Ultra-Area-Efficient 1024-Point In-Memory FFT Processor. *Micromachines* **2019**, *10*, 509. [CrossRef] [PubMed]

18. Gan, L.-R.; Wang, Y.-R.; Chen, L.; Zhu, H.; Sun, Q.-Q. A Floating Gate Memory with U-Shape Recessed Channel for Neuromorphic Computing and MCU Applications. *Micromachines* **2019**, *10*, 558. [CrossRef] [PubMed]

19. Yoon, J.; Lee, T.; Choi, J.-W. Development of Bioelectronic Devices Using Bionanohybrid Materials for Biocomputation System. *Micromachines* **2019**, *10*, 347. [CrossRef] [PubMed]

Review

Development of Bioelectronic Devices Using Bionanohybrid Materials for Biocomputation System

Jinho Yoon [1], Taek Lee [2] and Jeong-Woo Choi [1],*

[1] Department of Chemical & Biomolecular Engineering, Sogang University, 35 Baekbeom-Ro, Mapo-Gu, Seoul 04107, Korea; iverson0607@naver.com
[2] Department of Chemical Engineering, Kwangwoon University, Wolgye-dong, Nowon-gu, Seoul 01899, Korea; tlee@kw.ac.kr
*   Correspondence: jwchoi@sogang.ac.kr; Tel.: +82-2-705-8480

Received: 17 April 2019; Accepted: 22 May 2019; Published: 27 May 2019

Abstract: Bioelectronic devices have been researched widely because of their potential applications, such as information storage devices, biosensors, diagnosis systems, organism-mimicking processing system cell chips, and neural-mimicking systems. Introducing biomolecules including proteins, DNA, and RNA on silicon-based substrates has shown the powerful potential for granting various functional properties to chips, including specific functional electronic properties. Until now, to extend and improve their properties and performance, organic and inorganic materials such as graphene and gold nanoparticles have been combined with biomolecules. In particular, bionanohybrid materials that are composed of biomolecules and other materials have been researched because they can perform core roles of information storage and signal processing in bioelectronic devices using the unique properties derived from biomolecules. This review discusses bioelectronic devices related to computation systems such as biomemory, biologic gates, and bioprocessors based on bionanohybrid materials with a selective overview of recent research. This review contains a new direction for the development of bioelectronic devices to develop biocomputation systems using biomolecules in the future.

Keywords: bioelectronic devices; bionanohybrid material; biomemory; biologic gate; bioprocessor; protein; nucleic acid; nanoparticles

---

1. Introduction

Bioelectronics is defined as the combined field of biology and electronics that has recently been greatly developed to overcome the current limitation of silicon-based electronics and biology-based engineering [1]. By introducing biomolecules on the silicon-substrate, electrical functions have been demonstrated on the chip using the unique properties of biomolecules, such as specific target molecule detection and optoelectrical properties, that can be applied in bioelectronic devices such as biosensors, biophotodiodes, and biotransistors [2–5]. Various biomolecules including metalloprotein possess a metal ion at their core, and functional DNA with specific chemical group modifications such as amine and carboxyl groups have advantages for applications to develop bioelectronic devices because of their unique properties such as redox properties that are derived from the metal ion in the protein and the specific binding properties of DNA with its complementary DNA at the nanometer scale [6,7]. By fusing biomolecules with organic materials, electronic functions have been widely studied to develop bioelectronic devices with enhanced performance such as more sensitive target detection and increased signal [8–11]. Until now, many functional bioelectronic devices including protein-based bioelectronic chips that use the electron transfer mechanism of proteins and biophotodiode devices that use the photoelectric effect of rhodopsin have been reported [12–14]. However, current bioelectronic devices have certain critical limitations for practical application because the use of biomolecules inevitably accompanies limitations such as the low electrical/electrochemical signal-to-noise ratio derived from biomolecules, instability in harsh conditions, and narrow functionalization [15,16]. To overcome the limitations of biomolecules, innovative methods have been developed introducing nanoparticles to enhance the signal induced from biomolecules, combine biomolecules with carbon-based materials such as carbon nanotubes (CNT) or graphene for electrochemical signal increment and long-term stability using the biocompatibility of carbon-based materials, and the use of nanoscale-patterned chips as a platform for the extension of the functionality of bioelectronic devices such as by demonstrating nanoscale electronic functions and immobilizing different biomolecules independently at the nanometer scale to use these biomolecules simultaneously [17–22].

Recently, bionanohybrid materials composed of biomolecules and other nanomaterials have been developed widely for applications in bioelectronic devices. Bionanohybrid materials have received much attention for their wide application in developing delicate bioelectronic devices that accompany enhanced electronic functions or highly sensitive target detection for biosensors. As mentioned above, biomolecules have unique properties at the nanometer scale and nanomaterials such as nanoparticles, CNT, and biocompatible polymers that improve the properties of biomolecules can be hybridized precisely at the nanometer scale while retaining the properties of biomolecules and nanomaterials [23–25].

Among the various bioelectronic devices, certain bioelectronic devices that are capable of performing information storage or signal processing similar to memory or logic gates in conventional electronic devices have shown a new perspective and direction for the development of biocomputation systems [26,27]. Biomemory devices based on metalloprotein or redox-controllable linker have been reported [28–30] that can demonstrate the memory function using biomolecules through controlling two apparently distinguished biomolecular states reversibly. In addition, using the above-mentioned bionanohybrid materials as the core component, bioprocessor devices have been reported that can process the input signal to process the out signal using bionanohybrid materials as the processing platform [31]. In addition, to develop sophisticated and improved functional bioelectronic devices, various advanced materials have been studied and introduced to fabricate modern devices such as new functionalized structural graphene and two-dimensional materials [32–34]. Through these efforts, various bionanohybrid materials that are capable of performing information storage, logical functions, and information processing have been developed for the development of bioelectronic devices including biomemory, biologic gates, and bioprocessors. Such bioelectronic devices can be used as core components to develop a biocomputation system that is capable of performing computation similar to conventional computers that are common in our surroundings as depicted in Figure 1.

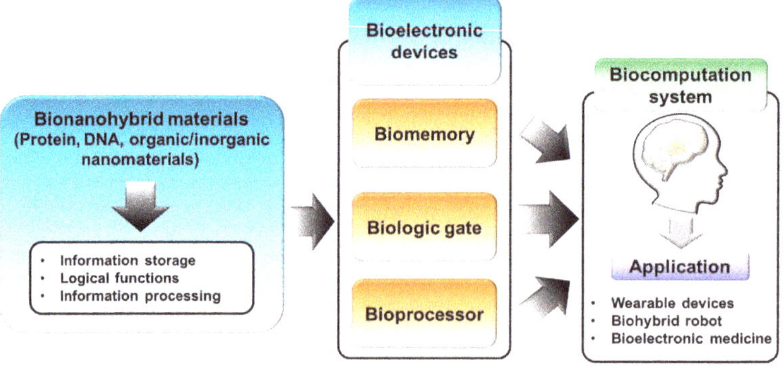

Figure 1. Bioelectronic devices based on bionanohybrid materials to develop biomemory, biologic gates, and bioprocessors for biocomputation systems.

In this review, bioelectronic devices based on bionanohybrid materials that are capable of performing information storage and signal processing for computation systems are discussed with a selective overview of recent research. Although there are many extensive reviews of bioelectronic devices, this review discusses in detail recent reports of various specific types of bioelectronic device for biocomputation systems. This review will suggest a new inspirable direction and aspects of bioelectronic devices to develop a biocomputation system [35,36].

2. Biomemory

Information storage is an important function for the operation of electronic devices. Until now, various information storage function devices have been developed in conventional silicon-based electronic devices through controlling two apparently distinguished states such as "1" and "0" states for the demonstration of conventional memory functions. From the bioelectronics perspective, some specific biomolecules have properties of existing in two distinguished states by external stimulation such as metal ion states that control metalloprotein, which can be utilized to develop biomolecular memory devices [37,38]. In addition, the hybridization of more than two types of biomolecule and bionanohybrid materials has been proposed to demonstrate multiple states control and increase the electrochemical signal derived from biomolecules for biomemory. In this chapter, we provide research related to biomemory devices including protein-based biomemory and resistive switching memory devices.

2.1. Multilevel Biomemory Devices

Metalloproteins have metal ions in their body that can be utilized for electrochemical investigation [39,40]. For example, the metal ion of a metalloprotein can be used to affect the redox reactions of specific materials, which can be measured using electrochemical techniques for developing biosensors [18,41]. In addition, this can be applied to develop biotransistors using redox properties [42]. This metal ion can exist in two different states like the Cu^+ and Cu^{2+} states of azurin, a metalloprotein that possesses copper ion, which shows the potential of metalloprotein-based biomemory devices [28] by controlling metal ions with distinguished states. Various research groups have developed metalloprotein-based biomemory devices [29,37,38]. Among them, our group developed various biomemory devices using metalloproteins such as azurin and cytochrome c, which have never been reported before. Beyond just controlling the ion states of one type of metalloprotein for biomemory, we suggested multilevel biomemory devices using two kinds of metalloproteins to achieve the incremental memory density [43]. By controlling isoelectric points of metalloproteins via pH control, we immobilized two different metalloproteins, recombinant azurin modified with cysteine group and cytochrome c, directly on to the gold substrate by self-assembly through the electrostatic bond without any chemical linkers for the control of multiple redox states [44,45]. This simple immobilization process could reduce the immobilizing time of biomolecules, and remove the introduction of the other chemical materials for immobilization. Figure 2A shows the schematic image and demonstration of multilevel biomemory using the direct immobilization of two kinds of metalloproteins. We confirmed multilevel memory device fabrication by surface plasmon resonance (SPR) and scanning tunneling microscopy (STM) to verify the metalloprotein double layer formation through morphological changes. Then, an electrochemical investigation was performed using cyclic voltammetry (CV) and chronoamperometry (CA). By introducing two different metalloproteins, this device showed the multiple redox states that could be derived from copper ions of azurin and iron ions of cytochrome c as shown in Figure 2B. This showed oxidation potential peaks at 0.294 V and 0.184 V that were derived from cytochrome c and azurin, respectively, and the reduction of potential peaks at 0.131 V and 0.062 V from cytochrome c and azurin, respectively. These potential values for each metalloprotein were used as input potentials to control the metal ion states of the two metalloproteins. Then, we estimated the memory performance for this device using the obtained redox potential peak values of two metalloproteins for the "writing step" and "erasing step" and obtained the open circuit potential (OCP) values of metalloproteins for the "reading step" for multilevel biomemory demonstration. As shown in Figure 2C, this device showed apparently distinguished states when applying a potential to the device following expected schematic images with two different forms of "writing step", "reading step", and "erasing step". From these results, we successfully developed new-concept multilevel biomemory devices using two different metalloproteins for multiple information storage biodevices.

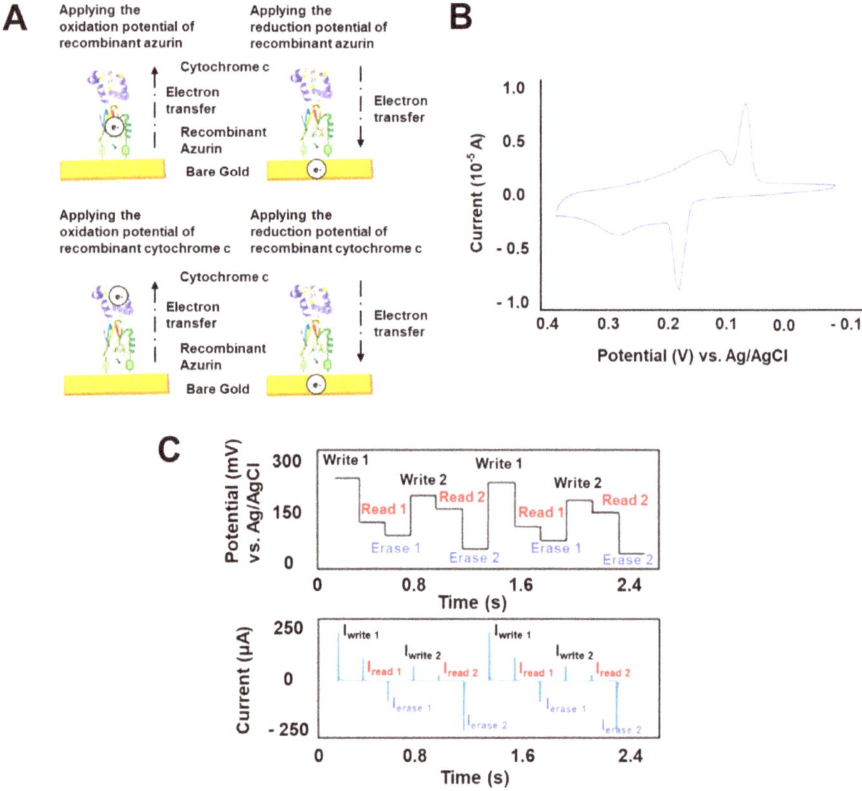

Figure 2. Multilevel biomemory device. (**A**) Schematic image demonstrating a multilevel biomemory device using metal ions states to control two different kinds of metalloprotein. (**B**) Cyclic voltammogram of a multilevel biomemory device composed of recombinant azurin and cytochrome c that shows two apparently distinguished reduction potential peaks and two oxidation potential peaks. (**C**) Memory performance of a multilevel biomemory device including writing, reading, and erasing steps by applying the potential values of reduction and oxidation potential peak values and the OCP values of metalloproteins. (Reproduced with permission from [43], published by John Wiley and Sons, 2010).

2.2. Electrochemical Signal-Enhanced Biomemory Device

As mentioned in the introduction, bioelectronic devices have certain limitations like the low electrical or electrochemical signal induced from biomolecules and low stability in harsh conditions [15,16]. To overcome these problems, various researchers have proposed the introduction of functional biocompatible nanomaterials for improved signal and stability [18–20]. Through these suggestions, biosensors and biofuel cells have been developed with advanced performance. In the case of biomemory devices, the extremely low electrochemical signal from biomolecules should be solved for application in practical applications. To achieve this, introducing metal nanoparticles can be a solution for signal enhancement. Gold nanoparticles (GNP) have been reported as an enhancer for the electron transfer reaction with metalloprotein [46]. Using the reported results, our group proposed a biomemory device using metalloprotein (azurin, Azu) and GNP to increase the electrochemical signal derived from metalloprotein (Figure 3A) [17]. To develop this electrochemical signal enhanced biomemory device, various GNP of nanometer size of diameter in the range 5–60 nm was introduced to find the optimized size for the GNP diameter. Based on the electrochemical signal increasing the redox potential peak values from CV results (Figure 3B), we found the optimized GNP size (5 nm) that showed smaller redox potential peak values compared to the results using the 60 nm GNP. However, in the case of the 60 nm GNP, the enhanced signal was not derived from Azu–GNP but directly induced from the immobilized GNP to the gold substrate without Azu. Therefore, the 5 nm GNP was chosen as the optimized size for biomemory fabrication. In addition, we assumed that the proposed increment of the electron transfer mechanism followed the equation below:

$$GNP \; \underset{k_{-1}}{\overset{k_1}{\leftrightarrow}} \; Protein \; \underset{k_{-2}}{\overset{k_2}{\leftrightarrow}} \; Electrode \tag{1}$$

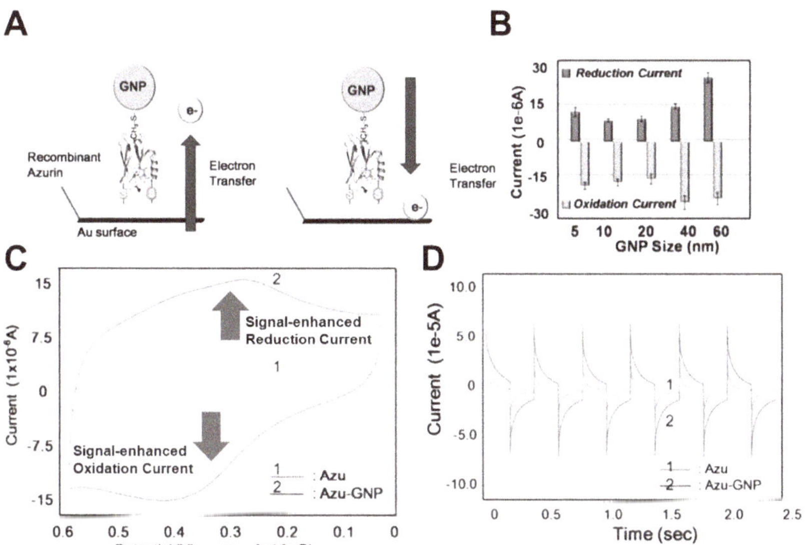

Figure 3. Electrochemical signal enhanced biomemory device. (**A**) Schematic image of the biomemory device composed of Azu and gold nanoparticles (GNP). (**B**) Redox potential peak values for optimizing the GNP diameter. (**C**) Cyclic voltammogram of Azu–GNP and Azu. (**D**) memory performance of Azu–GNP and Azu. (Reproduced with permission from [17], published by John Wiley and Sons, 2011).

In Equation (1), k_1 and k_{-1} are the electron transfer rate constants between the GNP and Azu and k_2 and k_{-2} are the electron transfer rate constants between Azu and the gold substrate. By introducing the GNP, the electrochemical signal from Azu could be enhanced through the better electric coupling between azurin and GNP and between Azu and the gold substrate. Furthermore, the better coupling between Azu and GNP compared to that between Azu and the gold substrate induced a remarkably enhanced signal. After verifying the signal enhancement, the biomemory function of the proposed device was estimated. As shown in Figure 3C,D, the biomemory device composed of Azu and GNP (Azu–GNP) showed enhanced memory function compared to biomemory prepared with only Azu. The stored charge amounts were calculated by the following equation,

$$Q = \int i \times dt \tag{2}$$

The current value (*i*) and time value (d*t*) were obtained by CA technique. To acquire the CA results, the redox potential peak values of Azu–GNP obtained by CV analysis were applied. From the calculation of the area underneath the CA graphs, the stored charge amounts of biomemory composed of Azu–GNP was about 4.503 μC, approximately four times higher than that of biomemory prepared with only Azu (about 1.1413 μC). This difference originated from the electric coupling between Azu and the GNP. Through this research, electrochemical signal-enhanced biomemory was developed for the first time, and this approach may demonstrate the possibility of developing accurate nanoscale biomemory devices that can overcome the problems associated with low electrochemical signals.

2.3. Resistive Biomemory Device

In conventional silicon-based electronic fields, huge attention has been paid to the development of resistive memory devices for resistive switching function demonstration. Resistive memory devices have been researched widely for commercialization due to their advantages such as fast processing and response and low energy requirement. In the case of the existence of metal–insulator–metal layers or semiconductor–insulator–metal layers on the substrate, there are specific unique hysteresis properties at some voltage range with two apparently different resistance values (extremely high resistance value and extremely low resistance value) following theories such as ohmic conduction, thermionic emission, Schottky emission, or tunneling current [47]. Various research groups have developed organic material-based resistive memory devices [48,49]. Biomolecules are suitable for demonstrating resistive switching functionality at the nanometer scale because they possess unique properties at such scale. Guo's group developed a resistive biomemory device using the RNA structure and quantum dot (QD) by collaboration with our group [50]. In previous research, they developed a packaging RNA (pRNA) three-way junction structure (pRNA-3WJ) that showed thermodynamically stability [51]. This pRNA-3WJ could overcome the critical limitations of conventional RNA such as extremely low stability even at room temperature. In resistive biomemory research, they introduced the developed pRNA-3WJ as a stable insulator to demonstrate the resistive switching function. Figure 4A shows the schematic images and resistive function in this device. Using the biological binding properties between streptavidin and biotin, they developed a nanoscale bionanohybrid material composed of pRNA-3WJ and QD. The conjugation of pRNA-3WJ and QD for bionanohybrid materials was verified by electrophoresis through the existence of the upper located band due to the increased total weight and size by QD introduction compared to the band in only pRNA-3WJ without QD. After immobilizing this bionanohybrid material on the gold substrate, pRNA-3WJ performed a role as an insulating layer and QD as the semiconducting layer on the conducting gold layer. Using a scanning tunneling spectroscopy (STS) technique, they estimated the resistive switching function of this device at the nanometer scale using the platinum tip as the probe located on this bionanohybrid material. Figure 4B displays the I–V curve of a bionanohybrid material on a gold substrate. Compared to the gold substrate alone, only pRNA-3WJ, and only QD on a gold substrate, a bionanohybrid material composed of pRNA-3WJ and QD on a gold substrate showed apparently distinguished resistance values with extremely high and low resistance at the voltage range of +3 to −3 V. This bistable behavior could be defined as "On state" and "Off state" for resistive memory applications.

Our group also developed a resistive biomemory device based on two-dimensional material. A bionanohybrid material composed of molybdenum disulfide nanoparticles (MoS_2) and a DNA layer on a gold substrate was developed to demonstrate resistive switching functionality at the nanometer scale [52]. To develop this resistive biomemory at the nanometer scale, we immobilized DNA and synthesized MoS_2 sequentially on a complementary DNA modified gold substrate. Then, a semiconductor (MoS_2)–insulator (DNA)–metal layer (gold substrate) was formed that could demonstrate resistive switching functionality through specific unique hysteresis properties at a certain voltage range with two apparently different resistance values. MoS_2 is a metal dichalcogenide material that has been widely used to develop bioelectronic devices because of its unique properties including biocompatibility, excellent semiconductivity, and its optical properties [53,54]. To demonstrate resistive switching functionality at the nanometer scale, MoS_2 nanoparticles with surface modification (carboxyl group) was synthesized for the first time to conjugate efficiently with amine-tagged DNA via EDC/NHS bonding. The synthesis of surface-modified MoS_2 nanoparticles and the fabrication of bionanohybrid materials were verified by transmission electron microscopy (TEM) for MoS_2 synthesized nanoparticles, energy-dispersive X-ray spectroscopy (EDS) for elemental analysis, electrophoresis for the conjugation of MoS_2 and DNA, and STM techniques to immobilize this bionanohybrid material on the gold substrate. Using STS analysis, the proposed resistive biomemory device based on MoS_2 and DNA showed the resistive switching function with bistable states at a wide voltage range (4 to −4 V) and long-term stability as shown in Figure 4C. Figure 4C shows that the resistance value dramatically decreased when the voltage reached 2.4 V; on the other hand, the resistance value abruptly increased when the voltage reached 0.01 V. In addition, by introducing DNA as the insulating layer, which is more stable than RNA, it showed resistive switching function for about 10 days. As with these studies, bionanohybrid material-based resistive biomemory devices have been researched to demonstrate resistive switching functionality at the nanometer scale, which suggests a future direction for the development of the next generation of memory devices using biomolecules.

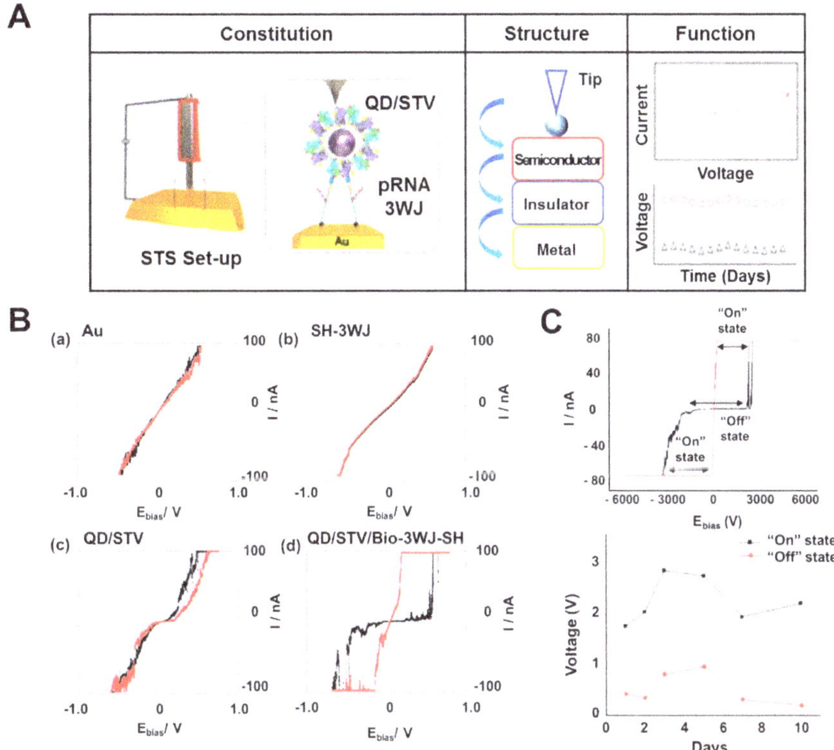

Figure 4. Resistive biomemory device. (**A**) Schematic image of a resistive biomemory device composed of pRNA-3WJ and quantum dot (QD) on a gold substrate, (**B**) I–V curves of bare Au, pRNA-3WJ, QD and pRNA-3WJ, and QD. (**C**) Resistive switching function and stability test for a resistive biomemory device composed of MoS_2 and DNA on a gold substrate with apparently distinguished resistance states and long-term stability for 10 days. (Reproduced with permission from [50], published by the American Chemical Society, 2015, and reproduced with permission from [52], published by Elsevier, 2019).

3. Biologic Gate

Among the various components of computing systems, logic gates are a core component in the computing process. These electrical circuits implement Boolean functions that can perform logical operations by converting more than two inputs to one binary output. Until now, many logic gates have been developed including AND (the gate that performing the logical conjugation), OR (the gate that performing the logical disjunction), NOT (the gate that performing the logical negation), XOR (the gate that giving the true outputted signal when the number of true inputs is odd), and NAND (the gate that giving the false outputted signal only when all inputs are true) gates [55,56]. In bioelectronics, some biomolecules can interact with specific chemical materials or biomolecules. For example, myoglobin can react with hydrogen peroxide [18] and glucose oxidase can react with glucose [57]. These properties can be utilized to demonstrate a logic gate using biomolecular interactions by controlling the input materials. In addition, conformational changes of biomolecules can be utilized to develop logic gates such as the conformational change of G-quadruplex DNA (G-rich DNA) as a bending shape and straightening shape that is dependent on the pH value [58]. Furthermore, these logic functions based on biomolecules can provide opportunities to mimic the analog human decision-making process [59] through controlling the combination of biomolecules and organic and inorganic materials. In this chapter, we will introduce research into biologic gates using biomolecules such as proteins and DNA and bioelectronic devices that mimic the analog human decision-making process.

3.1. DNA-Based Biologic Gate

DNA is the smallest level at which the composition of living organisms is developed. There have been reports related to DNA research such as DNA sequencing, immunoassay, and DNA structure formation for wider applications [60–62]. From the bioelectronics perspective, the unique properties of DNA have received attention for their granting of functionality to bioelectronic devices. DNA can specifically bind with complementary DNA and the structure of DNA can be controlled by external responses [58]. Until now, various DNA-based logic gates have been reported based on colorimetric or fluorescence investigations. However, electrochemical techniques are better suited to bioelectronic device fabrication due to their fast response, minimal required reagents, and simplified outputs compared to colorimetric-based bioelectronic devices. In addition, a report found that mismatched sequences of double strand DNA such as cytosine–cytosine (C–C) and thymine–thymine (T–T) mismatched pairs could possess metal ions in such locations [63]. From this perspective, Qiu's group developed a biologic gate using this unique property of DNA [64]. Figure 5A showed the schematic process and results of AND logic gates based on DNA mismatching. Silver ions (Ag^+) and mercury ions (Hg^{2+}) could enter the mismatched C–C and T–T pairs, respectively. Using these properties, they designed T- and C-rich DNA sequences with ferrocenecarboxylic acid (Fc) as the redox generator. In this device, metal ions were used as input molecules and the electrochemical signal from Fc was the output signal from the logic gate. By controlling the DNA sequences, they developed AND, NAND, and NOR logic gates through controlling the output signal using the unique electrochemical signals derived from the inserted Ag^+ and Hg^{2+} ions located inside the mismatched pairs in the DNA. From the results, in the case of the insertion of only both Ag^+ and Hg^{2+} ions, the electrochemical signal was detected by the differential pulse voltammetry (DPV), which was defined as "1" due to the co-existence of Ag^+ and Hg^{2+} ions as shown in Figure 5A. Furthermore, this logic gate based on DNA mismatching can be operated reversibly compared to DNA cleavage-based logic gates. This result showed the possibility of applying bionanohybrid materials based on specifically designed DNA sequences and metal ions for both bioelectronic devices and for the development of a multiplexed biosensing platform.

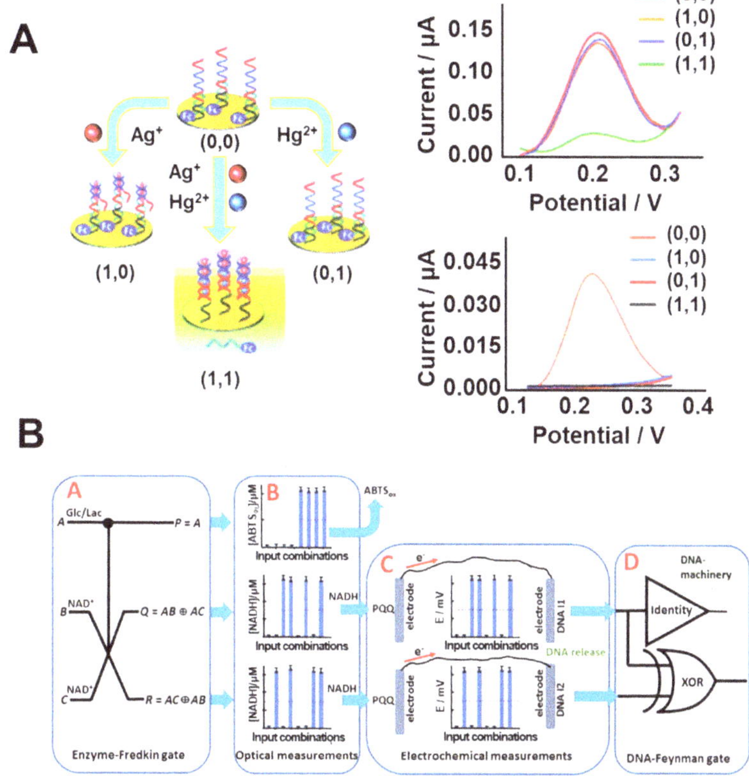

Figure 5. Biologic gates. (**A**) Schematic image of a DNA-based biologic gate based on metal ions inserted inside mismatched DNA pairs and differential pulse voltammetry (DPV) results of this device by controlling output signals through Ag^+ and Hg^{2+} ions inserted inside mismatched DNA pairs. (**B**) Schematic image of a protein/DNA-based biologic gate through the signal transduction of a protein-based biologic gate to a DNA-based biologic gate for the final outputted fluorescence signal (Reproduced with permission from [64], published by John Wiley and Sons, 2013 and reproduced with permission from [65], published by John Wiley and Sons, 2016).

3.2. Protein/DNA-Based Biologic Gate

Their distinct properties mean that proteins are widely utilized in bioelectronics. As mentioned in the introduction, proteins have certain unique advantages such as distinctive redox properties that can target specific molecules and reactions. Through these properties, protein-based biologic gates, particularly enzyme-based biologic gates, have been broadly developed [66,67]. Aida's group developed a biologic gate by protein folding [68] and Schöning's group proposed a biologic gate using a membrane composed of multiple enzymes [69]. Recently, Katz's group developed reversible biologic gates based on both enzymes and DNA [65]. They developed and combined an enzyme-based biologic gate with a reversible DNA-based biologic gate through a biomolecular electrode to create complex reversible logical computing systems. This proposed system was composed of an enzyme-based Fredkin gate that was capable of converting three input signals to three output signals and a DNA-based Feynman gate that was capable of converting two input signals to two output signals [70,71]. To demonstrate this complex biologic gate, they introduced the optical, electrochemical, and fluorescent measurement techniques. Figure 5B shows a schematic diagram of this complex biologic system that is composed of a protein-based biologic gate and a DNA-based biologic gate as demonstrated by the enzyme reaction and connected DNA reaction. Glucose (Glc), lactic acid (Lac), and β-nicotinamide adenine dinucleotide hydrate (NAD^+) were used as three input signals for the enzyme-based Fredkin biologic gate and glucose dehydrogenase (GDH), lactate dehydrogenase (LDH), glucose oxidase (GOx), and horseradish peroxidase (HRP) were utilized for enzyme-based biologic operation. After reacting in the first enzyme-based biologic gate, the generated signal that produced NADH through enzymatic reactions was measured by an optical technique and transferred to the electrochemical system for electrochemical enzymatic reaction. Then, in the final stage at the connected DNA-based biologic gate, the transferred signal was converted to fluorescent final outputs. Pyrroloquinoline quinone (PQQ)-modified electrode and iron ion (Fe^{3+}) crosslinked alginate-modified electrode with entrapped DNA were used for the electrochemical system and final DNA-based biologic system. By the electrochemical reaction, the Fe^{3+} ion of the crosslinked alginate-modified electrode with entrapped DNA was oxidized to Fe^{2+} and the entrapped DNA was released from the alginate-modified electrode to a DNA-based biologic gate for the final fluorescence output signal. Although many components and complex biological reactions were utilized for this biologic gate, they developed a complex biologic system that was composed of two different kinds of biomolecule-based biologic gates that were more complex than the reported biologic gate to accurately mimic a conventional silicon-based electronic logic system.

3.3. Analog Decision Mimicking Bioelectronic Device

In conventional silicon-based electronics, only digitalized processing, logic, and arithmetic operations have been developed and utilized in all devices [72,73]. These operations have certain advantages for the development of electronic devices in which the binary coded digital signals "1" and "0" can be distinguished, defined, and operated easily by converting input signals into an integrated single output signal. However, digital signal-based conventional electronic devices have limitations for the demonstration of human logic systems or other analog decision-making processes because these systems are not decided or operated by one simple and single digital input and output, but are instead affected by a myriad of complex factors such as personality, experience, and intelligence. The critical difference between conventional electronic devices and real human logic systems can hinder the development of biocomputation systems. Therefore, in bioelectronic fields, there have been studies to develop bioelectronic devices that are capable of mimicking analog decisions or analog logic systems [74] by considering various factors for analog calculation. Liu's group developed four analog computing systems and extended the range of computing to real numbers based on DNA by connecting DNA-based biologic gates using unique properties of DNA such as DNA strand displacement. This result showed the potential of DNA-based real number calculation such as a calculator and by extension a combination of various DNA-based biologic gates; this could demonstrate more complex number calculation. In addition, bioelectronic noses and tongues based on biomolecular receptors have been researched recently to mimic the processes of real living organisms [75,76]. To demonstrate the analog decision-making process on a bioelectronic chip, our group developed an electrochemical bioelectronic device based on a bionanohybrid material composed of metalloprotein and organic/inorganic nanomaterials or metal ions [59]. Figure 6A shows a conceptual image of their research for mimicking analog decision-making through the analogously processed output signals by inputting two different external factors (negative input and positive input) via electrochemical investigation. We defined specific regions of the acquired signal as the degree of confidence and reliability of a human following defined threshold values. Myoglobin (Mb) that is a metalloprotein used as signal generator and defined as an inherent human tendency, organic chemical linkers that are used as signal controllers and defined as experience-induced human tendencies, and inorganic materials that are used for signal modulation and defined as environment-dependent signal modulators were combined to demonstrate analog decision-making by signal control and modulation (Figure 6B). As shown in Figure 6C, the plotted results of analog decision-making based on the analysis of electrochemical signals by defined external factors showed the decision variation of 12 people based on defined threshold values. This research shows the conceptual potential for the development of analog-based bioelectronic devices which has never been reported to apply for biocomputation systems. Of course, many subjective definitions exist that can demonstrate the analog decision-making process on a bioelectronic chip. This research shows one potential development route for a human mimicking analog computation system.

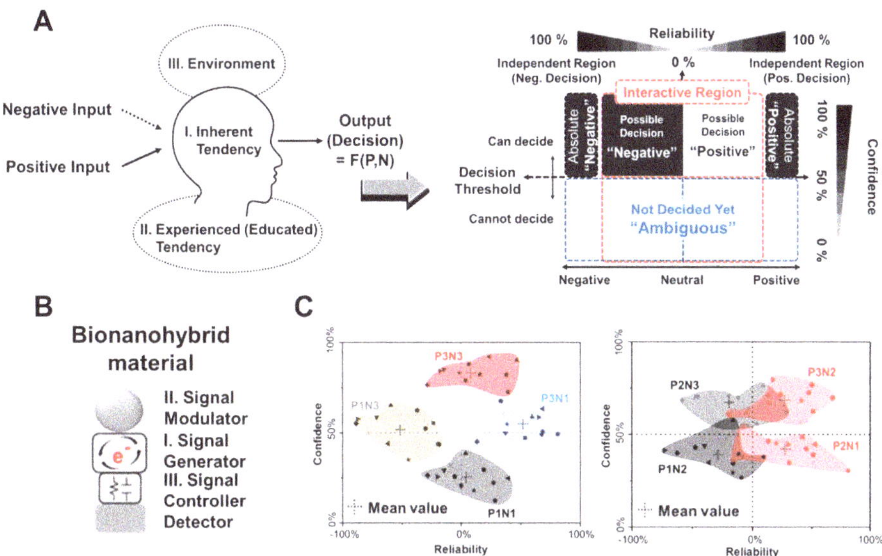

Figure 6. Analog decision mimicking bioelectronic device. (**A**) Schematic image and theory of this bioelectronic device through the analogously processed output signals by two different external factors inputted (negative input and positive input) by electrochemical investigation. (**B**) Bionanohybrid material used for this device composed of metalloprotein used as signal generator, organic chemical linkers as signal controller, and inorganic materials used for signal modulation. (**C**) The plotted results of analog decision-making based on the analysis of electrochemical signal by defined external factors showed the decision variation of 12 persons based on the defined threshold values. (Reproduced with permission from [59], the figures follow the terms of use under a Creative Commons Attribution 4.0 International License.).

4. Bioprocessor

Until now, numerous molecular electronic devices have been developed to miniaturize electronic devices at the molecular scale for overcoming the physical or technological limitations of conventional silicon-based electronic devices, such as difficulty to achieve compact integration at nanometer or molecular scale [77]. Biomolecules have unique properties even at the nanometer scale that are suitable to complement the molecular electronic devices with delicate functional processing properties. Accordingly, some researchers developed bioprocessors that could control the biological output signals, such as the expressed gene level, through the biological reaction process by specific inputted biomolecules [78,79]. To mimic the conventional silicon-based electronic processors, especially, the functional bionanohybrid material composed of biomolecules and various nanomaterials can used for processing the input signal converted to the processed output signal such as electrochemical signal. In addition, specifically designed microchips can be a powerful tool to control the biologically processed output signals through the control of biological reactions. In this chapter, we will provide the recently developed bioprocessor devices that could mimic the processing in electronic devices.

4.1. DNA-Based Bioprocessor

As mentioned in the above chapters, DNA is a suitable biomaterial for bioelectronic applications. Especially, the massive parallelism of DNA hybridization exhibits tremendous potential, which can be utilized to develop feasible electronic devices capable of performing processing or computing operation to fulfill the demands of monolithic parallel computing system with specific computational algorithm [80]. From this point of view, DNA-based bioprocessors or applications have been reported [81]. Lee's group proposed a novel programmable DNA-mediated processor to solve the optimal route planning problems [82]. To achieve this functional DNA-mediated bioprocessor, they fabricated the programmable optimal route planning apparatus comprising six stages, as shown in Figure 7A. Also, the routes shown in Figure 7A were defined following the specific DNA sequences (20mer single strand DNA) to find the optimal route based on DNA processing through the conventional PCR reaction using the inputted DNA sequences, which determine distance between specific locations as shown in the map of the right side of Figure 7A. To operate this PCR system, they defined the first stage as problem encoder for conversion of vertices and weighted edges of the designed route to DNA sequences, and all distances between each of the six locations were defined as specific DNA sequences (20mer single strand DNA) to apply for the program encoder. They defined the second stage as DNA solution bay for converted DNA preparation, the third as mixing controller for mixing and ligase of appropriate DNA sequences to make the template of DNA duplexes that represents the possible routes, the fourth as solution purifier for isolation of optimal DNA template from impurities such as the incompletely hybridized oligonucleotides or enzymes, and the fifth as PCR amplifier for amplification of optimal DNA template which is the optimal route, final as gel electrophoresis to acquire the final electrophoresis data for optimal DNA template as the find of the defined optimal route. Using these stages for optimal-route finding, they performed the DNA reactions at these mentioned stages by binding and amplification of the combined six defined DNA sequences for six locations. They obtained the results of electrophoresis to find the optimal route from home or company to the hospital as shown in Figure 7A using subjectively defined factors. Although there existed too many subjective definitions for operation, these results showed the possibility of DNA-based bioprocessing for solving the practical problems; this could be demonstrated with much fewer components and materials compared to the conventional silicon-based electronic devices. Until now, DNA-based bioprocessors remain at the early stage. However, due to the huge researches for DNA-based bioprocessors, the more sophisticate and functional processable bioprocessors will be developed.

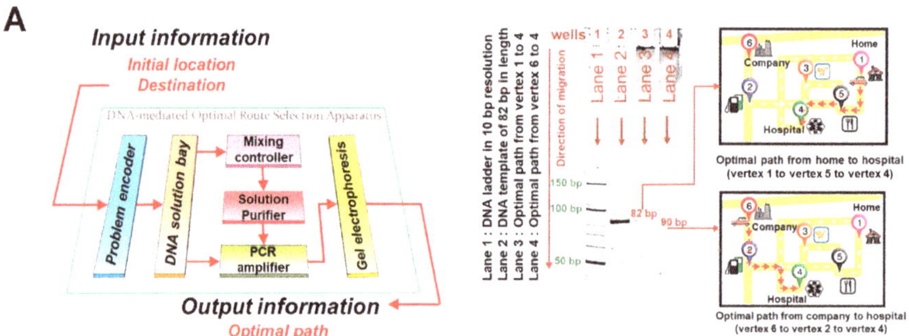

Figure 7. *Cont.*

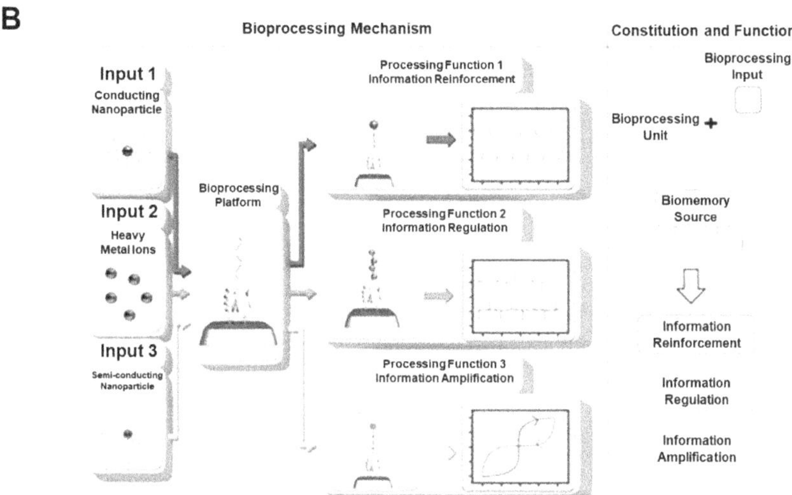

Figure 7. Bioprocessors. (**A**) Schematic image and electrophoresis results of DNA-based bioprocessor composed of six stages, including the first stage as problem encoder, the second stage as DNA solution bay for converted DNA preparation, the third as mixing controller for mixing and ligase of appropriate DNA sequences to make the template of DNA duplexes, the fourth as solution purifier for isolation of optimal DNA template from impurities such as the incompletely hybridized oligonucleotides or enzymes, the fifth as PCR amplifier for amplification of optimal DNA template which is the optimal route, and the sixth as gel electrophoresis to acquire the final electrophoresis data for solving optimal-route-planning problems, (**B**) Schematic image of bioprocessor based on bionanohybrid materials to demonstrate the specific processing functions including the electrochemical signal reinforcement, regulation, and amplification. (Reproduced with permission from [82], published by American Chemical Society, 2015, and reproduced with permission from [31], published by John Wiley and Sons, 2013).

4.2. Bioprocessor Based on Bionanohybrid Material

Advancing from only DNA-based bioprocessors, bionanohybrid material based on DNA can be used as the platform to develop the bioprocessor with more intuitive bioprocessing operation, without too much subjective definition seen in DNA-based bioprocessors, such as definition of the specific DNA sequences as the specific distance between home or company to hospital for solving the optimal-route finding. Our group developed the bioprocessing device based on bionanohybrid materials composed of protein, DNA, and inorganic nanomaterials to demonstrate the various bioprocessing functions using electrochemical/electrical investigation [31]. To develop this bioprocessor, the recombinant protein (azurin, Azu) and single-strand DNA were conjugated through the organic linker as the electrochemical signal generating bioprocessing unit (Azu/DNA hybrid) by the redox properties derived from recombinant azurin. Then, the complementary DNA (cDNA) and gold nanoparticle (GNP) hybrid (cDNA/GNP), heavy metal ions, and cDNA and quantum dot (QD) hybrid (cDNA/QD) were introduced to the bioprocessing unit as the input materials for electrochemical signal reinforcement, regulation, and amplification. Figure 7B showed the schematic image of this bioprocessor, which processed the three different outputs by introduced each input material. In the case of cDNA/GNP introduction, the electrochemical signal from bioprocessing unit was reinforced by the existence of conducting GNP. Moreover, in the case of introduction of heavy metal ions, the electrochemical signal was regulated by existed heavy metal ions such as Cu, Zn, Ni, Co, Fe, and Mn through the movement of redox peak values compared with the peak values of only Azu/DNA hybrid without metal ions. In addition, in the case of introduction of the cDNA/QD as the semiconducting nanoparticle to the bioprocessing unit, the processed electrical signal showed the electrical bistable properties as the resistive memory function by STS investigation compared to the result of only the bioprocessing unit without cDNA/QD. This developed bioprocessor device can process three different functions in the single bionanohybrid material using electrochemical and electrical signals intuitively compared to the bioprocessors demonstrated based on subjective definitions. It showed the possibility of the development of the biocomputation system in a single-biomolecular hybrid at nanometer scale.

5. Future Perspective

Since the 1960s, silicon-based electronic devices have been developed widely to demonstrate more complex functions with faster and more efficient processing on nanoscale-size chips. However, until now, the demonstration of a computation system on the single-molecular level has been impossible in the electronics field. To develop the single-molecular computation system, bioelectronic devices present new possibilities in the development of single biomolecular computation systems based on bionanohybrid materials. Bionanohybrid materials composed of biomolecules such as protein or DNA, and organic/inorganic nanomaterials can perform sophisticated functions at the single-biomolecular level to apply for bioelectronic devices. In this review, authors discussed the various research areas related to the bioelectronic devices including biomemory, biologic gates, and bioprocessors, which are the core components of the computation system. First, we discussed biomemory device based on the metalloprotein heterolayer, metalloprotein-nanoparticle hybrids, and nucleic acids-semiconducting nanoparticle hybrids. To achieve the memory function, developed bionanohybrid materials should demonstrate the two distinctive bistable states, which can be defined as '1' and '0' states for memory. As shown in results, those bionanohybrid materials showed apparently distinguished bistable states by electrochemical or electrical investigation. Next, we examined the studies about biologic gates based on the DNA–metal ion hybrids, protein–DNA connected reaction, and protein–organic/inorganic nanomaterial hybrids. Using these bionanohybrid material, various logic functions including the AND, NAND, Fredkin, or Feynman logic gates were demonstrated. Furthermore, the human analog decision-mimicking device was developed. In addition, we discussed about bioprocessors capable of processing of the inputted signals to the output signals such as finding of optimal routes and processing of different electrochemical signals through the DNA reactions and metalloprotein, DNA and inorganic nanomaterial hybrids. In addition to the results discussed in this review, many research groups have studied to develop the delicate functional bionanohybrid materials to apply for biomemory, biologic gates, and bioprocessors. The bioelectronic devices comprised with bionanohybrid materials would be a milestone for biomolecular-computation systems in the near future. Moreover, this will provide a useful way of bioelectronic devices to apply in development of wearable devices [83,84], biohybrid robots [85–87], and bioelectronic medicine [88,89].

Author Contributions: The manuscript was written through contributions of all authors. J.Y. reviewed and wrote the biologic gates and bioprocessors. T.L. reviewed and wrote the biomemory. J.-W.C. directed entire manuscript and contributed to this work as the corresponding author. All authors read and approved the submitted manuscript.

Funding: This research was supported by the Leading Foreign Research Institute Recruitment Program, through the National Research Foundation of Korea (NRF), funded by the Ministry of Science, ICT and Future Planning (MSIP) (2013K1A4A3055268), and the Basic Science Research Program through the National Research Foundation of Korea (NRF) funded by the Ministry of Education(2016R1A6A1A03012845).

Conflicts of Interest: The authors declare no conflict of interest.

References

1. Willner, I.; Willner, B. Biomaterials integrated with electronic elements: En route to bioelectronics. *Trends Biotechnol.* **2001**, *19*, 222–230. [CrossRef]
2. Goode, J.A.; Rushworth, J.V.H.; Millner, P.A. Biosensor Regeneration: A Review of Common Techniques and Outcomes. *Langmuir* **2015**, *31*, 6267–6276. [CrossRef]

3. Lee, T.; Lee, Y.; Park, S.Y.; Hong, K.; Kim, Y.; Park, C.; Chung, Y.-H.; Lee, M.-H.; Min, J. Fabrication of Electrochemical Biosensor Composed of Multi-functional DNA Structure/Au Nanospike on Micro-gap/PCB System for Detecting Troponin I in Human Serum. *Colloids Surf. B* **2019**, *175*, 343–350. [CrossRef] [PubMed]

4. Lee, T.; Park, S.Y.; Jang, H.; Kim, G.H.; Lee, Y.; Park, C.; Mohammadniaei, M.; Lee, M.-H.; Min, J. Fabrication of Electrochemical Biosensor consisted of Multi-functional DNA Structure/porous Au Nanoparticle for Avian Influenza Virus (H5N1) in Human Serum. *Mater. Sci. Eng. C* **2019**, *99*, 511–519. [CrossRef] [PubMed]

5. Brown, M.A.; Barker, L.; Semprini, L.; Minot, E.D. Graphene Biotransistor Interfaced with a Nitrifying Biofilm. *Environ. Sci. Technol. Lett.* **2015**, *2*, 118–122. [CrossRef]

6. Gorton, L.; Lindgren, A.; Larsson, T.; Munteanu, F.D.; Ruzgas, T.; Gazaryan, I. Direct electron transfer between heme-containing enzymes and electrodes as basis for third generation biosensors. *Anal. Chim. Acta* **1999**, *400*, 91–108. [CrossRef]

7. Akkilic, N.; Kamran, M.; Stan, R.; Sanghamitra, N.J.M. Voltage-controlled fluorescence switching of a single redox protein. *Biosens. Bioelectron.* **2015**, *67*, 747–751. [CrossRef]

8. Macchia, E.; Alberga, D.; Manoli, K.; Mangiatordi, G.F.; Magliulo, M.; Palazzo, G.; Giordano, F.; Lattanzi, G.; Torsi, L. Organic bioelectronics probing conformational changes in surface confined proteins. *Sci. Rep.* **2016**, *6*, 28085. [CrossRef]

9. Berggren, M.; Richter-Dahlfors, A. Organic Bioelectronics. *Adv. Mater.* **2007**, *19*, 3201–3213. [CrossRef]

10. Rivnay, J.; Owens, R.M.; Malliaras, G.G. The Rise of Organic Bioelectronics. *Chem. Mater.* **2014**, *26*, 679–685. [CrossRef]

11. Zhang, A.; Lieber, C.M. Nano-Bioelectronics. *Chem. Rev.* **2016**, *116*, 215–257. [CrossRef]

12. Bostick, C.D.; Mukhopadhyay, S.; Pecht, I. Protein bioelectronics: A review of what we do and do not know. *Rep. Prog. Phys.* **2018**, *81*, 026601. [CrossRef]

13. Kim, M.J.; Yang, M.-S.; Kwon, H.T.; Song, J.M. Low noise bipolar photodiode array protein chip based on on-chip bioassay for the detection of E. coli O157:H7. *Biomed. Microdevices* **2007**, *9*, 565–572. [CrossRef]

14. Vistas, C.R.; Soares, S.S.; Rodrigues, R.M.M.; Chu, V.; Conde, J.P.; Ferreira, G.N.M. An amorphous silicon photodiode microfluidic chip to detect nanomolar quantities of HIV-1 virion infectivity factor. *Analyst* **2014**, *139*, 3709–3713. [CrossRef]

15. Ingebrandt, S. Bioelectronics: Sensing beyond the limit. *Nat. Nanotechnol.* **2015**, *10*, 734–735. [CrossRef]

16. Thudi, L.; Jasti, L.S.; Swarnalatha, Y.; Fadnavis, N.W.; Mulani, K.; Deokar, S.; Ponrathnam, S. Adsorption induced enzyme denaturation: The role of protein surface in adsorption induced protein denaturation on allyl glycidyl ether (AGE)-ethylene glycol dimethacrylate (EGDM) copolymers. *Colloids Surf. B* **2012**, *90*, 184–190. [CrossRef]

17. Lee, T.; Yoo, S.-Y.; Chung, Y.-H.; Min, J.; Choi, J.-W. Signal Enhancement of Electrochemical Biomemory Device Composed of Recombinant Azurin/Gold Nanoparticle. *Electroanalysis* **2011**, *23*, 2023–2029. [CrossRef]

18. Yoon, J.; Lee, T.; Bapurao, G.B.; Jo, J.; Oh, B.-K.; Choi, J.-W. Electrochemical H_2O_2 biosensor composed of myoglobin on MoS_2 nanoparticle-graphene oxide hybrid structure. *Biosens. Bioelectron.* **2017**, *93*, 14–20. [CrossRef]

19. Zebda, A.; Gondran, C.; Goff, A.L.; Holzinger, M.; Cinquin, P.; Cosnier, S. Mediatorless high-power glucose biofuel cells based on compressed carbon nanotube-enzyme electrodes. *Nat. Commun.* **2011**, *2*, 370. [CrossRef]

20. Zuo, X.; He, S.; Li, D.; Peng, C.; Huang, Q.; Song, S.; Fan, C. Graphene Oxide-Facilitated Electron Transfer of Metalloproteins at Electrode Surfaces. *Langmuir* **2010**, *26*, 1936–1939. [CrossRef]

21. Kwak, S.K.; Lee, G.S.; Ahn, D.J.; Choi, J.-W. Pattern formation of cytochrome c by microcontact printing and dip-pen nanolithography. *Mater. Sci. Eng. C* **2004**, *24*, 151–155. [CrossRef]

22. Wu, X.; Xiao, T.; Luo, Z.; He, R.; Cao, Y.; Guo, Z.; Zhang, W.; Chen, Y. A micro-/nano-chip and quantum dots-based 3D cytosensor for quantitative analysis of circulating tumor cells. *J. Nanobiotechnol.* **2018**, *16*, 65. [CrossRef]

23. Mahyad, B.; Janfaza, S.; Hosseini, E.S. Bio-nano hybrid materials based on bacteriorhodopsin: Potential applications and future strategies. *Adv. Colloid Interface Sci.* **2015**, *225*, 194–202. [CrossRef]

24. Liu, Y.; Turner, A.P.F.; Zhao, M.; Mak, W.C. Processable enzyme-hybrid conductive polymer composites for electrochemical biosensing. *Biosens. Bioelectron.* **2018**, *100*, 374–381. [CrossRef]

25. Eguílaz, M.; Villalonga, R.; Rivas, G. Electrochemical biointerfaces based on carbon nanotubes-mesoporous silica hybrid material: Bioelectrocatalysis of hemoglobin and biosensing applications. *Biosens. Bioelectron.* **2018**, *111*, 144–151. [CrossRef]

26. Lv, Z.; Wang, Y.; Chen, Z.; Sun, L.; Wang, J.; Chen, M.; Xu, Z.; Liao, Q.; Zhou, L.; Chen, X.; et al. Phototunable Biomemory Based on Light-Mediated Charge Trap. *Adv. Sci.* **2018**, *5*, 1800714. [CrossRef]

27. Kramer, B.P.; Fischer, C.; Fussenegger, M. BioLogic gates enable logical transcription control in mammalian cells. *Biotechnol. Bioeng.* **2004**, *87*, 478–484. [CrossRef]

28. Choi, J.-W.; Oh, B.-K.; Kim, Y.J. Protein-based biomemory device consisting of the cysteine-modified azurin. *Appl. Phys. Lett.* **2007**, *91*, 263902. [CrossRef]

29. Güzel, R.; Ersöz, A.; Dolak, I.; Say, R. Multistate proteinous biomemory device based on redox controllable hapten cross-linker. *Mater. Sci. Eng. C* **2017**, *79*, 336–342. [CrossRef]

30. Min, J.; Kim, S.-U.; Kim, Y.J.; Yea, C.-H.; Choi, J.-W. Fabrication of Recombinant Azurin Self-assembled Layer for the Application of Bioelectronic Device. *J. Nanosci. Nanotechnol.* **2008**, *8*, 4982–4987. [CrossRef]

31. Lee, T.; Yagati, A.K.; Min, J.; Choi, J.-W. Bioprocessing Device Composed of Protein/DNA/Inorganic Material Hybrid. *Adv. Funct. Mater.* **2014**, *24*, 1781–1789. [CrossRef]

32. Kang, P.; Wang, M.C.; Nam, S.W. Bioelectronics with two-dimensional materials. *Microelectron. Eng.* **2016**, *161*, 18–35. [CrossRef]

33. Zhang, T.; Liu, J.; Wang, C.; Leng, X.; Xiao, Y.; Fu, L. Synthesis of graphene and related two-dimensional materials for bioelectronics devices. *Biosens. Bioelectron.* **2017**, *89*, 28–42. [CrossRef]

34. Choi, C.; Lee, Y.; Cho, K.W.; Koo, J.H.; Kim, D.-H. Wearable and Implantable Soft Bioelectronics Using Two-Dimensional Materials. *Acc. Chem. Res.* **2019**, *52*, 73–81. [CrossRef]

35. Dunn, K.E.; Trefzer, M.A.; Johnson, S.; Tyrrell, A.M. Towards a Bioelectronic Computer: A Theoretical Study of a Multi-Layer Biomolecular Computing System That Can Process Electronic Inputs. *Int. J. Mol. Sci.* **2018**, *19*, 2620. [CrossRef]

36. Katz, E. Biocomputing- tools, aims, perspectives. *Curr. Opin. Biotechnol.* **2015**, *34*, 202–208. [CrossRef]

37. Choi, J.-W.; Kim, J.S.; Kim, S.-U.; Min, J. Charge Storage in Redox-active Azurin Monolayer on 11-MUA Modified Gold Surface. *Biochip J.* **2009**, *3*, 157–163.

38. Güzel, R.; Ersöz, A.; Ziyadanoğulları, R.; Say, R. Nano-hemoglobin film based sextet state biomemory device by cross-linked photosensitive hapten monomer. *Talanta* **2018**, *176*, 85–91. [CrossRef]

39. Hwang, H.J.; Carey, J.R.; Brower, E.T.; Gengenbach, A.J.; Abramite, J.A.; Lu, Y. Blue Ferrocenium Azurin: An Organometalloprotein with Tunable Redox Properties. *J. Am. Chem. Soc.* **2005**, *127*, 15356–15357. [CrossRef]

40. Liu, J.; Chakraborty, S.; Hosseinzadeh, P.; Yu, Y.; Tian, S.; Petrik, I.; Bhagi, A.; Lu, Y. Metalloproteins Containing Cytochrome, Iron–Sulfur, or Copper Redox Centers. *Chem. Rev.* **2014**, *114*, 4366–4469. [CrossRef]

41. Suárez, G.; Santschi, C.; Martin, O.J.F.; Slaveykova, V.I. Biosensor based on chemically-designed anchorable cytochrome c for the detection of H_2O_2 released by aquaticcells. *Biosens. Bioelectron.* **2013**, *42*, 385–390. [CrossRef]

42. Artés, J.M.; Díez-Pérez, I.; Gorostiza, P. Transistor-like Behavior of Single Metalloprotein Junctions. *Nano Lett.* **2012**, *12*, 2679–2684. [CrossRef]

43. Lee, T.; Kim, S.-U.; Min, J.; Choi, J.-W. Multilevel Biomemory Device Consisting of Recombinant Azurin/Cytochrome c. *Adv. Mater.* **2010**, *22*, 510–514. [CrossRef]

44. Balabushevich, N.G.; Sholina, E.A.; Mikhalchik, E.V.; Filatova, L.Y.; Vikulina, A.S.; Volodkin, D. Self-Assembled Mucin-Containing Microcarriers via Hard Templating on $CaCO_3$ Crystals. *Micromachines* **2018**, *9*, 307. [CrossRef]

45. Norris, K.; Mishukova, O.I.; Zykwinska, A.; Colliec-Jouault, S.; Sinquin, C.; Koptioug, A.; Cuenot, S.; Kerns, J.G.; Surmeneva, M.A.; Surmenev, R.A.; Douglas, T.E.L. Marine Polysaccharide-Collagen Coatings on Ti6Al4V Alloy Formed by Self-Assembly. *Micromachines* **2019**, *10*, 68. [CrossRef]

46. Jensen, P.S.; Chi, Q.; Zhang, J.; Ulstrup, J. Long-Range Interfacial Electrochemical Electron Transfer of Pseudomonas aeruginosa Azurin−Gold Nanoparticle Hybrid Systems. *J. Phys. Chem. C* **2009**, *113*, 13993–14000. [CrossRef]

47. Cho, B.; Song, S.; Ji, Y.; Kim, T.-W.; Lee, T. Organic Resistive Memory Devices: Performance Enhancement, Integration, and Advanced Architectures. *Adv. Funct. Mater.* **2011**, *21*, 2806–2829. [CrossRef]

48. Li, L.; Li, G. High-Performance Resistance-Switchable Multilayers of Graphene Oxide Blended with 1,3,4-Oxadiazole Acceptor Nanocomposite. *Micromachines* **2019**, *10*, 140. [CrossRef]

49. Li, L. Tunable Memristic Characteristics Based on Graphene Oxide Charge-Trap Memory. *Micromachines* **2019**, *10*, 151. [CrossRef]

50. Lee, T.; Yagati, A.K.; Pi, F.; Sharma, A.; Choi, J.-W.; Guo, P. Construction of RNA-Quantum Dot Chimera for Nanoscale Resistive Biomemory Application. *ACS Nano* **2015**, *9*, 6675–6682. [CrossRef]

51. Shu, D.; Shu, Y.; Haque, F.; Abdelmawla, S.; Guo, P. Thermodynamically Stable RNA three-way junctions as platform for constructing multi-functional nanoparticles for delivery of therapeutics. *Nat. Nanotechnol.* **2011**, *6*, 658–667. [CrossRef]

52. Yoon, J.; Mohammadniaei, M.; Choi, H.K.; Shin, M.; Bharate, B.G.; Lee, T.; Choi, J.-W. Resistive switching biodevice composed of MoS_2-DNA heterolayer on the gold electrode. *Appl. Surf. Sci.* **2019**, *478*, 134–141. [CrossRef]

53. Lembke, D.; Bertolazzi, S.; Kis, A. Single-Layer MoS_2 Electronics. *Acc. Chem. Res.* **2015**, *48*, 100–110. [CrossRef]

54. Barua, S.; Dutta, H.S.; Gogoi, S.; Devi, R.; Khan, R. Nanostructured MoS2-Based Advanced Biosensors: A Review. *ACS Appl. Nano Mater.* **2018**, *1*, 2–25. [CrossRef]

55. Andersson, M.; Sinks, L.E.; Hayes, R.T.; Zhao, Y.; Wasielewski, M.R. Bio-Inspired Optically Controlled Ultrafast Molecular AND Gate. *Angew. Chem. Int. Ed.* **2003**, *42*, 3139–3143. [CrossRef]

56. Sivan, S.; Tuchman, S.; Lotan, N. A biochemical logic gate using an enzyme and its inhibitor. Part II: The logic gate. *BioSystems* **2003**, *70*, 21–33. [CrossRef]

57. Dung, N.Q.; Patil, D.; Duong, T.-T.; Jung, H.; Kim, D.; Yoon, S.-G. An amperometric glucose biosensor based on a GOx-entrapped TiO_2–SWCNT composite. *Sens. Actuator B-Chem.* **2012**, *166–167*, 103–109. [CrossRef]

58. Chen, Q.; Yoo, S.-Y.; Chung, Y.-H.; Lee, J.-Y.; Min, J.; Choi, J.-W. Control of electrochemical signals from quantum dots conjugated to organic materials by using DNA structure in an analog logic gate. *Bioelectrochemistry* **2016**, *111*, 1–6. [CrossRef]

59. Chung, Y.-H.; Lee, T.; Yoo, S.-Y.; Min, J.; Choi, J.-W. Electrochemical Bioelectronic Device Consisting of Metalloprotein for Analog Decision Making. *Sci. Rep.* **2015**, *5*, 14501. [CrossRef]

60. Gurunathan, S.; Klinman, D.M.; Seder, R.A. DNA Vaccines: Immunology, Application, and Optimization. *Annu. Rev. Immunol.* **2000**, *18*, 927–974. [CrossRef]

61. França, L.T.; Carrilho, E.; Kist, T.B. A review of DNA sequencing techniques. *Q. Rev. Biophys.* **2002**, *35*, 169–200. [CrossRef]

62. Pei, H.; Lu, N.; Wen, Y.; Song, S.; Liu, Y.; Yan, H.; Fan, C. A DNA Nanostructure-based Biomolecular Probe Carrier Platform for Electrochemical Biosensing. *Adv. Mater.* **2010**, *22*, 4754–4758. [CrossRef]

63. Ono, A.; Cao, S.; Togashi, H.; Tashiro, M.; Fujimoto, T.; Machinami, T.; Oda, S.; Miyake, Y.; Okamotoa, I.; Tanaka, Y. Specific interactions between silver(I) ions and cytosine–cytosine pairs in DNA duplexes. *Chem. Commun.* **2008**, *39*, 4825–4827. [CrossRef]

64. Zhang, Y.M.; Zhang, L.; Liang, R.P.; Qiu, J.D. DNA Electronic: Logic Gates Based on Metal-Ion-Dependent Induction of Oligonucleotide Structural Motifs. *Chem. Eur. J.* **2013**, *19*, 6961–6965. [CrossRef]

65. Guz, N.; Fedotova, T.A.; Fratto, B.E.; Schlesinger, O.; Alfonta, L.; Kolpashchikov, D.M.; Katz, E. Bioelectronic Interface Connecting Reversible Logic Gates Based on Enzyme and DNA Reactions. *ChemPhysChem* **2016**, *17*, 2247–2255. [CrossRef]

66. Katz, E.; Minko, S. Enzyme-based logic systems interfaced with signal-responsive materials and electrodes. *Chem. Commun.* **2015**, *51*, 3493–3500. [CrossRef]

67. Katz, E.; Poghossian, A.; Schöning, M.J. Enzyme-based logic gates and circuits-analytical applications and interfacing with electronics. *Anal. Bioanal. Chem.* **2017**, *409*, 81–94. [CrossRef]

68. Muramatsu, S.; Kinbara, K.; Taguchi, H.; Ishii, N.; Aida, T. Semibiological Molecular Machine with an Implemented "AND" Logic Gate for Regulation of Protein Folding. *J. Am. Chem. Soc.* **2006**, *128*, 3764–3769. [CrossRef]

69. Poghossian, A.; Katz, E.; Schöning, M.J. Enzyme logic AND-Reset and OR-Reset gates based on a field-effect electronic transducer modified with multi-enzyme membrane. *Chem. Commun.* **2015**, *51*, 6564–6567. [CrossRef]

70. Fredkin, E.; Toffoli, T. Conservative Logic. In *Collision-Based Computing*; Adamatzky, A., Ed.; Springer: London, UK, 2002; pp. 47–81.

71. O'Brien, J.L.; Pryde, G.J.; White, A.G.; Ralph, T.C.; Branning, D. Demonstration of an all-optical quantum controlled-NOT gate. *Nature* **2003**, *426*, 264–267. [CrossRef]

72. Sun, D.-M.; Timmermans, M.Y.; Kaskela, A.; Nasibulin, A.G.; Kishimoto, S.; Mizutani, T.; Kauppinen, E.I.; Ohno, Y. Mouldable all-carbon integrated circuits. *Nat. Commun.* **2013**, *4*, 2302. [CrossRef] [PubMed]

73. Kelley, T.W.; Baude, P.F.; Gerlach, C.; Ender, D.E.; Muyres, D.; Haase, M.A.; Vogel, D.E.; Theiss, S.D. Recent Progress in Organic Electronics: Materials, Devices, and Processes. *Chem. Mater.* **2004**, *16*, 4413–4422. [CrossRef]

74. Zou, C.; Wei, C.; Zhang, Q.; Liu, C.; Zhou, C.; Liu, Y. Four-Analog Computation Based on DNA Strand Displacement. *ACS Omega* **2017**, *2*, 4143–4160. [CrossRef]

75. Son, M.; Park, T.H. The bioelectronic nose and tongue using olfactory and taste receptors: Analytical tools for food quality and safety assessment. *Biotechnol. Adv.* **2018**, *36*, 371–379. [CrossRef]

76. Lim, J.-H.; Park, J.; Ahn, J.H.; Jin, H.J.; Hong, S.; Park, T.H. A peptide receptor-based bioelectronic nose for the real-time determination of seafood quality. *Biosens. Bioelectron.* **2013**, *39*, 244–249. [CrossRef] [PubMed]

77. Szaciłowski, K. Digital Information Processing in Molecular Systems. *Chem. Rev.* **2008**, *108*, 3481–3548. [CrossRef] [PubMed]

78. Toriello, N.M.; Douglas, E.S.; Thaitrong, N.; Hsiao, S.C.; Francis, M.B.; Bertozzi, C.R.; Mathies, R.A. Integrated microfluidic bioprocessor for single-cell gene expression analysis. *Proc. Natl. Acad. Sci. USA* **2008**, *105*, 20173–20178. [CrossRef]

79. Evans, A.C.; Thadani, N.N.; Suh, J. Biocomputing nanoplatforms as therapeutics and diagnostics. *J. Control. Release* **2016**, *240*, 387–393. [CrossRef]

80. Yurke, B.; Turberfield, A.J.; Mills, A.P., Jr.; Simmel, F.C.; Neumann, J.L. A DNA-fuelled molecular machine made of DNA. *Nature* **2000**, *406*, 605–608. [CrossRef]

81. MacConnell, A.B.; Price, A.K.; Paegel, B.M. An Integrated Microfluidic Processor for DNA-Encoded Combinatorial Library Functional Screening. *ACS Comb. Sci.* **2017**, *19*, 181–192. [CrossRef]

82. Shu, J.-J.; Wang, Q.-W.; Yong, K.-Y.; Shao, F.; Lee, K.J. Programmable DNA-Mediated Multitasking Processor. *J. Phys. Chem. B* **2015**, *119*, 5639–5644. [CrossRef]

83. Kim, J.; Jeerapan, I.; Sempionatto, J.R.; Barfidokht, A.; Mishra, R.K.; Campbell, A.S.; Hubble, L.J.; Wang, J. Wearable Bioelectronics: Enzyme-Based Body-Worn Electronic Devices. *Acc. Chem. Res.* **2018**, *51*, 2820–2828. [CrossRef]

84. Lopes, P.A.; Paisana, H.; De Almeida, A.T.; Majidi, C.; Tavakoli, M. Hydroprinted Electronics: Ultrathin Stretchable Ag-In-Ga E-Skin for Bioelectronics and Human-Machine Interaction. *ACS Appl. Mater. Interfaces* **2018**, *10*, 38760–38768. [CrossRef]

85. Nawroth, J.C.; Lee, H.; Feinberg, A.W.; Ripplinger, C.M.; McCain, M.L.; Grosberg, A.; Dabiri, J.O.; Parker, K.K. A tissue-engineered jellyfish with biomimetic propulsion. *Nat. Biotechnol.* **2012**, *30*, 792–797. [CrossRef]

86. Williams, B.W.; Anand, S.V.; Rajagopalan, J.; Saif, M.; Taher, A. A self-propelled biohybrid swimmer at low Reynolds number. *Nat. Commun.* **2014**, *5*, 3081. [CrossRef]

87. Shin, S.R.; Zihlmann, C.; Akbari, M.; Assawes, P.; Cheung, L.; Zhang, K.; Manoharan, V.; Zhang, Y.S.; Yüksekkaya, M.; Wan, K.-T.; Nikkhah, M.; Dokmeci, M.R.; Tang, X.S.; Khademhosseini, A. Reduced Graphene Oxide-GelMA Hybrid Hydrogels as Scaffolds for Cardiac Tissue Engineering. *Small* **2016**, *12*, 27. [CrossRef]

88. Olofsson, P.S.; Tracey, K.J. Bioelectronic medicine: Technology targeting molecular mechanisms for therapy. *J. Intern. Med.* **2017**, *282*, 3–4. [CrossRef]

89. Löffler, S.; Melican, K.; Nilsson, K.P.R.; Richter-Dahlfors, A. Organic bioelectronics in medicine. *J. Intern. Med.* **2017**, *282*, 24–36. [CrossRef]

 *micromachines*

Review

Recent Progress in the Voltage-Controlled Magnetic Anisotropy Effect and the Challenges Faced in Developing Voltage-Torque MRAM

Takayuki Nozaki [1,*], Tatsuya Yamamoto [1], Shinji Miwa [2,3], Masahito Tsujikawa [4], Masafumi Shirai [4], Shinji Yuasa [1] and Yoshishige Suzuki [1,3]

[1] National Institute of Advanced Industrial Science and Technology (AIST), Spintronics Research Center, Tsukuba, Ibaraki 305-8568, Japan; yamamoto-t@aist.go.jp (T.Y.); yuasa-s@aist.go.jp (S.Y.); suzuki-y@mp.es.osaka-u.ac.jp (Y.S.)
[2] The Institute of Solid State Physics, The University of Tokyo, Kashiwa, Chiba 277-8531, Japan; miwa@issp.u-tokyo.ac.jp
[3] Graduate School of Engineering Science, Osaka University, Toyonaka, Osaka 560-8531, Japan
[4] Research Institute of Electrical Communication, Tohoku University, Sendai, Miyagi 980-8577, Japan; t-masa@riec.tohoku.ac.jp (M.T.); shirai@riec.tohoku.ac.jp (M.S.)
* Correspondence: nozaki-t@aist.go.jp

Received: 24 April 2019; Accepted: 12 May 2019; Published: 15 May 2019

Abstract: The electron spin degree of freedom can provide the functionality of "nonvolatility" in electronic devices. For example, magnetoresistive random access memory (MRAM) is expected as an ideal nonvolatile working memory, with high speed response, high write endurance, and good compatibility with complementary metal-oxide-semiconductor (CMOS) technologies. However, a challenging technical issue is to reduce the operating power. With the present technology, an electrical current is required to control the direction and dynamics of the spin. This consumes high energy when compared with electric-field controlled devices, such as those that are used in the semiconductor industry. A novel approach to overcome this problem is to use the voltage-controlled magnetic anisotropy (VCMA) effect, which draws attention to the development of a new type of MRAM that is controlled by voltage (voltage-torque MRAM). This paper reviews recent progress in experimental demonstrations of the VCMA effect. First, we present an overview of the early experimental observations of the VCMA effect in all-solid state devices, and follow this with an introduction of the concept of the voltage-induced dynamic switching technique. Subsequently, we describe recent progress in understanding of physical origin of the VCMA effect. Finally, new materials research to realize a highly-efficient VCMA effect and the verification of reliable voltage-induced dynamic switching with a low write error rate are introduced, followed by a discussion of the technical challenges that will be encountered in the future development of voltage-torque MRAM.

Keywords: voltage-controlled magnetic anisotropy; magnetoresistive random access memory; magnetic tunnel junction

---

1. Introduction

The evolving information society has triggered the rapid spread of advanced technologies, such as Artificial Intelligence (AI), Advanced Safety Vehicle (ASV), and IoT (Internet of Things), and this has led to further industrial innovation. In the society of the future, Big-Data collected from physical space will be stored and analyzed in cyber space, which creates new social values. Such a data-driven society can only be sustained by the high-speed processing of Big-Data; therefore, reducing the power consumption of nano-electronic devices is becoming increasingly crucial. One promising approach is the introduction of nonvolatile computation.

It is expected that the stand-by power of future computing systems will be reduced by utilizing the nonvolatile features of spintronic devices, such as a magnetoresistive random-access memory (MRAM) while using magnetic tunnel junctions (MTJ). An MTJ consists of two ferromagnetic layers that are separated by an ultrathin insulating layer, such as magnesium oxide (MgO) [1,2]. Electrons can tunnel through the barrier when a bias voltage is applied between the two ferromagnetic layers due to the ultrathin thickness of the insulating layer. The amplitude of the tunneling current depends on the relative angle between the magnetizations in each ferromagnetic layer through a spin-dependent tunneling process, which is called the tunnel magnetoresistance (TMR) effect. The direction of the magnetizations of one of the ferromagnetic layers is fixed (reference layer), typically by exchange coupling with an antiferromagnetic material. An external field (free layer), using an electric-current, can control the direction in the other, as discussed below. In this way information is written to the memory device. Then, the information can be stored by controlling the magnetization configuration between parallel and anti-parallel states, exhibiting two resistance states, in a nonvolatile manner.

MRAM has great potential to be a fast, high write endurance, and CMOS-compatible nonvolatile memory, which is suitable for embedded as well as standalone memory applications. However, one of the significant remaining challenges is to reduce the energy that is needed to write information, that is, to switch the magnetization. In the long history of magnetism, magnetic fields that are produced by electric-current have been used for magnetization reversal. This indirect approach is inefficient and not scalable. Spintronics has brought us a new way of switching the magnetization through the *s-d* exchange interaction between the conduction electron spin and localized spin, called the spin-transfer torque (STT) effect [3–8]. The spin angular momentum that is carried by conduction electrons can be transferred to localized electrons and can induce magnetization reversal. Recently, an alternative technique for magnetization switching using the spin Hall effect, which is called the spin-orbit torque (SOT) switching [9–12], has also been attracting attention. A typical SOT device comprises a bilayer that consists of a non-magnetic heavy metal layer, such as Ta or W, and a ferromagnetic layer capped by an oxide. A transverse pure-spin current is generated when an in-plane electric-current is injected into the bilayer due to the spin Hall effect. The accumulation of spin at the heavy metal/ferromagnet interface exerts a torque and induces magnetization switching. In this switching scheme, high write endurance can be realized, even with high speed switching of the order of a few nanoseconds, because the read and write passes are separate.

STT-based switching (STT-MRAM) has brought a drastic reduction in writing energy and expanded potential for applications; STT-MRAM [13–15]. Figure 1 summarizes the reported writing energies for a MRAM (red dots) and STT-MRAM (blue dots) as a function of the MTJ cell size. For example, recent developments in STT-MRAMs have achieved writing energies of approximately 100 fJ/bit in perpendicularly magnetized MTJs [13], which is close to the writing energy for a dynamic-RAM (DRAM). However, it is still much higher than that of a static-RAM (SRAM), which is made up of several MOSFETs that an electric-field operates. Furthermore, a writing energy of 100 fJ/bit corresponds to $10^7 \, k_B T$ (k_B is the Boltzmann constant and T is the temperature, assumed to be 300 K). On the other hand, the energy that is required to maintain magnetic information, i.e. the thermal stability, is about $60 \, k_B T$ (green line in Figure 1), which means that we have a large energy gap between data writing and retention, in the order of 10^5. This difference mainly comes from unwanted energy consumption due to ohmic dissipation of the electric-current flow. Overcoming this fundamental issue using a novel way of electric-field based spin manipulation is strongly desired. Not only for MRAMs, but all of the nonvolatile memories that have been proposed so far have a dilemma of choosing between stable nonvolatility and high operating energy. Therefore, the development of a novel type of memory having low operating energy as well as low stand-by energy can have great impact on the design of future memory hierarchy.

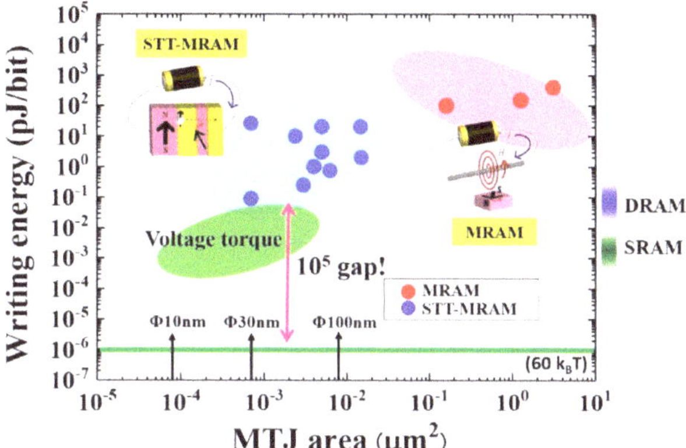

Figure 1. Reported writing energy for toggle magnetoresistive random-access memory (MRAM) (red dots) and spin-transfer torque-based switching (STT-MRAM) (blue dots) as a function of magnetic tunnel junctions (MTJ) cell size and the target area for voltage-torque MRAM.

Various kinds of approaches to the electric-field manipulation of spin have been proposed and experimentally demonstrated, such as using the inverse magnetostriction effect in a multilayered stack with piezoelectric materials [16–18], the gate-controlled Curie temperature in ferromagnetic semiconductors [19–21] or even in an ultrathin ferromagnetic metal layer [22], magnetoelectric switching of exchange bias [23–26], electric polarization induced control in magnetic anisotropy at the ferromagnetic/ferroelectric interface [27,28], electric-field induced magnetic phase transition through structural phase transition [29], and the utilization of multiferroic materials [30,31]. However, each of these approaches have the drawbacks of limited operation temperature or low write endurance or difficulty in the introduction to magnetoresistive devices, although these requirements should be simultaneously satisfied for memory applications. We have focused on the voltage-controlled magnetic anisotropy (VCMA) effect in an ultrathin ferromagnetic layer [32,33] to overcome this problem.

This paper reviews recent progress in the research of the VCMA effect and the challenges that are faced in developing new types of MRAM controlled by voltage, called voltage-torque MRAM (also called Magnetoelectric (ME)-RAM) [34–39]. Section 2 presents an overview of the early experimental observations of the VCMA effect in all solid state devices and the concept of voltage-induced dynamic switching, with a discussion of the technical challenges. In Section 3, the current understanding of the physical origin of the VCMA effect is discussed through experimental investigations while using X-ray absorption spectroscopy (XAS) and magnetic circular dichroism (XMCD) analyses with first-principles calculation. Section 4 presents the materials research being done to enhance the VCMA effect, especially focusing on the heavy metal doping technique. Finally, in Section 5, experimental demonstrations of reliable voltage-induced dynamic switching and an understanding of the voltage-induced spin dynamics are discussed, together with a discussion on the theoretical investigations being made.

2. Overview of the VCMA Effect and Voltage-Induced Dynamic Switching

Weisheit et al. first reported the VCMA effect in a *3d* transition ferromagnetic layer in 2007 [32]. They observed a coercivity change of a few % in 2–4 nm-thick FePt(Pd) films immersed in a liquid electrolyte. Opposing trends in the change in coercivity in FePt and FePd, depending on the applied voltage, were observed. An electric double layer is effective for applying a large electric-field at the interface; however, the operating speed is limited and we need to take care of the influence of chemical reactions. The voltage control of in-plane magnetic anisotropy was also found in ferromagnetic semiconductors at low temperature [40]. Theoretical attempts to understand the physical origin of the VCMA effect in metal started around the same time. Duan et al. proposed that spin-dependent screening of the electric-field can induce modification in the surface magnetization and magnetic anisotropy [41]. Nakamura et al. calculated the VCMA effect in a freestanding Fe(001) monolayer and pointed out that electric-field induced changes in the band structure, especially the *p* orbitals near the Fermi level, which are coupled to the *d* states, play an important role [42]. Tsujikawa et al. studied the VCMA effect in a Pt/Fe/Pt/vacuum system and found that relative modification in the electron filling of the 3*d* orbital induced by the accumulated charges at the interface causes a change in the perpendicular magnetic anisotropy (PMA) [43]. Other possible mechanisms have also been discussed, such as electric-field induced modification in Rashba spin-orbit anisotropy [44,45] and electric-field induced atomic displacement at the interface between ferromagnetic oxide and dielectric layers [46].

We attempted to apply the VCMA effect in an all solid state structure, which consisted of epitaxial Au/ultrathin Fe(Co)/MgO/polyimide/ITO junctions grown on MgO(001) substrates (see Figure 2a) to investigate the feasibility for practical applications [33,47]. Figure 2b shows an example of polar magneto-optical Kerr effect (MOKE) hysteresis curves that were measured under the application of a voltage. The thickness of the $Fe_{80}Co_{20}$ layer is fixed at 0.58 nm. The bias direction is defined with respect to the top ITO electrode. A clear change in the saturation field in the out-of-plane direction can be seen, which suggests a modification in the PMA. Under the application of a positive bias, the PMA is suppressed and the in-plane anisotropy becomes more stable. On the other hand, the application of a negative voltage enhances the PMA and even the transition of the magnetic easy axis can be realized from the in-plane to the out-of-plane direction.

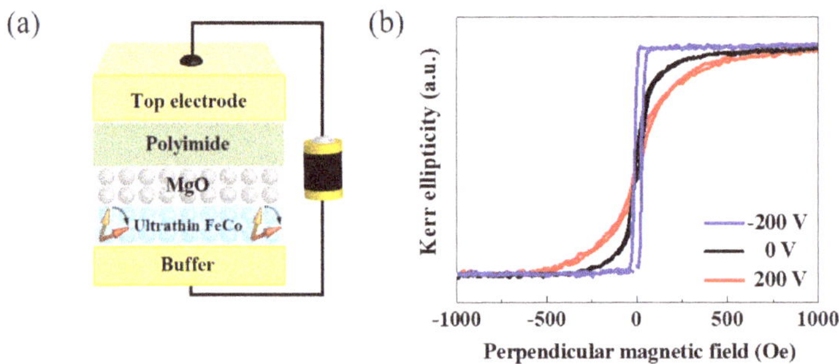

Figure 2. (a) Schematic illustration of sample stack used for the first demonstration of the voltage-controlled magnetic anisotropy (VCMA) effect in an all-solid state structure, and (**b**) applied bias voltage dependence of the polar-magneto-optical Kerr effect (MOKE) hysteresis curves for a 0.58 nm-thick $Fe_{80}Co_{20}$ layer.

Due to screening by free electrons, the penetration of the electric-field into a metal is limited to the surface, unlike in the case of a semiconductor; however, if the thickness of the ferromagnetic layer is thin enough, e.g. several monoatomic layers, the modulation in the electronic structure at the interface can make a sizable impact on the magnetic properties. Details of an experimental verification for the physical origin of the VCMA effect are discussed in Section 2.

One great advantage of the VCMA effect is its high applicability in a MTJ structure, which is the most important practical devices in spintronics. Figure 3 exhibits the first demonstration of the VCMA effect that was observed in a MTJ structure, which consisted of Cr/ Au/ultrathin $Fe_{80}Co_{20}(0.5 \text{ nm})/MgO(t_{MgO})/Fe$ grown on a MgO(001) substrate [48]. Here, we made electrical ferromagnetic resonance (FMR) measurements through the TMR effect. The PMA energy, K_{PMA}, was evaluated from the resonant frequency of the free layer at each applied voltage. In addition to FMR measurements, the effect of a bias voltage on normalized TMR curves has also often been used for the quantitative evaluation of the VCMA effect, as discussed later [49]. Generally, the PMA energy linearly changes as a function of the applied electric field, E, which is defined as the applied bias voltage, V_{bias}, divided by the MgO thickness, t_{MgO}. The slope of the linear relationship represents the VCMA coefficient in units of J/Vm, *e.g.* −37 fJ/Vm for the case in Figure 3. The VCMA coefficient is one of the most important parameters for demonstrating scalability and also in the reliable switching of the magnetization and thus the development of voltage-torque MRAM.

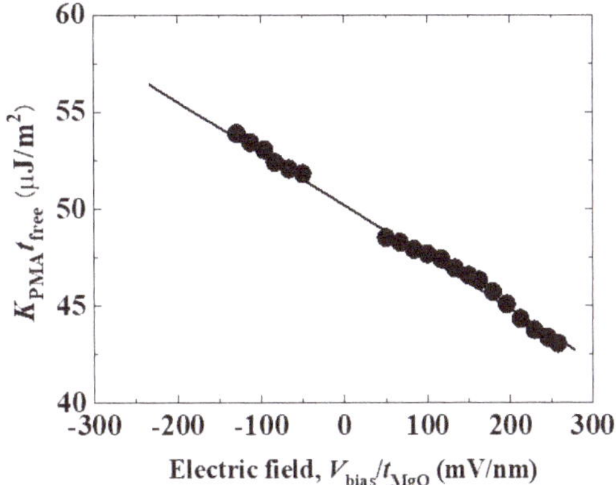

Figure 3. Example of applied electric-field dependence of $K_{PMA}t_{free}$ observed in an MgO-based MTJ structure. Reprinted figure with permission from [48], Copyright 2010 by the AIP Publishing LLC.

The realization of the VCMA effect in all-solid state devices, including a MTJ structure, made it possible for us to demonstrate the high speed response of this effect, such as in voltage-induced ferromagnetic resonance excitation [50–54], dynamic magnetization switching driven solely by the application of a voltage [55], and spin wave excitation [56–58].

In addition to ultrathin epitaxial films with large PMA [35,59–67], VCMA effects have been observed in various materials systems, for example, in sputter-deposited CoFeB [68–81], which is an important practical material that is used in the mass production of MTJs, and in self-assembled nano-islands [82], nanocomposite structures [83], and ultrathin layers with quantum well states [84]. The VCMA effect can also be applied for the control of domain wall motion [85–87] and magnetic skyrmions [88–90]. In addition, voltage control of the magnetic properties has been expanded not only for the PMA, but also for the Curie temperature [22], Dzyaloshinskii-Moriya interactions [91], interlayer exchange coupling [92], and proximity induced magnetism in non-magnetic metal thin films [93–95].

The VCMA effect can induce a transition of the magnetic easy axis between the in plane and out-of-plane directions by the application of a static voltage; however, bi-stable switching is not easily attained, because the VCMA effect does not break the time reversal symmetry. One possible way is to use the VCMA effect to assist other external fields. For example, the coercivity of the perpendicularly magnetized film can be reduced by the application of dc voltage [47,96,97] or of voltage-induced FMR [98], just as in the microwave-assisted magnetization reversal (MAMR) technique. Moreover, the combination of STT [99,100] or SOT [101] and the VCMA effect has also been experimentally demonstrated. These approaches are effective in reducing the energy that is required for writing by electric-current based manipulation; however, the realization of magnetization switching solely by a voltage effect is much more preferable.

We proposed pulse voltage-induced dynamic switching to overcome this problem (see Figure 4). This technique was first demonstrated in in-plane magnetized MTJs [55,102] and it was then applied in perpendicularly-magnetized MTJs [103–109]. For example, we assume the initial state (Figure 4a) to be the perpendicularly magnetized "up" state under the application of an in-plane bias magnetic field, H_{bias}. When a short pulse voltage is applied to eliminate the PMA completely, the magnetization starts to precess around the H_{bias} (Figure 4b). If the voltage pulse is turned off at the timing of half turn precession, then the magnetization can be stabilized in the opposite "down" direction (Figure 4c). H_{bias}

is required to determine the axis of magnetization precession. The effective field, such as crystalline anisotropy field and the exchange bias field, is also applicable.

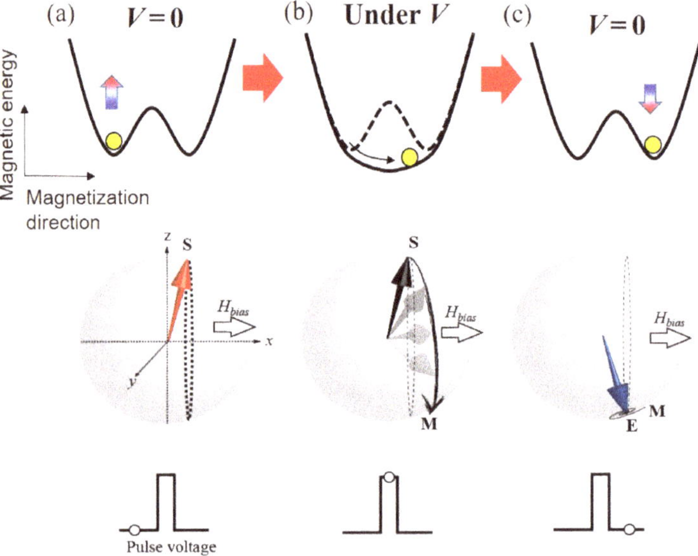

Figure 4. Conceptual diagram of voltage-induced dynamic switching for a perpendicularly-magnetized film. The in-plane bias magnetic field, H_{bias}, which determines the axis of the precessional dynamics, is applied in the $+x$ direction. (**a**) initial state (point S), (**b**) precessional switching process induced by an application of pulse voltage (from point S to point M), and (**c**) relaxation process (from point M to point E).

Figure 5a shows an example of an experimental demonstration of voltage-induced dynamic switching being observed in perpendicularly magnetized MTJs [105]. The top FeB layer with a W cap is the voltage-driven free layer. Under an optimized applied magnetic field, we achieved the stable toggle switching by the successive application of voltage pulses with a width of 1 ns and amplitude of −1.2 V. The precessional dynamics of the magnetization are reflected in the oscillation of the switching probability (P_{SW}) as a function of pulse width, as shown in Figure 5b. A high P_{SW} is obtained at the timing of half turn precession; however, when the pulse width is twice this, one turn precession results in low P_{SW}. From a practical point of view, the first half turn precession is effective in obtaining a low WER with fast switching speed. Under the condition that the PMA is completely eliminated, the amplitude of H_{bias} determines the precession frequency, and then the switching time, t_{SW} for the half turn precession is expressed as

$$t_{SW} \sim \frac{\pi\left(1 - \alpha^2\right)}{\gamma \mu_0 H_{bias}} \tag{1}$$

where α, γ, and μ_0 are the magnetic damping constant, the gyromagnetic ratio, and the permeability of vacuum, respectively.

The possible advantages of voltage-induced dynamic switching are as follows. (i) Fast switching (~1 nanosecond) can be induced with an ultralow switching power of the order of a few fJ/bit. (ii) The switching transistor can be downsized, because we do not need to apply a large electric-current. (iii) Unipolar switching can separate the polarity of voltages for writing and reading. In addition, the VCMA-induced enhancement in PMA has been used to propose a unique approach to reduce the read disturbance [110].

On the other hand, the following technical challenges remain. Firstly, the realization of a large VCMA effect is the most important issue to show the scalability of the voltage-torque MRAM, as discussed in Section 4. Furthermore, as seen in Figure 5b, the switching probability is sensitive to the writing pulse width, due to the precession-mediated switching process. Therefore, we need verification as to whether a sufficiently-low WER can be achieved by the voltage-induced dynamic switching technique. In addition, this is a toggle switching technique, so pre-read and read-verify processes are always required for writing. These reading processes dominate the total write time, and it can be critical when the resistance of the MTJ cell increases. In addition, the removal of the external magnetic field is also an important issue for practical applications.

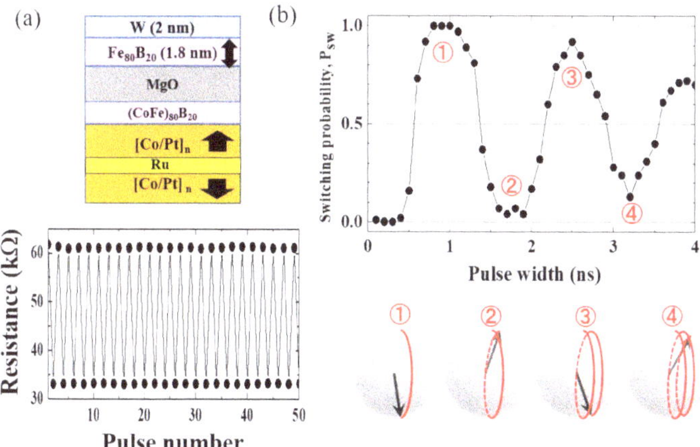

Figure 5. Experimental demonstration of voltage-induced dynamic switching. (**a**) Schematic of the sample structure of a voltage-controlled perpendicularly-magnetized MTJ and observed bi-stable switching between parallel and antiparallel magnetization configurations induced by successive pulse voltage applications. (**b**) Pulse width dependence of switching probability, P_{SW}. Due to the precessional dynamics, P_{SW} exhibits oscillatory behavior depending on the pulse width.

3. Physical Origin of the VCMA Effect

In this section, recent experimental trials conducted to understand the physical origin of the VCMA effect are introduced [111]. The following two mechanisms account for the purely electronic VCMA effect. The first mechanism comes from the charge-doping-induced anisotropy in the orbital angular momentum, as shown in Figure 6a. As each electron orbital in the vicinity of the Fermi level has a different density of states, selective charge doping may induce anisotropy in the orbital angular momentum. This effect changes the PMA energy through spin-orbit interactions from the spin-conserved virtual excitation processes [112,113], as expressed by the first term in Equation (2) [114].

$$-\frac{1}{4}\frac{\lambda}{\hbar}\left(\langle \Delta L_{\xi,\,\downarrow\downarrow}\rangle - \langle \Delta L_{\xi,\,\uparrow\uparrow}\rangle\right) + \frac{7}{2}\frac{\lambda}{\hbar}\left(\langle \Delta T'_{\zeta,\,\downarrow\uparrow}\rangle - \langle \Delta T'_{\zeta,\,\uparrow\downarrow}\rangle\right) \tag{2}$$

Here, λ is the spin-orbit interaction coefficient. L and T' are the orbital angular momentum and part of the magnetic dipole operator, respectively. Here, $\langle\Delta L_{\xi}\rangle \equiv \langle L_z\rangle - \langle L_x\rangle$ and $\langle\Delta T'_{\zeta}\rangle \equiv \langle T'_z\rangle - \langle T'_x\rangle$ are used. $\langle L_z\rangle$ and $\langle L_x\rangle$ are evaluated for the z- and x- components of the spin angular momentum, respectively. The same is the case for $\langle T'_z\rangle$ and $\langle T'_x\rangle$. $\uparrow$ and $\downarrow$ denote the contributions from the majority and minority spin-bands, respectively. We call the first mechanism the orbital magnetic moment mechanism. The second mechanism is the VCMA effect from the induction of an electric quadrupole (Figure 6b). An electric-field applied to the metal/dielectric interface is inhomogeneous, owing to the strong electrostatic screening effect in the metal, such as electric-field, including higher-order quadratic components, can couple with the electric quadrupole correlated with the magnetic dipole operator in an electron shell of the metal layer. The induced energy split of each orbital changes the magnetic anisotropy through spin-orbit interactions from spin-flip virtual excitation processes [115,116], as shown in Figure 6c. The latter mechanism corresponds to the second term in Equation (2). We call this the electric quadrupole mechanism. As the expectation values for the orbital angular momentum and the magnetic dipole operator can be measured as the orbital magnetic moment and the magnetic dipole T_z term (m_T), respectively, the aforementioned two mechanisms can be validated by X-ray absorption spectroscopy (XAS) and X-ray magnetic circular dichroism (XMCD) spectroscopy.

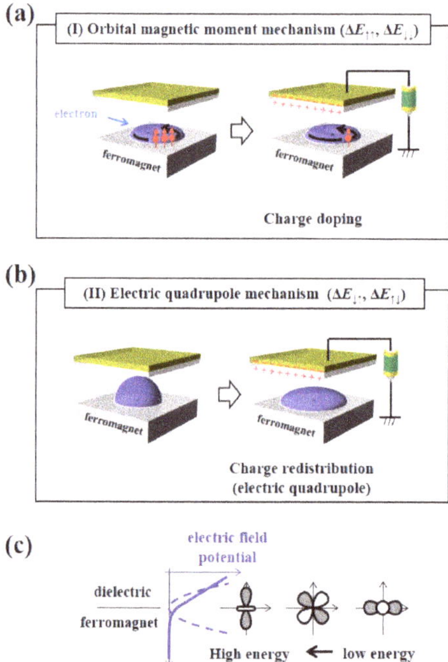

Figure 6. Microscopic origin of the VCMA effect. (**a**) Orbital magnetic moment mechanism. (**b**) Electric quadrupole mechanism. (**c**) Schematic of the nonlinear electric field at the interface between the dielectrics and the ferromagnet, which induces a charge redistribution-induced VCMA effect.

The XAS/XMCD experiments provide element-specific information on the electronic structure via the optical transition from the core level to unoccupied states in the valence band. Based on the use of circularly polarized X-rays, X-ray absorption techniques provide interesting features for the study of magnetic materials. Figure 7 shows a schematic diagram of the electronic states that are involved in an optical transition from the $2p$ core to d valence states, which is related to XMCD at the L edges of transition metals. The dichroic signal directly reflects the difference in the density of the states near the Fermi level between the up and down spin sub-bands. From the XMCD results with sum-rule analysis [117,118], the magnetic moments (spin magnetic moment: m_S, m_L, and m_T) can be determined from the measured XAS/XMCD spectra. Here, the measured orbital magnetic moments and magnetic dipole T_z term have the following relationships;

$$\Delta m_{\rm L} = -\frac{\mu_{\rm B}\left(\langle \Delta L_{\downarrow\downarrow}\rangle+\langle \Delta L_{\downarrow\downarrow}\rangle\right)}{\hbar}, \text{ and}$$

$$-7\Delta m_{\rm T} = -\mu_{\rm B}\left(\langle \Delta L^2_{\uparrow\uparrow}\rangle - \langle \Delta L^2_{\downarrow\downarrow}\rangle\right) - 7\mu_{\rm B}\left(\langle \Delta T_{\downarrow\uparrow}\rangle + \langle \Delta T_{\uparrow\downarrow}\rangle\right)/\hbar \tag{3}$$

It should be noted that the PMA energy from the spin-conserved virtual excitation process (first term in Equation (2)) is related to the orbital magnetic moment and the PMA energy from the spin-flip virtual excitation process (second term in Equation (2)) is related to the magnetic dipole T_z term.

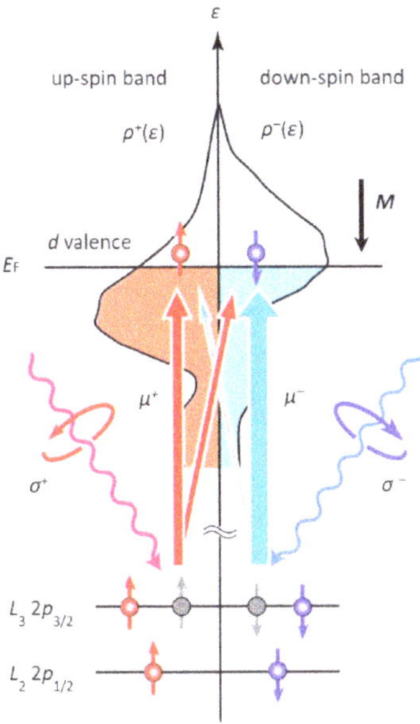

Figure 7. Diagram of the electronic states related to X-ray absorption spectroscopy and X-ray magnetic circular dichroism (XAS/XMCD) measurements at the *L*-edges of transition metals.

A Fe/Co (1 ML)/MgO multilayer was employed to see the changes in the orbital magnetic moment in XAS/XMCD experiments [113]. The sample stack is depicted in Figure 8a. A multilayered structure, consisting of bcc-V(001) (30 nm)/bcc-Fe(001) (0.4 nm)/Co (0.14 nm)/MgO(001) (2 nm)/SiO$_2$ (5 nm)/Cr (2 nm)/Au (5 nm), was deposited on a MgO(001) substrate. Figure 8b shows the typical XAS/XMCD results around the L_3 and L_2 edges of Co with a magnetic field of 1.9 T ($\theta = 20°$) to saturate the magnetization of the Fe/Co layer. The changes in the orbital magnetic moment and effective spin magnetic moment ($m_S - 7m_T$) of Co were determined while using sum-rule analysis, and they are summarized in Figure 8c,d. We can see that m_L of Co with an electric-field of −0.2 V/nm is larger than that corresponding to +0.2 V/nm. Moreover, the induced change in m_L with $\theta = 20°$ is larger than that with $\theta = 70°$. The experiment demonstrates that an orbital magnetic moment anisotropy change of $(0.013 \pm 0.008)\mu_B$ between the magnetization angles of $\theta = 20°$ and 70° was generated in the presence of applied electric fields of ±0.2 V/nm. Figure 8d shows the electric-field-induced change in $m_S - 7m_T$ of Co. As with m_L, $m_S - 7m_T$ is enhanced under the application of a negative electric-field. Moreover, the electric-field-induced change in the magnetic moment is anisotropic. In contrast to m_T, it is known that m_S is not sensitive to the magnetization direction. Hence, the anisotropic part of the induced change in the magnetic moment should be attributed to m_T.

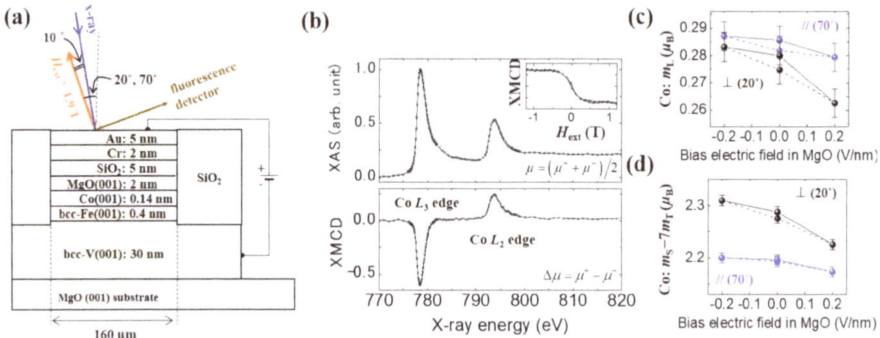

Figure 8. Voltage-induced changes to the magnetic moment of Co in the Fe/Co/MgO system. (**a**) Schematic of the sample structure. (**b**) Typical XAS/XMCD results around the Co-absorption edges. (**c**) Voltage-induced change to the orbital magnetic moment in Co. (**d**) Voltage-induced changes to the effective spin magnetic moment ($m_S - 7m_T$) in Co. Reprinted figure with permission from [113], Copyright 2017 by the American Physical Society.

As discussed in the previous section, Equation (2) can be used to analyze the VCMA effect. If we employ the spin-orbit interaction coefficient of Co, $\lambda_{Co} = 5$ meV, then the induced change in the PMA energy is estimated to be 0.039 ± 0.023 mJ/m^2 when the applied electric-field is switched from $+0.2$ V/nm to -0.2 V/nm. Here, the experimentally obtained $\Delta m_L = (0.017 \pm 0.010)\mu_B$ was used. From the VCMA coefficient in the Fe/Co/MgO system (-82 fJ/Vm), the PMA energy change at ±0.2 V/nm is 0.03 mJ/m^2, which is in good agreement with the PMA energy change that was obtained using the first term of Equation (2). From the discussion above, the change in the orbital magnetic moment anisotropy in Co seems to explain the VCMA effect. However, the impact of the change in the magnetic dipole T_z term (m_T) that is shown in Figure 8d on the VCMA effect remains to be seen. In Ref. 113, a first principles study was employed to clarify this point. As a result, the VCMA effect from the spin-flip terms ($\Delta E_{\downarrow\uparrow} + \Delta E_{\uparrow\downarrow}$) is found to be negligible and that from the spin-conserved terms ($\Delta E_{\uparrow\uparrow} + \Delta E_{\downarrow\downarrow}$) appeared to be dominant. Therefore, the change in orbital magnetic moment is responsible for the VCMA effect. Due to the large exchange splitting for Co, the observed changes in m_T do not contribute to the VCMA effect, as described by the second term in Equation (2).

It has been reported that the spin-orbit interaction energy from a spin-flip virtual excitation process makes a significant contribution to the VCMA effect when $3d/5d$-layered transition metals are employed [116]. Figure 9a shows an experimental design and a high-angle annular dark-field scanning transmission electron microscopy (HAADF-STEM) image of the device. Figure 9b shows the typical results of the polarization-averaged XAS and its XMCD around the L_3 and L_2 energy edges of Pt. A perpendicular magnetic field of ±60 mT was applied to saturate the magnetization of FePt. Figure 9c,d show electric-field-induced changes in the magnetic moments of Pt. The results confirm a clear bias voltage inductions of $m_S - 7m_T$, while there is no significant change to m_L under voltage applications.

In general, in low-symmetry systems, such as interfaces, the atomic electron orbital may possess an electric quadrupole moment. If the atom is also spin-polarized, the electric quadrupole moment induces the anisotropic part of the spin-density distribution, i.e., the magnetic dipole T_z term (m_T) [114–116,118]. In contrast to m_T, m_S is not sensitive to the magnetization direction. In Ref. 116, the voltage-induced change in $m_S - 7m_T$ shows large magnetization direction dependence. Thus, the observations indicate the significant induction of m_T in Pt by an external voltage. A first-principles study was also conducted for the FePt/MgO system, similar to the Fe/Co/MgO study. As a result, firstly, the monoatomic Pt layer at the interface with MgO makes the dominant contribution to the VCMA effect. Moreover, while the VCMA effect from the spin-conserved terms ($\Delta E_{\uparrow\uparrow} + \Delta E_{\downarrow\downarrow}$) decreases the PMA energy, the VCMA effect that is induced by the applied voltage from the spin-flip terms of interfacial Pt increases the PMA energy ($\Delta E_{\downarrow\uparrow} + \Delta E_{\uparrow\downarrow}$). The total PMA energy in the FePt/MgO system increases under the condition of electron depletion at the Pt/MgO interface, as the PMA energy increase by the spin-flip terms is greater than the PMA energy decrease by the spin-conserved terms.

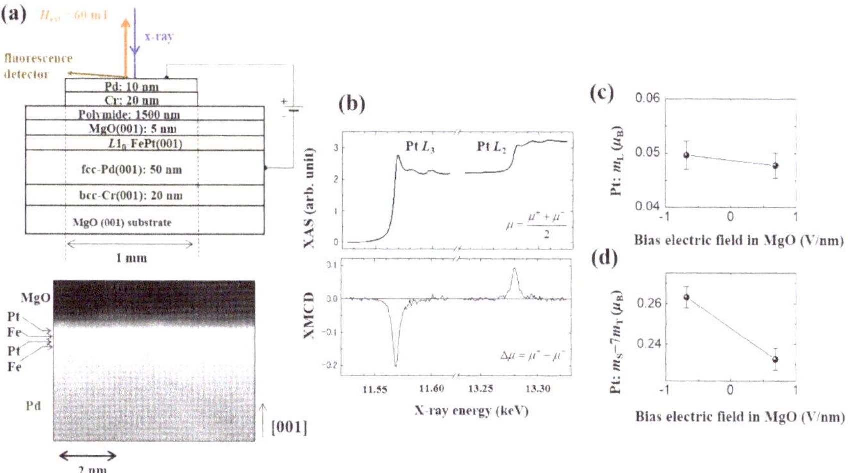

Figure 9. Voltage-induced changes to the magnetic moment of Pt in the Fe/Pt/MgO system. (**a**) Schematic of the sample structure and its high-angle annular dark-field scanning transmission electron microscopy (HAADF-STEM) image. (**b**) Typical XAS/XMCD results around the Pt-absorption edges. (**c**) Voltage-induced change to the orbital magnetic moment in Pt. (**d**) Voltage-induced changes to the effective spin magnetic moment ($m_S - 7m_T$) in Pt. Reproduced from [116]. CC BY 4.0.

To conclude, for the 3*d*-transition ferromagnetic metals, it is important to consider the orbital magnetic moment anisotropy. The validity of the Bruno model [112] (first term of Equation (2) and Figure 6a) has been experimentally demonstrated in Ref. 113. For the 3*d*/5*d*-multilayered ferromagnetic metals, the orbital magnetic moment anisotropy in 3*d*-metals cannot completely explain the VCMA effect. In addition to the magnetic moments in 3*d* metals, those in 5*d* metals should be considered in treating the total PMA energy in the system. Moreover, both the orbital magnetic moments and the electric quadrupole mechanisms (second term of Equation (2) and Figure 6b) of Pt dominate the VCMA in the case of $L1_0$-FePt, as shown in Ref. 116. As discussed in the recent review paper [111], it has been widely recognized that the XAS/XMCD spectroscopy is a powerful tool to investigate the voltage-induced effects in spintronic devices [28,113,116,119–124].

A much larger VCMA coefficient can be obtained when compared with that of purely electronic origin if we use a chemical reaction [122,125]. For example, a VCMA coefficient exceeding 10,000 fJ/Vm originating from reversible oxygen ion migration has been demonstrated in the Co/GdO_x system. In Ref. 122, XAS/XMCD spectroscopy at the Co absorption edge was employed to a Ta (4 nm)/Pt (3 nm)/Co (0.9 nm)/GdO_x (33 nm)/Ta (2 nm)/Au (12 nm) multilayer and found that an applied voltage changes the oxidation state and magnetization of the Co. Ref. 125 also reports real-time measurements of such an electrochemical VCMA effect. The operating speed strongly depends on the applied voltage and temperature, which strongly indicates that the electrochemical VCMA requires a thermal activation process. The reported maximum speed was in the sub-millisecond range. Therefore, such large values of the electrochemical VCMA seem attractive, but lie beyond the scope of VCMA studies for working memory applications. A similarly large VCMA effect with limited operating speed has been observed in many systems with electrochemical reactions [28,126,127] and/or charge traps [128,129].

Recently, strain-induced modulation of electronic structures and its influence on the VCMA effect has attracted attention [130,131]. For example, Hibino et al. reported a high VCMA coefficient of +1600 fJ/Vm in a Pt/Co/Pd/MgO structure at 10 K [95]. Here, the thin Pd layer possesses a magnetic moment that is induced by the proximity effect from the adjacent Co layer. At room temperature, a conventional linear VCMA effect with an efficiency of −90 fJ/Vm was observed. On the other hand, at lower temperatures below 100 K, a strong nonlinear VCMA effect appeared with the sign reversal. They explained that the observed effect can be attributed to the temperature dependence of the strain in the Pd. Similarly, Kato et al. reported a VCMA coefficient of over +1000 fJ/Vm at room temperature in an Ir/tetragonal FeCo/MgO structure [132]. So far, only static measurements have been done in these experiments. A demonstration of a high speed response is required to confirm whether they actually originate from the purely-electronic VCMA effect or not.

4. Materials Research for a Large VCMA Effect

The VCMA coefficient is one of the most important parameters for the scalability design of voltage-torque MRAM. When the cell size is reduced, we need to increase the PMA of the free layer to maintain the target thermal stability. As described in Section 2, voltage-induced dynamic switching requires the elimination of the PMA during the precessional dynamics.

Figure 10 shows a simple estimate of the PMA and VCMA coefficient required to consider the scalability [34,35]. As the simplest example, if we assume a free layer whose PMA is only determined by the interface magnetic anisotropy at the interface with the dielectric layer, the effective PMA energy is expressed as

$$K_{\text{PMA}}(E) = \frac{K_i(E)}{t_{\text{free}}} - \frac{1}{2}\mu_0 M_{\text{S}}^2 \tag{4}$$

Here, t_{free} and M_{S} are the thickness and saturation magnetization of the free layer. $K_i(E)$ is the PMA under application of the electric-field (E), and it is given by

$$K_i(E) = K_i(E = 0) - \eta E \tag{5}$$

where η is the VCMA coefficient. The thermal stability $\Delta(E)$ of the free layer under the application of the electric-field can be expressed by

$$\Delta(E) = \frac{K_{\text{PMA}}(E)At_{\text{free}}}{k_{\text{B}}T} = \Delta_0 - \frac{\eta A}{k_{\text{B}}T}E \tag{6}$$

Here, A and Δ_0 are the area of the free layer and the thermal stability under zero electric-field, respectively.

Consequently, the VCMA coefficient, η, which is required to eliminate Δ_0 can be expressed as,

$$\eta = \frac{k_{\text{B}}T\Delta_0}{AE_{\text{SW}}} \tag{7}$$

where E_{SW} is the amplitude of the switching electric-field.

For the curves in Figure 10, it was assumed that $t_{\text{free}} = 1$ nm and $E_{\text{SW}} = 1$ V/nm for each value of Δ_0. If we take cache memory applications as an example, the required $K_{\text{PMA}}t_{\text{free}}$ values range from 0.2 mJ/m^2 to 0.5 mJ/m^2, depending on the target Δ_0 values; consequently, the required VCMA coefficient is estimated to be from 200 fJ/Vm to 500 fJ/Vm. The main memory applications need higher $K_{\text{PMA}}t_{\text{free}}$ values in the range from 0.6 mJ/m^2 to 1.5 mJ/m^2. As a result, the required VCMA coefficient is in the range from 600 fJ/Vm to 1500 fJ/Vm. However, in experiments that have only focused on the purely-electronic VCMA effect, the achieved VCMA coefficient that is demonstrated in practical materials, such as CoFeB, has been limited to about 100 fJ/Vm [71,78,81,98].

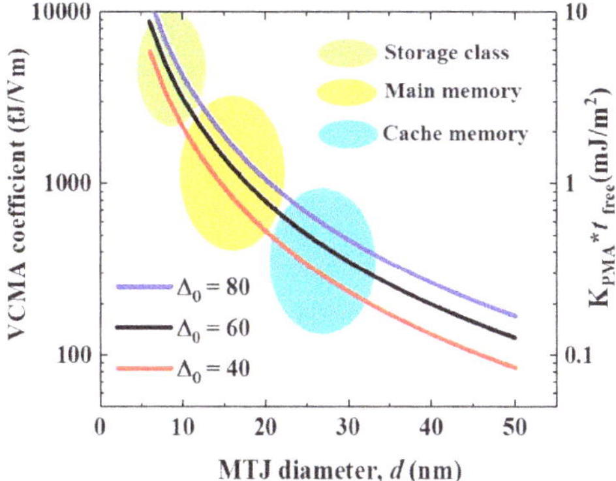

Figure 10. Scalability issue for voltage-torque MRAMs. The dependence of the required $K_{PMA}t_{free}$ and VCMA coefficient on the diameter of the MTJ was estimated for each thermal stability factor (Δ_0).

We employed a fully epitaxial Cr/ultrathin Fe/MgO system as a standard system for the materials research of VCMA effect [133], because large interface magnetic anisotropy can be obtained due to the flat and well-defined Fe/MgO interface [134–136] when compared to MTJs with noble metal buffers, which can have the problem of surface segregation [137]. To evaluate the VCMA properties, we used molecular beam epitaxy to prepare orthogonally-magnetized MTJ structures that consisted of a MgO seed (3 nm)/Cr buffer (30 nm)/ultrathin Fe (t_{Fe})/MgO (t_{MgO} = 2.3 nm)/Fe(10 nm) on MgO(001) substrates. Here, the bottom ultrathin Fe layer is the voltage-controlled free layer with perpendicular magnetic easy axis and the top 10 nm-thick Fe is the in-plane magnetized reference layer. Figure 11a shows an example of the applied bias voltage, V_{bias}, and dependence of the half-MR loop measured under an in-plane magnetic field, H_{ex}. The vertical axis is normalized using the maximum (H_{ex} = 0 Oe) and minimum (H_{ex} = −20 kOe) resistances. The Fe thickness is fixed at t_{Fe} = 0.44 nm.

The application of an in-plane magnetic field tilts the magnetization of the ultrathin Fe layer into the magnetic hard axis, while that of the reference layer remains in the film plane (see the drawings in Figure 11a). Therefore, the effective perpendicular anisotropy field is reflected in the saturation behavior of tunneling resistance. The tunneling conductance, G, depends on the relative angle (θ) between the magnetizations of the free and reference layers, i.e. $G(\theta) = G_{90} + (G_P - G_{90})\cos\theta$. Here, G_{90} and G_P are the conductance under the orthogonal and parallel magnetization configurations. Therefore, the ratio of the in-plane component of the magnetization of the free layer, $M_{in\text{-}plane}$, to its saturation magnetization, M_S, is expressed as

$$\frac{M_{in-plane}}{M_S} = \cos\theta = \frac{R_{90} - R(\theta)}{R(\theta)} \frac{R_P}{R_{90} - R_P} \tag{8}$$

where R_P is the MTJ resistance in the parallel magnetization configuration, R_{90} is the MTJ resistance in the orthogonal magnetization configuration, and $R(\theta)$ is the MTJ resistance when the magnetization of the ultrathin Fe layer is tilted towards the in-plane direction at angle θ under the application of an in-plane magnetic field. Using Equation (8), we can evaluate the normalized in-plane magnetization versus the applied magnetic field. The inset in Figure 11b shows an example of a normalized *M-H* curve measured under V_{bias} = 10 mV. The PMA energy, K_{PMA} can be calculated from $M_{in\text{-}plane}$ (H) with the saturation magnetization value evaluated by SQUID measurements (yellow area in the inset of Figure 11b). Figure 11b summarizes the applied electric-field, V_{bias}/t_{MgO}, dependence of $K_{PMA}t_{Fe}$. With ultrathin layers of Fe, an unexpected nonlinear VCMA effect was observed. Under the application of negative voltages, the PMA monotonically increases with a large VCMA coefficient of −290 fJ/Vm. On the other hand, the PMA deviates from a linear relationship under the application of positive voltages. Figure 12 summarizes the Fe thickness dependence of the VCMA coefficient. This nonlinear

VCMA effect was only observed with ultrathin layers of Fe, $t_{Fe} < 0.6$ nm (blue dots), and the usual linear VCMA effect appears for thicker layers (red dots). Xiang et al. systematically investigated the tunneling conductance, the PMA, and the VCMA effect in a similar system to determine the origin of the nonlinear VCMA effect, but the MgO was replaced by a $MgAl_2O_4$ barrier, which has smaller lattice mismatch with Fe. Interestingly, they found strong correlation between the VCMA effect and the quantum well states of Δ_1 band formed in an ultrathin Fe layer that is sandwiched between the Cr and MgO layers [138]. These results may indicate that artificial control of the electronic states in an ultrathin ferromagnetic layer may provide a new approach for designing the VCMA properties. In addition to the influence of quantum well states, we found that intentional Cr doping at the Fe/MgO interface can enhance the PMA and the VCMA effect [62]. Therefore, intermixing with the bottom Cr buffer may also have an influence on the observed large VCMA effect. A theoretical investigation to understand the role of the inter-diffused Cr atoms has been proceeded [139,140].

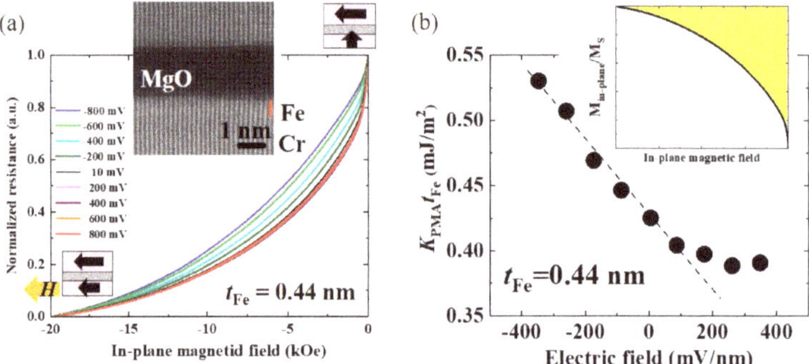

Figure 11. (**a**) Bias voltage dependence of normalized tunnel magnetoresistance (TMR) curves measured under in-plane magnetic fields for an orthogonally magnetized MTJ consisting of Cr/ultrathin Fe (0.44 nm)/MgO/Fe (10 nm). The inset shows a cross-sectional TEM image of the MTJ. (**b**) Applied electric-field dependence of $K_{PMA} t_{Fe}$ values. The inset displays an example of a normalized *M-H* curve. K_{PMA} was evaluated from the yellow-colored area with the saturation magnetization value that was obtained by a SQUID measurement. Reprinted figure with permission from [133], Copyright 2017 by the American Physical Society.

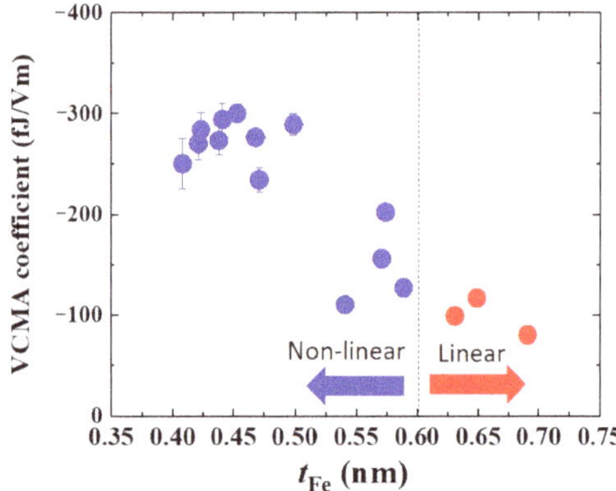

Figure 12. Fe thickness dependence of the VCMA coefficient observed in a Cr/ultrathin Fe(t_{Fe})/MgO/Fe structure. A large VCMA coefficient with nonlinear behavior was found in the thinner Fe thickness range, $t_{Fe} < 0.6$ nm (blue dots). Reprinted figure with permission from [133], Copyright by the American Physical Society.

A large VCMA effect can be obtained with the Cr/ultrathin Fe/MgO system; however, we can only induce an enhancement in the PMA. As explained in Section 2, reduction in the PMA is required for voltage-induced dynamic switching of the perpendicularly-magnetized free layer.

Nakamura et al. proposed inserting a heavy metal monolayer at the Fe/MgO interface to improve the VCMA properties, and found using first-principles calculations that $5d$ transition metals, such as Ir and Os, would be effective in enhancing the VCMA coefficient [141]. A few experimental trials of interface engineering that included the insertion of a heavy metal layer at a CoFe-based film/MgO interface have been reported [81,142]; however, the VCMA coefficient was still less than 100 fJ/Vm. Ir seems to be a promising candidate for this purpose due to its huge spin-orbit coupling constant, which is more than 10 times larger than that of $3d$ transition ferromagnets [141].

We prepared multilayer structures consisting of Cr (30 nm)/ultrathin Fe(t_{Fe})/Ir(t_{Ir})/MgO (2.5 nm) with indium-tin oxide (ITO) or Fe (10 nm) top electrodes to investigate the impact of the introduction of Ir on the interfacial PMA and the VCMA effect [35]. The ultrathin Ir layer was inserted between the Fe and MgO layers; however, we found that the Ir atoms were dispersed inside the Fe layer during the post-annealing process, as seen in the HAADF-STEM images in Figure 13a. Atomic-scale Z-contrast HAADF-STEM imaging enabled the identification of inter-diffused Ir atoms as bright spots that are indicated by yellow arrows. The first-principles calculation predicts strong in-plane anisotropy at the Ir/MgO interface [141]; however, we observed an unexpected enhancement in the PMA. Figure 13b shows a comparison between the polar MOKE hysteresis curves of a single Fe layer (t_{Fe} = 1.0 nm) and an Ir-doped Fe layer formed the bilayer structure consisting of Fe (1.0 nm)/Ir (0.1 nm)). The pure Fe layer exhibits large saturation fields of about 7 kOe, which indicated an in-plane magnetic easy axis. On the other hand, the introduction of the quite thin Ir doping layer resulted in transition of the magnetic easy axis from the in-plane to the out-of-plane direction. Figure 13c summarizes the dependence of the intrinsic interfacial magnetic anisotropy, $K_{i,0}$, on the thickness of the Ir layer. With appropriate Ir doping, $K_{i,0}$ reaches 3.7 mJ/m^2, which is about 1.8 times that observed at the Fe/MgO interface (2.0 mJ/m^2) [35,134].

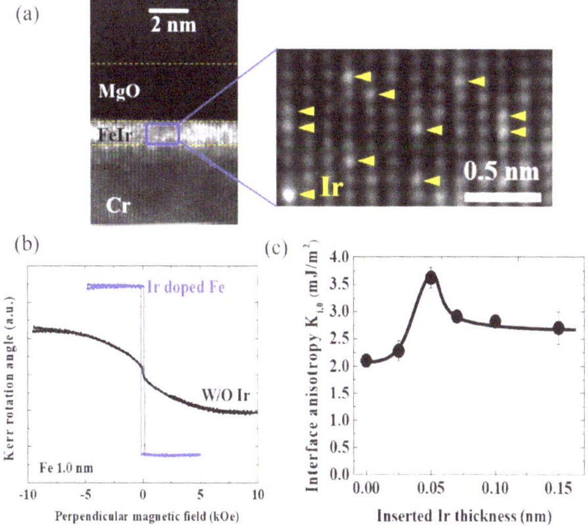

Figure 13. (a) HAADF-STEM images of a multilayer structure of Cr/ultrathin Ir-doped Fe/MgO. Inter-diffused Ir atoms can be identified by atomic-scale Z-contrast HAADF-STEM imaging as indicated by the yellow arrows. (b) Comparison of the polar MOKE hysteresis curves for pure Fe (1 nm)/MgO and Fe (1 nm)/Ir (0.1 nm)/MgO structures. (c) Dependence of the intrinsic interface magnetic anisotropy energy, K$_{i,0}$, on the thickness of the Ir layer. Reproduced from [35]. CC BY 4.0.

The Ir doping also has an effect on the VCMA. Figure 14a shows an example of the bias voltage effect on the TMR curves that were measured under in-plane magnetic fields for an orthogonally-magnetized MTJ with an Ir-doped Fe free layer ($t_{\text{FeIr}} = 0.82$ nm; formed from Fe (0.77 nm)/Ir (0.05 nm)). The saturation field shifts with changes in the applied voltage, as is the case in a pure Fe/MgO structure. However, the applied electric-field dependence of $K_{\text{PMA}}t_{\text{FeIr}}$ exhibits a completely different trend when compared with that observed in the Fe/MgO structure. We observed a large reduction in PMA with a VCMA coefficient of -320 fJ/Vm under positive voltages (see Figure 14b). It is interesting that such a low doping concentration of Ir, which is even thinner than one monolayer, can have a drastic effect on the VCMA properties. In addition, voltage-induced FMR measurements confirmed the high speed response of the VCMA effect, as shown in the inset in Figure 14b. Thus, the observed large VCMA comes from purely-electronic origin.

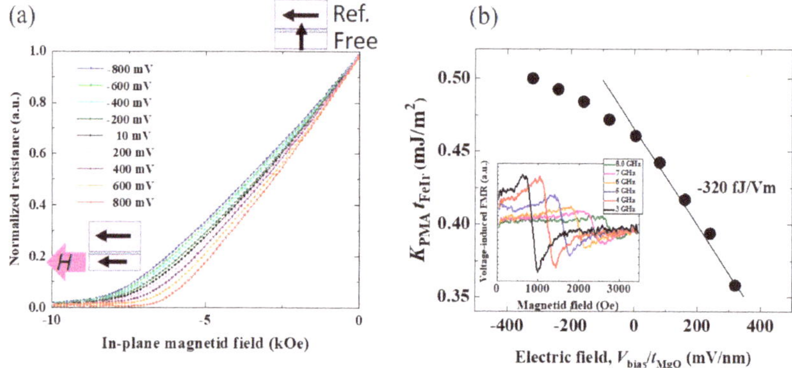

Figure 14. (**a**) Bias voltage dependence of normalized TMR curves measured under in-plane magnetic fields for an orthogonally-magnetized MTJ consisting of Cr/Ir-doped Fe(0.82 nm)/MgO/Fe(10 nm). (**b**) Applied electric-field dependence of $K_{\text{PMA}}t_{\text{FeIr}}$. The inset shows an example of voltage-induced FMR excitation measured by a homodyne detection technique, which proves the high speed responsiveness of the observed VCMA effect. Reproduced from [35]. CC BY 4.0.

A theoretical analysis using first-principles calculation was performed in Cu(5ML)/Fe$_{94}$Ir$_6$(5ML)/MgO(5ML) structures to discuss the physical origin of the large VCMA effect in Ir-doped Fe. The Ir-doped bcc Fe was modeled by a supercell consisting of 4×4 unit cells as shown in Figure 15a. Figure 15b depicts the atomic-resolved electric-field induced magnetic anisotropy energies (MAE) for the Fe and Ir atoms. The variation in the MAE for the Ir atoms is more than five times greater than that for the Fe atoms. Interestingly, MAE change in the second layer (layer 2 in Figure 15b) from the interface with the MgO layer is larger than that of the layer 1, contrary to expectations.

We also attempted to divide the MAE into contributions from the spin-flip and spin-conserved terms between the occupied and unoccupied states. Figure 15c shows the voltage-induced changes in MAE that arise from second-order perturbation of the Ir sites in layers 1 and 2. The electric-field modulation of the spin-conserved term for the majority spin occupied and unoccupied states $\delta E_{\uparrow\uparrow}$ is larger than that for the minority spin states $\delta E_{\downarrow\downarrow}$. On the other hand, the spin-flip terms that are by the electric-field, $\delta E_{\uparrow\downarrow}$ and $\delta E_{\downarrow\uparrow}$ have almost the same absolute value, but with opposite sign, so the VCMA effect that arises from the spin-flip term is small. Therefore, the large VCMA effect in Ir-doped Fe is mainly caused by the electric-field effect on the majority spin Ir-5d states and it can be interpreted by the modulation in the first term of Equation (2), i.e. the orbital magnetic moment mechanism.

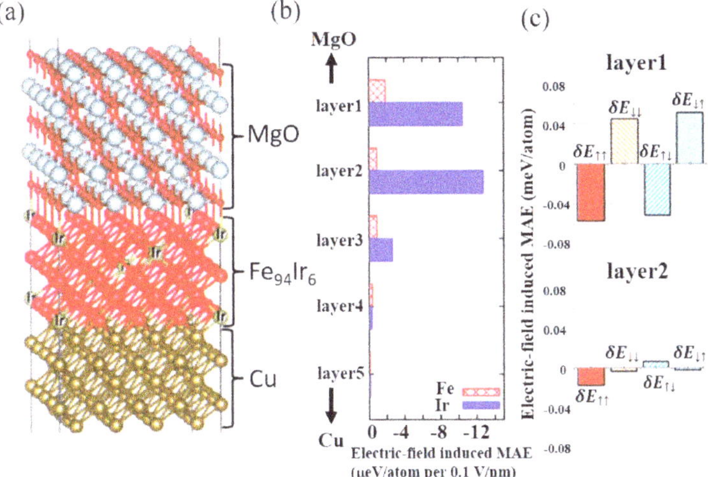

Figure 15. First principles calculations of the electric-field induced magnetic anisotropy energy change in an Ir-doped Fe/MgO system. (**a**) Supercell structure used for the calculation, consisting of MgO (5 ML)/FeIr (5 ML)/MgO (5 ML). (**b**) Atomic-resolved magnetic anisotropy energies (MAE) change induced by an electric-field of 0.1 V/nm in MgO. The Ir concentration was maintained at about 6% in the FeIr layer. (**c**) The electric-field induced MAE arising from second-order perturbation of the spin-orbit coupling for Ir atoms in layers 1 and 2. Reproduced from [35]. CC BY 4.0.

Figure 16 shows the density of states for Ir atoms in layer 2. The majority spin $5d$ states are dominant near the Fermi level, since the minority spin $5d$ states near the Fermi level form bonding and anti-bonding states by hybridization with the minority spin Fe-$3d$ states. On the other hand, the majority spin $5d$ states are well-localized when compared with the minority spin states near the Fermi level. Figure 16 also shows the MAE as a function of the Fermi energy shift (black line). The PMA energy is drastically modified by a small shift in Fermi energy reflecting the localized majority spin states and the large spin-orbit coupling of the Ir atoms. As a result, a large VCMA is obtained for the charge-doping effect even in layer 2.

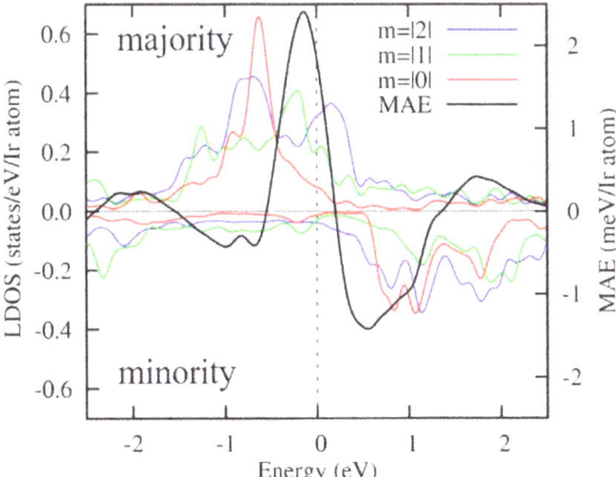

Figure 16. Spin polarized local density of states of Ir-$5d$ orbitals and magnetic anisotropy energy as a function of the band energy in layer 2.

The theoretical calculations predict the larger VCMA effect exceeding a few thousand fJ/Vm by inserting a monolayer of Ir at the Fe/MgO interface; however, such a structure can drastically degrade the TMR properties in the MTJ device, in addition to the strong in-plane anisotropy. On the other hand, if Ir doping can improve both the PMA and the VCMA effect while minimizing degradation in TMR, the MTJs should be much more manufacturable, even by sputtering processes. In fact, the enhancement of the PMA and the VCMA effect by Ir doping has also been confirmed in polycrystalline MTJs that are mainly prepared by sputtering [143]. We still have numerous choices for the 4d and 5d elements, therefore materials engineering using heavy metal doping has enormous possibilities for further improvement in the interfacial PMA and VCMA properties.

5. Towards Reliable Voltage-Induced Dynamic Switching

In this section, recent experimental trials for reliable voltage-induced dynamic switching are discussed. As shown in Figure 4, voltage-driven magnetization switching is initiated by precession of the magnetization that is induced by the VCMA effect and the associated voltage-torque, which is proportional to the time derivative of the applied voltage. During the application of a voltage, the magnetization precesses around the effective field while undergoing magnetization damping. Once the voltage is turned off, the magnetic anisotropy immediately recovers as the ferromagnetic layer/dielectric layer junction discharges, and the magnetization relaxes into one of two polarities. We can achieve bipolar magnetization switching using a unipolar voltage pulse with a controlled duration since the polarity of the final state can be controlled by the voltage pulse width. In the absence of thermal fluctuations, the magnetization trajectory during the switching process is uniquely determined for a given initial state and voltage pulse shape, and therefore error-free magnetization switching can be achieved by choosing the appropriate voltage pulse width. However, in practice, the magnetization inevitably suffers thermal fluctuations and that results in the stochastic generation of write errors. Special care must be taken when attempting to reduce the write errors in voltage-torque MRAM cells. In the case of STT, the current polarity determines the polarity of magnetization switching, and a longer pulse may be used to reduce write errors. On the other hand, in the case of voltage-induced dynamic switching, a longer pulse dampens the magnetization along the effective field direction, and this degrades the switching accuracy.

Although earlier experiments have characterized the basics of voltage-driven magnetization switching, it was only in 2016 that the WER in a practical MTJ was quantitatively evaluated for the first time [105]. Figure 17a shows a schematic illustration of an experimental setup for evaluating the WER of an MTJ. Voltage pulses that were generated by the pulse generator are fed to the MTJ and these switch the free layer magnetization. The free layer magnetization direction, either parallel or antiparallel with respect to the reference layer magnetization, can be monitored via the TMR effect.

Figure 17b displays the typical behavior of voltage-driven magnetization switching; P_{sw} is the switching probability, t_{pulse} is the pulse width; and, V_{pulse} is the voltage amplitude. When V_{pulse} is small, the VCMA effect cannot completely eliminate the magnetic energy barrier; therefore, the magnetization switching in this region is dominated by thermal activation. As V_{pulse} is increased, well-defined oscillation of P_{sw} appears, which is a signature of precession-mediated switching induced by the VCMA effect. As discussed in Section 2, the highest P_{sw} is obtained at t_{pulse} that corresponds to one-half the magnetization precession cycle, and then P_{sw} gradually moves toward 0.5 while undergoing damped oscillations. This behavior can be understood as the combined action of magnetization damping and thermal fluctuations.

In Ref. 105, Shiota et al. employed perpendicularly magnetized MTJ (p-MTJ) that consisted of a reference layer/MgO/Fe$_{80}$B$_{20}$/W cap and experimentally demonstrated a WER of 4×10^{-3}. They also demonstrated in numerical simulations that the WER could be reduced by improving the thermal stability factor, Δ and by reducing the magnetic damping, α of the free layer, as shown in Figure 17c. An improved Δ effectively reduces the thermal fluctuations in the initial state and in the relaxation process after switching. Moreover, a lower α can reduce the influence of thermal fluctuations during the switching process, which leads to more accurate writing. However, it should be noted that, the larger the value of Δ, the larger the VCMA efficiency required, otherwise the magnetization switching is dominated by thermal activation, and well-controlled magnetization switching cannot be obtained. By using CoFeB/MgO/CoFeB p-MTJs, Grezes et al. experimentally investigated the WER and the read disturbance rate as a function of read/write pulse width and amplitude, and examined the compatibility of the bit-level device performance for integration with CMOS processes [110]. They also simulated the performance of a 256 kbit voltage-torque MRAM block in a 28 nm CMOS process, and showed the capability of the MTJs for delivering WERs below 10^{-9} for 10 ns total write time by introducing the read verify processes. The introduction of read verify processes makes it possible to reduce the effective WER, however it causes an increase in the total writing time. Therefore, we need further effort to reduce the essential WER that is induced by single pulse switching. Recently, Shiota et al. showed that improvement in the PMA and VCMA properties can be achieved in the MTJ consisting of Ta/$(Co_{30}Fe_{70})_{80}B_{20}$/MgO/reference layer, and demonstrated a WER of 2×10^{-5} without the read verify process [106]. Further optimization of the composition of the CoFeB alloy and the device structure allowed for a WER lower than 10^{-6} to be achieved, as shown in Figure 18 [109]. In this case, the introduction of a once read verify process enables a practical WER of the order of 10^{-12}.

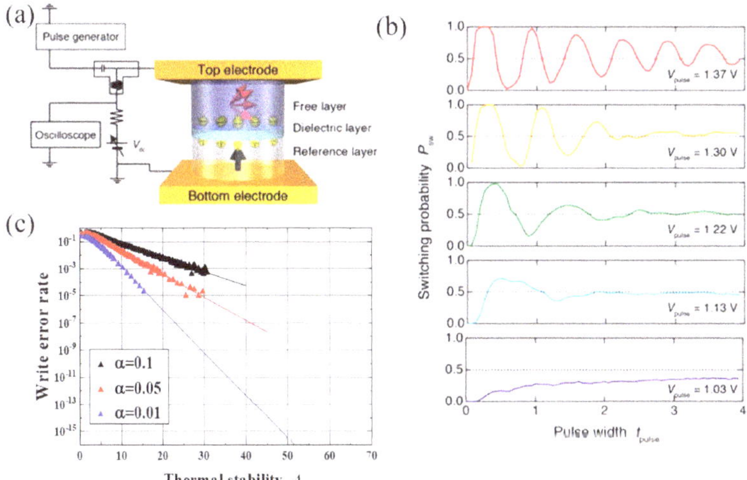

Figure 17. (a) Experimental setup for evaluating the WER of an MTJ. (b) Pulsed-voltage-driven magnetization switching in a p-MTJ. (c) WER as a function of Δ obtained from numerical simulations.

In addition to materials engineering, a physical understanding of the voltage-driven magnetization dynamics is also needed in order to facilitate reductions in the WER. Recent studies [107,108] showed that numerical simulations that are based on the macrospin approximation could well reproduce the experimental data by taking into account thermal fluctuations and magnetization damping. In macrospin approximation, the free layer spins are represented by a magnetic moment $\mathbf{M}$, and its time evolution can be obtained by numerically solving the Landau-Lifshitz-Gilbert equation:

$$\frac{d\mathbf{M}}{dt} = \gamma \mathbf{M} \times \mathbf{H}_{\text{eff}} + \frac{\alpha \mathbf{M}}{M_s} \times \frac{d\mathbf{M}}{dt} \tag{9}$$

where M_s is the saturation magnetization, t is the time, α is the damping constant, and $\mathbf{H}_{\text{eff}}$ is the effective field given by

$$\mathbf{H}_{\text{eff}} = -\frac{dE}{d\mathbf{M}} \tag{10}$$

and E is the energy density expressed as

$$E = K_{\text{PMA}}\left(1 - m_z^2\right) - M_s H_x m_x \tag{11}$$

where $m = (m_x, m_y, m_z)$ is the magnetization unit vector and H_x is an in-plane bias magnetic field. As displayed in Figure 19a, without the VCMA effect, the magnetization has two energy equilibrium at $\widetilde{m}_\pm = \left(\widetilde{m}_x, 0, \pm\sqrt{1 - \widetilde{m}_x^2}\right)$, where $\widetilde{m}_x = M_s H_x/(2K_{\text{PMA}})$, one maximum at $m_x = -1$, and one saddle point at $m_x = 1$. By letting K_{PMA} fall to zero, the magnetization precesses around H_x associated with damping, and the appropriate duration can switch the magnetization direction.

Figure 19b displays a typical plot of the dependence of WER on t_{pulse} that was observed in an MTJ consisting of a Ta/(Co$_{30}$Fe$_{70}$)$_{80}$B$_{20}$ (1.1 nm)/MgO/reference layer. The amplitude of the in-plane component of the bias magnetic field is 890 Oe. The filled circles and the line denote data were obtained from experiments and numerical simulations, respectively [107]. Good agreement with the experimental data suggests the validity of the model used for the numerical simulations. It is noteworthy that the WER exhibits a local maximum at a certain t_{pulse}, which cannot be explained just by considering the VCMA effect. A detailed analysis of the magnetization trajectory revealed that thermal agitation during the relaxation process (*i.e.*, after the pulse application) induces the transition of the magnetization between the precession orbits surrounding the energy minima and that the precession-orbit transition enhances the WER. The numerical simulations also revealed that the probability of the precession-orbit transition depends on t_{pulse} (see Ref. 107 for more details). In the present case, the probability is maximized at around $t_{\text{pulse}} = 0.12$ ns. This results in the appearance of a local maximum in the WER, and it narrows the operating t_{pulse} range for which reliable magnetization switching is assured. As the appearance of the WER local maximum is related to magnetization fluctuations during the relaxation process, we need to reduce its influence by improving the PMA and VCMA properties in order to achieve a wide operating t_{pulse} range.

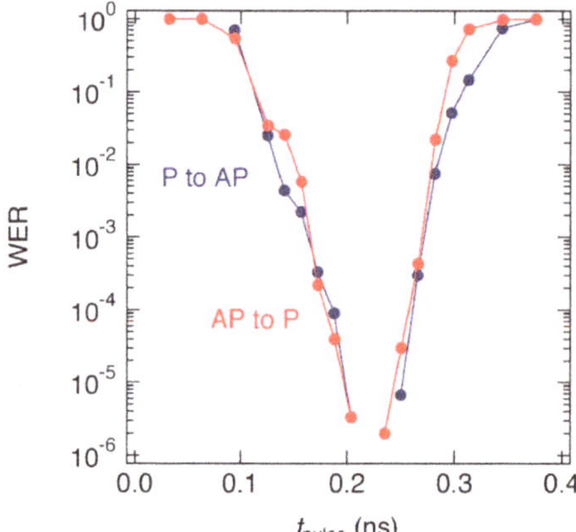

Figure 18. Example of the optimized WER as a function of t_{pulse} observed in a perpendicularly-magnetized MTJ consisting of Ta/(Co$_{50}$Fe$_{50}$)$_{80}$B$_{20}$/MgO/reference layer. The blue and red symbols represent the WER of parallel (P) to antiparallel (AP) and AP to P switching, respectively. Reprinted figure with permission from [109], Copyright 2019 by the IOP Publishing Ltd.

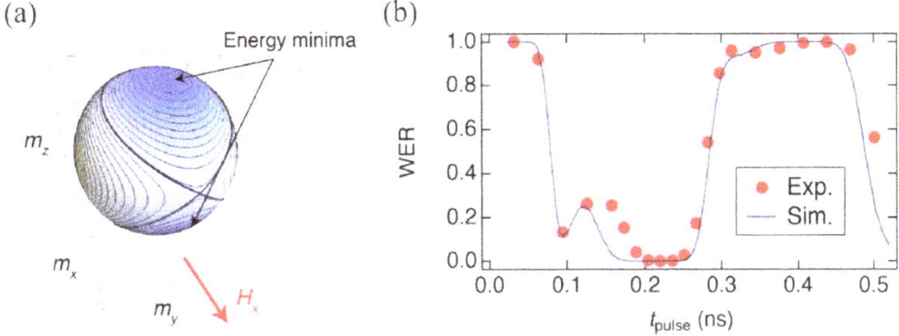

Figure 19. (**a**) Contour plot of energy density in the absence of a bias voltage. (**b**) Appearance of a local peak in the WER observed in an MTJ consisting of Ta/(Co$_{30}$Fe$_{70}$)$_{80}$B$_{20}$ (1.1 nm)/MgO/reference layer. The filled circles and the lines represent experimental data and numerical simulations, respectively. Reprinted figure with permission from [107], Copyright 2018 by the American Physical Society.

In addition to t_{pulse}, a recent study revealed that the WER depends in a unique manner on the rise time (t_{rise}) and fall time (t_{fall}) [108]. Figure 20 displays the magnetization trajectories that were obtained by using three different waveforms. When a pulsed voltage is applied, the magnetization rotates from $m+$ towards $m-$ (red line) and, after the pulse, the magnetization relaxes to either $\widetilde{m}_+$ or $\widetilde{m}_-$, depending on t_{pulse} (green line). An important thing is that, due to the nonzero magnetization damping, the magnetization direction at the end of the voltage pulse (m') never reaches $\widetilde{m}_+$ or $\widetilde{m}_-$ whatever t_{pulse} is chosen as long as one uses square pulses (Figure 20a). Therefore, it takes some time before the magnetization settles down to the energy minimum. During that time, the magnetization is subjected to thermal agitation, and a finite number of write errors will be counted. When a nonzero t_{rise} and/or nonzero t_{fall} is introduced, the magnetization is subjected not only to H_x, but also to the anisotropy field due to the uncompensated PMA $K_{PMA}'(V,t)$, which is given by

$$H_{ani} = -\frac{2K_{PMA}'(V,t)m_z}{M_s} \tag{12}$$

Since the polarity of H_{ani} switches according to the polarity of m_z, it applies additional torque to the magnetization that tilts the magnetization to H_x during t_{rise} (Figure 20b), and it pulls the magnetization away from H_x during t_{fall} (Figure 20c). As a result, for $t_{rise} = 0.085$ ns, m' comes closer to the saddle point, whereas, for $t_{fall} = 0.085$ ns, m' almost overlaps with $\widetilde{m}_-$ and thereby one can minimize the time that is required for relaxation. This suggests that there is a certain t_{fall} which can minimize the WER. Indeed, such WER reduction is experimentally obtained and the numerical simulations reproduce it, as shown in Figure 20d,e.

The inverse bias method is another unique technique for reducing the WER. Figure 21a illustrates the write sequence of the conventional and inverse bias methods. In the inverse bias method, a bias voltage with a negative polarity is applied before and after the write pulse. If the system exhibits a linear VCMA effect, then the inverse bias enhances the K_{PMA} of the free layer, and thereby reduces the thermal fluctuations in the initial state and during the relaxation process. It should be noted that inverse biases can also be used for the pre-read and read verify processes, which thereby offers a read-disturbance-free operation as well as WER reduction. Noguchi et al. first proposed the inverse bias method [37] and the effectiveness was later studied using numerical simulations [144]. In Ref. 144, a substantial reduction in WER was confirmed by introducing inverse biases, whose absolute intensity was the same as that of the write pulse, but with opposite sign (see Figure 21b).

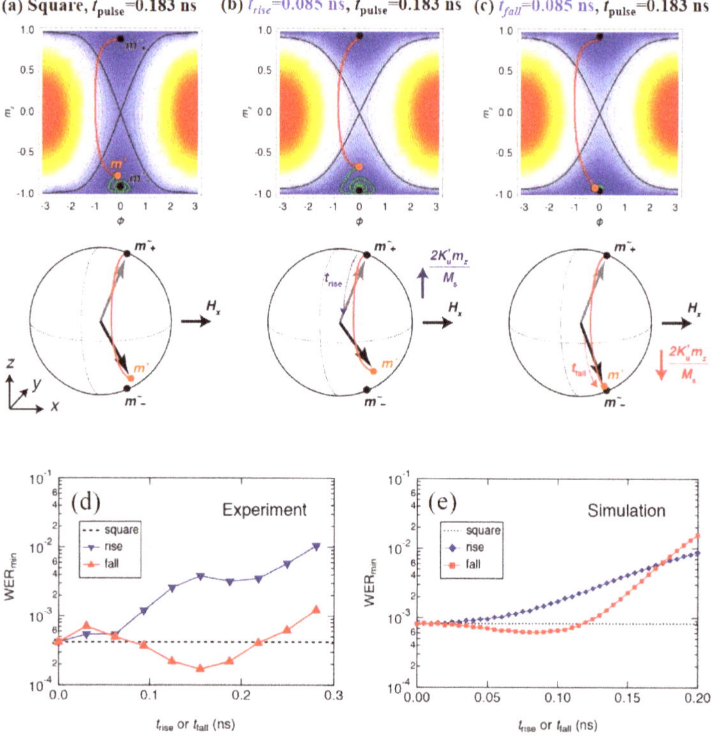

Figure 20. (**a**)–(**c**) Effects of pulse shaping on magnetization trajectory. The red and green lines represent the magnetization trajectory during and after application of the pulse, t_{pulse}, respectively. (**d**), (**e**) WER minimum as a function of rise time (blue symbols) and fall time (red symbols). (**d**) experimental results; (**e**) numerical simulation results. Reprinted figure with permission from [108], Copyright 2019 by the American Physical Society.

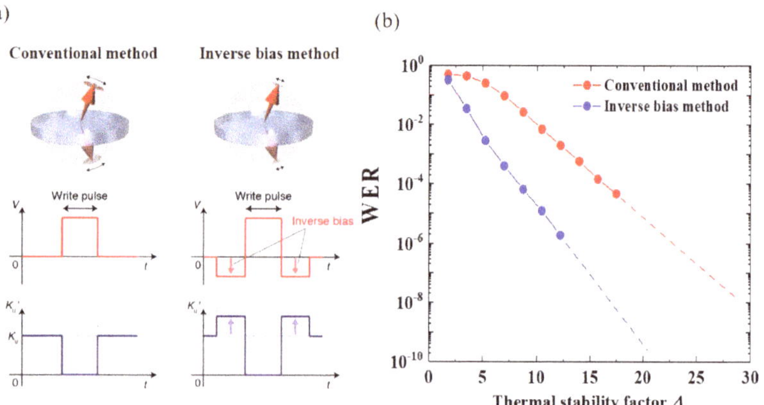

Figure 21. (**a**) Comparison of write pulse sequence in conventional and inverse bias methods. (**b**) Numerically obtained WER as a function of Δ using two different methods.

Since precise control of voltage-driven magnetization switching relies on the precise control of the voltage pulse shape, accurate calculation and shaping of the voltage pulse waveform [38,145] are also an important technique for studying the voltage-driven magnetization dynamics in detail. The procedure that is presented in Ref. 145 allows for one to accurately analyze and control the voltage waveform applied to an MTJ. This is especially important in the development of voltage-torque MRAM, because the MTJ resistance becomes much higher than 50 Ω to suppress the flow of charge current, whereas nearly all microwave interconnects have a characteristic impedance of 50 Ω. This impedance mismatch gives rise to multiple reflections between the signal source and the MTJ, and/or the deformation of the waveform, and this obscures the correlation between the applied voltage waveform and the induced magnetization dynamics.

An external bias magnetic field has been used to determine the axis for magnetization precession in most experimental demonstrations of voltage-induced dynamic switching. However, the application of a magnetic field is not suitable for practical circuits. Therefore, we also need efforts to replace the external bias field by an effective field, such as through crystalline anisotropy and exchange bias fields. Matsumoto et al. proposed using a combination of a conical magnetization state and shape anisotropy to induce precessional switching under zero-bias magnetic field [146]. Conical magnetization states have been mainly studied in multilayer structures containing Co, such as Co/Pt and Co/Pd [147–149], however recently it can be realized, even in a practical CoFeB/MgO structure [150–152]. Therefore, the above proposed structure might be applicable if we can realize a sufficiently-high thermal stability while keeping the conical states.

6. Conclusions

Electric-field control of spin has the potential to make substantial impact on the development of novel nonvolatile memory with ultra-low operating power, as well as the expected zero stand-by power. The utilization of the voltage-controlled magnetic anisotropy (VCMA) effect is a promising approach to realizing voltage-torque MRAMs. Bi-stable magnetization switching has been demonstrated while using precessional dynamics that are induced by the VCMA effect. The purely-electronic VCMA effect originates from electric-field induced modification of the electronic structure at the interface between an ultrathin ferromagnet and a dielectric layer, such as MgO. In a $3d$ transition ferromagnet, e.g. Fe and Co, the voltage-induced change in the orbital magnetic moment plays an important role in the origin of the VCMA effect through the carrier accumulation/depletion effect at the interface. On the other hand, in a $3d/5d$ composite system, e.g. L1$_0$-FePt film, an electric quadrupole mechanism also has significant influence on the VCMA effect. To increase of the VCMA coefficient, the utilization of proximity-induced magnetism in a $5d$ transition metal, which has large spin-orbit coupling, is promising. A large VCMA coefficient of -320 fJ/Vm has been achieved in an Ir-doped ultrathin Fe layer with a demonstration of high-speed responsiveness. As for the reliability of writing while using voltage-induced dynamic switching, low write error rates of the order of 10^{-6} have been realized by improving the thermal stability and the VCMA effect in practical perpendicularly-magnetized MTJs. Further enhancement in the VCMA coefficient is the key to demonstrating the potential for scalability and realizing more reliable switching for voltage-torque MRAM. A novel nonvolatile memory maintaining low operating power as well as zero stand-by power can provide a broader option for the design of memory hierarchy in future data-driven society. We expect that the voltage-torque MRAM has the potential to be applied in IoT edge devices and wearable/implantable computing systems, in which, ultimately, low power consumption is strongly demanded. Furthermore, the voltage-control of spin may also lead to the improvement in other spintronic devices, such as a voltage-tuned magnetic sensor, spin-torque oscillator, and spin-based neuromorphic devices.

Author Contributions: T.N. wrote Section 1 "Introduction", Section 2 "Overview of the VCMA effect and voltage-induced dynamic switching", and Section 4 "Materials research for large VCMA effect". T.Y. wrote Section 5 "Towards the reliable voltage-induced dynamic switching". S.M. wrote Section 3 "Physical origin of the

VCMA effect". M.T. wrote part of the theoretical discussion in Section 4. M.S., Y.S. and S.Y. supervised the work and edited the manuscript.

Acknowledgments: This work was supported by the ImPACT Program of the Council for Science. We thank Y. Shiota, A. Kozioł-Rachwał, W. Skowroński, X. Xu, T. Ikeura, T. Ohkubo, T. Tsukahara, M. Suzuki, S. Tamaru, H. Kubota, A. Fukushima, K. Hono, K. Nakamura, T. Oda, R. Matsumoto, H. Imamura, Y. Miura, T. Taniguchi, T. Yorozu, Y. Kotani, T. Nakamura and M. Sahashi for fruitful discussions. The XAS and XMCD measurements were performed in SPring-8 with the approval of the Japan Synchrotron Radiation Research Institute (Proposal Nos. 2016B1017).

Conflicts of Interest: The authors declare no conflict of interest.

References

1. Yuasa, S.; Nagahama, T.; Fukushima, A.; Suzuki, Y.; Ando, K. Giant room-temperature magnetoresistance in single-crystal Fe/MgO/Fe magnetic tunnel junctions. *Nat. Mater.* **2004**, *3*, 868–871. [CrossRef]
2. Parkin, S.S.; Kaiser, C.; Panchula, A.; Rice, P.M.; Hughes, B.; Samant, M.; Yang, S.H. Giant tunnelling magnetoresistance at room temperature with MgO (100) tunnel barriers. *Nat. Mater.* **2004**, *3*, 862–867. [CrossRef]
3. Slonczewski, J.C. Current-driven excitation of magnetic multilayers. *J. Magn. Magn. Mater.* **1996**, *159*, L1. [CrossRef]
4. Berger, L. Emission of spin waves by a magnetic multilayer traversed by a current. *Phys. Rev. B* **1996**, *54*, 9353–9358. [CrossRef]
5. Myers, E.B.; Ralph, D.C.; Katine, J.A.; Louie, R.N.; Buhrman, R.A. Current-induced switching of domains in magnetic multilayer devices. *Science* **1999**, *285*, 867–870. [CrossRef] [PubMed]
6. Katine, J.A.; Albert, F.J.; Buhrman, R.A.; Myers, E.B.; Ralph, D.C. Current-driven magnetization reversal and spin-wave excitation in Co/Cu/Co pillars. *Phys. Rev. Lett.* **2000**, *84*, 3149–3152. [CrossRef]
7. Huai, Y.; Albert, F.; Nguyen, P.; Pakala, M.; Valet, T. Observation of spin-transfer switching in deep submicron-sized and low-resistance magnetic tunnel junctions. *Appl. Phys. Lett.* **2004**, *84*, 3118–3120. [CrossRef]
8. Kubota, H.; Fukushima, A.; Ootani, Y.; Yuasa, S.; Ando, K.; Maehara, H.; Tsunekawa, K.; Djayaprawira, D.D.; Watanabe, N.; Suzuki, Y. Evaluation of Spin-Transfer Switching in CoFeB/MgO/CoFeB Magnetic Tunnel Junctions. *Jpn. J. Appl. Phys.* **2005**, *44*, L1237. [CrossRef]
9. Chernyshov, A.; Overby, M.; Liu, X.; Furdyna, J.K.; Lyanda-Geller, Y.; Rokhinson, L.P. Evidence for reversible control of magnetization in a ferromagnetic material by means of spin–orbit magnetic field. *Nat. Phys.* **2009**, *5*, 656–659. [CrossRef]
10. Miron, I.M.; Gaudin, G.; Auffret, S.; Rodmacq, B.; Schuhl, A.; Pizzini, S.; Vogel, J.; Gambardella, P. Current-driven spin torque induced by the Rashba effect in a ferromagnetic metal layer. *Nat. Mater.* **2010**, *9*, 230–234. [CrossRef]
11. Miron, I.M.; Garello, K.; Gaudin, G.; Zermatten, P.J.; Costache, M.V.; Auffret, S.; Bandiera, S.; Rodmacq, B.; Schuhl, A.; Gambardella, P. Perpendicular switching of a single ferromagnetic layer induced by in-plane current injection. *Nature* **2011**, *476*, 189–193. [CrossRef] [PubMed]
12. Liu, L.; Pai, C.-F.; Li, Y.; Tseng, H.W.; Ralph, D.C.; Buhrman, R.A. Spin-torque switching with the giant spin Hall effect of tantalum. *Science* **2012**, *336*, 555–558. [CrossRef] [PubMed]
13. Ando, K.; Fujita, S.; Ito, J.; Yuasa, S.; Suzuki, Y.; Nakatani, Y.; Miyazaki, T.; Yoda, H. Spin-transfer torque magnetoresistive random-access memory technologies for normally off computing (invited). *J. Appl. Phys.* **2014**, *115*, 172607. [CrossRef]
14. Khvalkovskiy, A.V.; Apalkov, D.; Watts, S.; Chepulskii, R.; Beach, R.S.; Ong, A.; Tang, X.; Driskill-Smith, A.; Butler, W.H.; Visscher, B.P.; et al. Basic principles of STT-MRAM cell operation in memory arrays. *J. Phys. D Appl. Phys.* **2013**, *46*, 074001. [CrossRef]
15. Rizal, C.; Moa, B.; Niraula, B. Ferromagnetic Multilayers: Magnetoresistance, Magnetic Anisotropy, and Beyond. *Magnetochemistry* **2016**, *2*, 22. [CrossRef]
16. Novosad, V.; Otani, Y.; Ohsawa, A.; Kim, S.G.; Fukamichi, K.; Koike, J.; Maruyama, K.; Kitakami, O.; Shimada, Y. Novel magnetostrictive memory device. *J. Appl. Phys.* **2000**, *87*, 6400–6402. [CrossRef]
17. Hu, J.M.; Li, Z.; Chen, L.Q.; Nan, C.W. High-density magnetoresistive random access memory operating at ultralow voltage at room temperature. *Nat. Commun.* **2011**, *2*, 553. [CrossRef] [PubMed]

18. Wu, T.; Bur, A.; Wong, K.; Zhao, P.; Lynch, C.S.; Amiri, P.K.; Wang, K.L.; Carman, G.P. Electrical control of reversible and permanent magnetization reorientation for magnetoelectric memory devices. *Appl. Phys. Lett.* **2011**, *98*, 262504.

19. Ohno, H.; Chiba, D.; Matsukura, F.; Omiya, T.; Abe, E.; Dietl, T.; Ohno, Y.; Ohtani, K. Electric-field control of ferromagnetism. *Nature* **2000**, *408*, 944–946. [CrossRef] [PubMed]

20. Chiba, D.; Yamanouchi, M.; Matsukura, F.; Ohno, H. Electrical manipulation of magnetization reversal in a ferromagnetic semiconductor. *Science* **2003**, *301*, 943–945. [CrossRef] [PubMed]

21. Yamada, Y.; Ueno, K.; Fukumura, T.; Yuan, H.T.; Shimotani, H.; Iwasa, Y.; Gu, L.; Tsukimoto, S.; Ikuhara, Y.; Kawasaki, M. Electrically induced ferromagnetism at room temperature in cobalt-doped titanium dioxide. *Science* **2011**, *332*, 1065–1067. [CrossRef] [PubMed]

22. Chiba, D.; Fukami, S.; Shimamura, K.; Ishiwata, N.; Kobayashi, K.; Ono, T. Electrical control of the ferromagnetic phase transition in cobalt at room temperature. *Nat. Mater.* **2011**, *10*, 853–856. [CrossRef] [PubMed]

23. Borisov, P.; Hochstrat, A.; Chen, X.; Kleemann, W.; Binek, C. Magnetoelectric switching of exchange bias. *Phys. Rev. Lett.* **2005**, *94*, 117203. [CrossRef] [PubMed]

24. He, X.; Wang, Y.; Wu, N.; Caruso, A.N.; Vescovo, E.; Belashchenko, K.D.; Dowben, P.A.; Binek, C. Robust isothermal electric control of exchange bias at room temperature. *Nat. Mater.* **2010**, *9*, 579–585. [CrossRef] [PubMed]

25. Ashida, T.; Oida, M.; Shimomura, N.; Nozaki, T.; Shibata, T.; Sahashi, M. Observation of magnetoelectric effect in Cr_2O_3/Pt/Co thin film system. *Appl. Phys. Lett.* **2014**, *104*, 152409. [CrossRef]

26. Toyoki, K.; Shiratsuchi, Y.; Kobane, A.; Mitsumata, C.; Kotani, Y.; Nakamura, T.; Nakatani, R. Magnetoelectric switching of perpendicular exchange bias in Pt/Co/α-Cr_2O_3/Pt stacked films. *Appl. Phys. Lett.* **2015**, *106*, 162404. [CrossRef]

27. Duan, C.G.; Jaswal, S.S.; Tsymbal, E.Y. Predicted magnetoelectric effect in Fe/$BaTiO_3$ multilayers: Ferroelectric control of magnetism. *Phys. Rev. Lett.* **2006**, *97*, 047201. [CrossRef]

28. Radaelli, G.; Petti, D.; Plekhanov, E.; Fina, I.; Torelli, P.; Salles, B.R.; Cantoni, M.; Rinaldi, C.; Gutierrez, D.; Panaccione, G.; et al. Electric control of magnetism at the Fe/$BaTiO_3$ interface. *Nat. Commun.* **2014**, *5*, 3404. [CrossRef]

29. Gerhard, L.; Yamada, T.K.; Balashov, T.; Takacs, A.F.; Wesselink, R.J.; Dane, M.; Fechner, M.; Ostanin, S.; Ernst, A.; Mertig, I.; et al. Magnetoelectric coupling at metal surfaces. *Nat. Nanotechnol.* **2010**, *5*, 792–797. [CrossRef]

30. Heron, J.T.; Bosse, J.L.; He, Q.; Gao, Y.; Trassin, M.; Ye, L.; Clarkson, J.D.; Wang, C.; Liu, J.; Salahuddin, S.; et al. Deterministic switching of ferromagnetism at room temperature using an electric field. *Nature* **2014**, *516*, 370–373. [CrossRef] [PubMed]

31. Tokunaga, Y.; Taguchi, Y.; Arima, T.-H.; Tokura, Y. Electric-field-induced generation and reversal of ferromagnetic moment in ferrites. *Nat. Phys.* **2012**, *8*, 838–844. [CrossRef]

32. Weisheit, M.; Fahler, S.; Marty, A.; Souche, Y.; Poinsignon, C.; Givord, D. Electric field-induced modification of magnetism in thin-film ferromagnets. *Science* **2007**, *315*, 349–351. [CrossRef] [PubMed]

33. Maruyama, T.; Shiota, Y.; Nozaki, T.; Ohta, K.; Toda, N.; Mizuguchi, M.; Tulapurkar, A.A.; Shinjo, T.; Shiraishi, M.; Mizukami, S.; et al. Large voltage-induced magnetic anisotropy change in a few atomic layers of iron. *Nat. Nanotechnol.* **2009**, *4*, 158–161. [CrossRef]

34. Amiri, P.K.; Alzate, J.G.; Cai, X.Q.; Ebrahimi, F.; Hu, Q.; Wong, K.; Grezes, C.; Lee, H.; Yu, G.; Li, X.; et al. Electric-Field-Controlled Magnetoelectric RAM: Progress, Challenges, and Scaling. *IEEE Trans. Magn.* **2015**, *51*, 3401507.

35. Nozaki, T.; Kozioł-Rachwał, A.; Tsujikawa, M.; Shiota, Y.; Xu, X.; Ohkubo, T.; Tsukahara, T.; Miwa, S.; Suzuki, M.; Tamaru, S.; et al. Highly efficient voltage control of spin and enhanced interfacial perpendicular magnetic anisotropy in iridium-doped Fe/MgO magnetic tunnel junctions. *NPG Asia Mater.* **2017**, *9*, e451. [CrossRef]

36. Wang, K.L.; Lee, H.; Amiri, P.K. Magnetoelectric Random Access Memory-Based Circuit Design by Using Voltage-Controlled Magnetic Anisotropy in Magnetic Tunnel Junctions. *IEEE Trans. Nanotechnol.* **2015**, *14*, 992–997. [CrossRef]

37. Noguchi, H.; Ikegami, K.; Abe, K.; Fujita, S.; Shiota, Y.; Nozaki, T.; Yuasa, S.; Suzuki, Y. Novel Voltage Controlled MRAM(VCM) with Fast Read/Write Circuits for Ultra Large Last Level Cache. In Proceedings of the 2016 IEEE International Electron Devices Meeting (IEDM), San Francisco, CA, USA, 3–7 Decembder 2016.

38. Lee, H.; Lee, A.; Wang, S.; Ebrahimi, F.; Gupta, P.; Amiri, P.K.; Wang, K.L. Analysis and Compact Modeling of Magnetic Tunnel Junctions Utilizing Voltage-Controlled Magnetic Anisotropy. *IEEE Trans. Magn.* **2018**, *54*, 4400209. [CrossRef]

39. Long, M.; Zeng, L.; Gao, T.; Zhang, D.; Qin, X.; Zhang, Y.; Zhao, W. Self-Adaptive Write Circuit for Magnetic Tunneling Junction Memory With Voltage-Controlled Magnetic Anisotropy Effect. *IEEE Trans. Nanotechnol.* **2018**, *17*, 492–499. [CrossRef]

40. Chiba, D.; Sawicki, M.; Nishitani, Y.; Nakatani, Y.; Matsukura, F.; Ohno, H. Magnetization vector manipulation by electric fields. *Nature* **2008**, *455*, 515–518. [CrossRef]

41. Duan, C.G.; Velev, J.P.; Sabirianov, R.F.; Zhu, Z.; Chu, J.; Jaswal, S.S.; Tsymbal, E.Y. Surface magnetoelectric effect in ferromagnetic metal films. *Phys. Rev. Lett.* **2008**, *101*, 137201. [CrossRef]

42. Nakamura, K.; Shimabukuro, R.; Fujiwara, Y.; Akiyama, T.; Ito, T.; Freeman, A.J. Giant modification of the magnetocrystalline anisotropy in transition-metal monolayers by an external electric field. *Phys. Rev. Lett.* **2009**, *102*, 187201. [CrossRef] [PubMed]

43. Tsujikawa, M.; Oda, T. Finite electric field effects in the large perpendicular magnetic anisotropy surface Pt/Fe/Pt(001): A first-principles study. *Phys. Rev. Lett.* **2009**, *102*, 247203. [CrossRef] [PubMed]

44. Xu, L.; Zhang, S. Electric field control of interface magnetic anisotropy. *J. Appl. Phys.* **2012**, *111*, 07C501. [CrossRef]

45. Barnes, S.E.; Ieda, J.; Maekawa, S. Rashba spin-orbit anisotropy and the electric field control of magnetism. *Sci. Rep.* **2014**, *4*, 4105. [CrossRef] [PubMed]

46. Nakamura, K.; Akiyama, T.; Ito, T.; Weinert, M.; Freeman, A.J. Role of an interfacial FeO layer in the electric-field-driven switching of magnetocrystalline anisotropy at the Fe/MgO interface. *Phys. Rev. B* **2010**, *81*, 220409(R). [CrossRef]

47. Shiota, Y.; Maruyama, T.; Nozaki, T.; Shinjo, T.; Shiraishi, A.M.; Suzuki, Y. Voltage-Assisted Magnetization Switching in Ultrathin $Fe_{80}Co_{20}$ Alloy Layers. *Appl. Phys. Exp.* **2009**, *2*, 063001. [CrossRef]

48. Nozaki, T.; Shiota, Y.; Shiraishi, M.; Shinjo, T.; Suzuki, Y. Voltage-induced perpendicular magnetic anisotropy change in magnetic tunnel junctions. *Appl. Phy. Lett.* **2010**, *96*, 022506. [CrossRef]

49. Shiota, Y.; Murakami, S.; Bonell, F.; Nozaki, T.; Shinjo, T.; Suzuki, Y. Quantitative Evaluation of Voltage-Induced Magnetic Anisotropy Change by Magnetoresistance Measurement. *Appl. Phys. Exp.* **2011**, *4*, 043005. [CrossRef]

50. Nozaki, T.; Shiota, Y.; Miwa, S.; Murakami, S.; Bonell, F.; Ishibashi, S.; Kubota, H.; Yakushiji, K.; Saruya, T.; Fukushima, A.; et al. Electric-field-induced ferromagnetic resonance excitation in an ultrathin ferromagnetic metal layer. *Nat. Phys.* **2012**, *8*, 492–496. [CrossRef]

51. Zhu, J.; Katine, J.A.; Rowlands, G.E.; Chen, Y.J.; Duan, Z.; Alzate, J.G.; Upadhyaya, P.; Langer, J.; Amiri, P.K.; Wang, K.L.; et al. Voltage-induced ferromagnetic resonance in magnetic tunnel junctions. *Phys Rev Lett* **2012**, *108*, 197203. [CrossRef]

52. Shiota, Y.; Miwa, S.; Tamaru, S.; Nozaki, T.; Kubota, H.; Fukushima, A.; Suzuki, Y.; Yuasa, S. High-output microwave detector using voltage-induced ferromagnetic resonance. *Appl. Phys. Lett.* **2014**, *105*, 192408. [CrossRef]

53. Kanai, S.; Gajek, M.; Worledge, D.C.; Matsukura, F.; Ohno, H. Electric field-induced ferromagnetic resonance in a CoFeB/MgO magnetic tunnel junction under dc bias voltages. *Appl. Phys. Lett.* **2014**, *105*, 242409. [CrossRef]

54. Rana, B.; Fukuma, Y.; Miura, K.; Takahashi, H.; Otani, Y. Effect of excitation power on voltage induced local magnetization dynamics in an ultrathin CoFeB film. *Sci. Rep.* **2017**, *7*, 2318. [CrossRef] [PubMed]

55. Shiota, Y.; Nozaki, T.; Bonell, F.; Murakami, S.; Shinjo, T.; Suzuki, Y. Induction of coherent magnetization switching in a few atomic layers of FeCo using voltage pulses. *Nat. Mater.* **2012**, *11*, 39–43. [CrossRef] [PubMed]

56. Verba, R.; Carpentieri, M.; Finocchio, G.; Tiberkevich, V.; Slavin, A. Excitation of Spin Waves in an In-Plane-Magnetized Ferromagnetic Nanowire Using Voltage-Controlled Magnetic Anisotropy. *Phys. Rev. Appl.* **2017**, *7*, 064023. [CrossRef]

57. Chen, Y.J.; Lee, H.K.; Verba, R.; Katine, J.A.; Barsukov, I.; Tiberkevich, V.; Xiao, J.Q.; Slavin, A.N.; Krivorotov, I.N. Parametric Resonance of Magnetization Excited by Electric Field. *Nano Lett.* **2017**, *17*, 572–577. [CrossRef] [PubMed]

58. Rana, B.; Fukuma, Y.; Miura, K.; Takahashi, H.; Otani, Y. Excitation of coherent propagating spin waves in ultrathin CoFeB film by voltage-controlled magnetic anisotropy. *Appl. Phys. Lett.* **2017**, *111*, 052404. [CrossRef]

59. Bonell, F.; Murakami, S.; Shiota, Y.; Nozaki, T.; Shinjo, T.; Suzuki, Y. Large change in perpendicular magnetic anisotropy induced by an electric field in FePd ultrathin films. *Appl. Phys. Lett.* **2011**, *98*, 232510. [CrossRef]

60. Seki, T.; Kohda, M.; Nitta, J.; Takanashi, K. Coercivity change in an FePt thin layer in a Hall device by voltage application. *Appl. Phys. Lett.* **2011**, *98*, 212505. [CrossRef]

61. Kikushima, S.; Seki, T.; Uchida, K.; Saitoh, E.; Takanashi, K. Electric field effect on magnetic anisotropy for Fe-Pt-Pd alloys. *AIP Adv.* **2017**, *7*, 085210. [CrossRef]

62. Kozioł-Rachwał, A.; Nozaki, T.; Freindl, K.; Korecki, J.; Yuasa, S.; Suzuki, Y. Enhancement of perpendicular magnetic anisotropy and its electric field-induced change through interface engineering in Cr/Fe/MgO. *Sci. Rep.* **2017**, *7*, 5993. [CrossRef] [PubMed]

63. Xiang, Q.; Wen, Z.; Sukegawa, H.; Kasai, S.; Seki, T.; Kubota, T.; Takanashi, K.; Mitani, S. Nonlinear electric field effect on perpendicular magnetic anisotropy in Fe/MgO interfaces. *J. Phys. D Appl. Phys.* **2017**, *50*, 40LT04. [CrossRef]

64. Wen, Z.; Sukegawa, H.; Seki, T.; Kubota, T.; Takanashi, K.; Mitani, S. Voltage control of magnetic anisotropy in epitaxial Ru/Co$_2$FeAl/MgO heterostructures. *Sci. Rep.* **2017**, *7*, 45026. [CrossRef] [PubMed]

65. Miwa, S.; Fujimoto, J.; Risius, P.; Nawaoka, K.; Goto, M.; Suzuki, Y. Strong Bias Effect on Voltage-Driven Torque at Epitaxial Fe-MgO Interface. *Phys. Rev. X* **2017**, *7*, 031018. [CrossRef]

66. Shukla, A.K.; Goto, M.; Xu, X.; Nawaoka, K.; Suwardy, J.; Ohkubo, T.; Hono, K.; Miwa, S.; Suzuki, Y. Voltage-Controlled Magnetic Anisotropy in Fe1-xCox/Pd/MgO system. *Sci. Rep.* **2018**, *8*, 10362. [CrossRef] [PubMed]

67. Suzuki, K.Z.; Kimura, S.; Kubota, H.; Mizukami, S. Magnetic Tunnel Junctions with a Nearly Zero Moment Manganese Nanolayer with Perpendicular Magnetic Anisotropy. *ACS Appl. Mater. Interfaces* **2018**, *10*, 43305. [CrossRef] [PubMed]

68. Endo, M.; Kanai, S.; Ikeda, S.; Matsukura, F.; Ohno, H. Electric-field effects on thickness dependent magnetic anisotropy of sputtered MgO/Co$_{40}$Fe$_{40}$B$_{20}$/Ta structures. *Appl. Phys. Lett.* **2010**, *96*, 212503. [CrossRef]

69. Kita, K.; Abraham, D.W.; Gajek, M.J.; Worledge, D.C. Electric-field-control of magnetic anisotropy of Co0.6Fe0.2B0.2/oxide stacks using reduced voltage. *J. Appl. Phys.* **2012**, *112*, 033919. [CrossRef]

70. Skowronński, W.; Wiśniowski, P.; Stobiecki, T.; Cardoso, S.; Freitas, P.P.; van Dijken, S. Magnetic field sensor with voltage-tunable sensing properties. *Appl. Phys. Lett.* **2012**, *101*, 192401. [CrossRef]

71. Nozaki, T.; Yakushiji, K.; Tamaru, S.; Sekine, M.; Matsumoto, R.; Konoto, M.; Kubota, H.; Fukushima, A.; Yuasa, S. Voltage-Induced Magnetic Anisotropy Changes in an Ultrathin FeB Layer Sandwiched between Two MgO Layers. *Appl. Phys. Exp.* **2013**, *6*, 073005. [CrossRef]

72. Shiota, Y.; Bonell, F.; Miwa, S.; Mizuochi, N.; Shinjo, T.; Suzuki, Y. Opposite signs of voltage-induced perpendicular magnetic anisotropy change in CoFeB|MgO junctions with different underlayers. *Appl. Phys. Lett.* **2013**, *103*, 082410. [CrossRef]

73. Alzate, J.G.; Amiri, P.K.; Yu, G.; Upadhyaya, P.; Katine, J.A.; Langer, J.; Ocker, B.; Krivorotov, I.N.; Wang, K.L. Temperature dependence of the voltage-controlled perpendicular anisotropy in nanoscale MgO|CoFeB|Ta magnetic tunnel junctions. *Appl. Phys. Lett.* **2014**, *104*, 112410. [CrossRef]

74. Meng, H.; Naik, V.B.; Liu, R.; Han, G. Electric field control of spin re-orientation in perpendicular magnetic tunnel junctions—CoFeB and MgO thickness dependence. *Appl. Phys. Lett.* **2014**, *105*, 042410. [CrossRef]

75. Okada, A.; Kanai, S.; Yamanouchi, M.; Ikeda, S.; Matsukura, F.; Ohno, H. Electric-field effects on magnetic anisotropy and damping constant in Ta/CoFeB/MgO investigated by ferromagnetic resonance. *Appl. Phys. Lett.* **2014**, *105*, 052415. [CrossRef]

76. Li, X.; Yu, G.; Wu, H.; Ong, P.V.; Wong, K.; Hu, Q.; Ebrahimi, F.; Upadhyaya, P.; Akyol, M.; Kioussis, N.; et al. Thermally stable voltage-controlled perpendicular magnetic anisotropy in Mo|CoFeB|MgO structures. *Appl. Phys. Lett.* **2015**, *107*, 142403. [CrossRef]

77. Skowroński, W.; Nozaki, T.; Lam, D.D.; Shiota, Y.; Yakushiji, K.; Kubota, H.; Fukushima, A.; Yuasa, S.; Suzuki, Y. Underlayer material influence on electric-field controlled perpendicular magnetic anisotropy in CoFeB/MgO magnetic tunnel junctions. *Phys. Rev. B* **2015**, *91*, 184410. [CrossRef]

78. Skowroński, W.; Nozaki, T.; Shiota, Y.; Tamaru, S.; Yakushiji, K.; Kubota, H.; Fukushima, A.; Yuasa, S.; Suzuki, Y. Perpendicular magnetic anisotropy of Ir/CoFeB/MgO trilayer system tuned by electric fields. *Appl. Phys. Exp.* **2015**, *8*, 053003. [CrossRef]

79. Piotrowski, S.K.; Bapna, M.; Oberdick, S.D.; Majetich, S.A.; Li, M.; Chien, C.L.; Ahmed, R.; Victora, R.H. Size and voltage dependence of effective anisotropy in sub-100-nm perpendicular magnetic tunnel junctions. *Phys. Rev. B* **2016**, *94*, 014404. [CrossRef]

80. YLau, -C.; Sheng, P.; Mitani, S.; Chiba, D.; Hayashi, M. Electric field modulation of the non-linear areal magnetic anisotropy energy. *Appl. Phys. Lett.* **2017**, *110*, 022405.

81. Li, X.; Fitzell, K.; Wu, D.; Karaba, C.T.; Buditama, A.; Yu, G.; Wong, K.L.; Altieri, N.; Grezes, C.; Kioussis, N.; et al. Enhancement of voltage-controlled magnetic anisotropy through precise control of Mg insertion thickness at CoFeB|MgO interface. *Appl. Phys. Lett.* **2017**, *110*, 052401. [CrossRef]

82. Sonntag, A.; Hermenau, J.; Schlenhoff, A.; Friedlein, J.; Krause, S.; Wiesendanger, R. Electric-field-induced magnetic anisotropy in a nanomagnet investigated on the atomic scale. *Phys. Rev. Lett.* **2014**, *112*, 017204. [CrossRef] [PubMed]

83. Zhou, T.; Leong, S.H.; Yuan, Z.M.; Hu, S.B.; Ong, C.L.; Liu, B. Manipulation of magnetism by electrical field in a real recording system. *Appl. Phys. Lett.* **2010**, *96*, 012506. [CrossRef]

84. Bauer, U.; Przybylski, M.; Beach, G.S.D. Voltage control of magnetic anisotropy in Fe films with quantum well states. *Phys. Rev. B* **2014**, *89*, 174402. [CrossRef]

85. Chiba, D.; Kawaguchi, M.; Fukami, S.; Ishiwata, N.; Shimamura, K.; Kobayashi, K.; Ono, T. Electric-field control of magnetic domain-wall velocity in ultrathin cobalt with perpendicular magnetization. *Nat. Commun.* **2012**, *3*, 888. [CrossRef] [PubMed]

86. Schellekens, A.J.; van den Brink, A.; Franken, J.H.; Swagten, H.J.; Koopmans, B. Electric-field control of domain wall motion in perpendicularly magnetized materials. *Nat. Commun.* **2012**, *3*, 847. [CrossRef] [PubMed]

87. Bauer, U.; Emori, S.; Beach, G.S.D. Voltage-controlled domain wall traps in ferromagnetic nanowires. *Nat. Nanotechnol.* **2013**, *8*, 411–416. [CrossRef] [PubMed]

88. Schott, M.; Bernand-Mantel, A.; Ranno, L.; Pizzini, S.; Vogel, J.; Bea, H.; Baraduc, C.; Auffret, S.; Gaudin, G.; Givord, D. The Skyrmion Switch: Turning Magnetic Skyrmion Bubbles on and off with an Electric Field. *Nano Lett.* **2017**, *17*, 3006–3012. [CrossRef] [PubMed]

89. Srivastava, T.; Schott, M.; Juge, R.; Krizakova, V.; Belmeguenai, M.; Roussigne, Y.; Bernand-Mantel, A.; Ranno, L.; Pizzini, S.; Cherif, S.M.; et al. Large-Voltage Tuning of Dzyaloshinskii-Moriya Interactions: A Route toward Dynamic Control of Skyrmion Chirality. *Nano Lett.* **2018**, *18*, 4871–4877. [CrossRef] [PubMed]

90. Nozaki, T.; Jibiki, Y.; Goto, M.; Tamura, E.; Nozaki, T.; Kubota, H.; Fukushima, A.; Yuasa, S.; Suzuki, Y. Brownian motion of skyrmion bubbles and its control by voltage applications. *Appl. Phys. Lett.* **2019**, *114*, 012402. [CrossRef]

91. Nawaoka, K.; Miwa, S.; Shiota, Y.; Mizuochi, N.; Suzuki, Y. Voltage induction of interfacial Dzyaloshinskii–Moriya interaction in Au/Fe/MgO artificial multilayer. *Appl. Phys. Express* **2015**, *8*, 063004. [CrossRef]

92. Newhouse-Illige, T.; Liu, Y.; Xu, M.; Hickey, D.R.; Kundu, A.; Almasi, H.; Bi, C.; Wang, X.; Freeland, J.W.; Keavney, D.J.; et al. Voltage-controlled interlayer coupling in perpendicularly magnetized magnetic tunnel junctions. *Nat. Commun.* **2017**, *8*, 15232. [CrossRef]

93. Hibino, Y.; Koyama, T.; Obinata, A.; Miwa, K.; Ono, S.; Chiba, D. Electric field modulation of magnetic anisotropy in perpendicularly magnetized Pt/Co structure with a Pd top layer. *Appl. Phys. Express* **2015**, *8*, 113002. [CrossRef]

94. Obinata, A.; Hibino, Y.; Hayakawa, D.; Koyama, T.; Miwa, K.; Ono, S.; Chiba, D. Electric-field control of magnetic moment in Pd. *Sci. Rep.* **2015**, *5*, 14303. [CrossRef]

95. Hibino, T.K.Y.; Obinata, A.; Harai, T.; Ota, S.; Miwa, K.; Ono, S.; Matsukura, F.; Ohno, H.; Chiba, D. Peculiar temperature dependence of electric-field effect on magnetic anisotropy in Co/Pd/MgO system. *Appl. Phys. Lett.* **2016**, *109*, 082403. [CrossRef]

96. Amiri, P.K.; Upadhyaya, P.; Alzate, J.G.; Wang, K.L. Electric-field-induced thermally assisted switching of monodomain magnetic bits. *J. Appl. Phys.* **2013**, *113*, 013912. [CrossRef]

97. Han, G.; Huang, J.; Chen, B.; Lim, S.T.; Tran, M. Electric Field Assisted Switching in Magnetic Random Access Memory. *IEEE Trans. Magn.* **2015**, *51*, 3401207. [CrossRef]

98. Nozaki, T.; Arai, H.; Yakushiji, K.; Tamaru, S.; Kubota, H.; Imamura, H.; Fukushima, A.; Yuasa, S. Magnetization switching assisted by high-frequency-voltage-induced ferromagnetic resonance. *Appl. Phys. Express* **2014**, *7*, 073002. [CrossRef]

99. Wang, W.G.; Li, M.; Hageman, S.; Chien, C.L. Electric-field-assisted switching in magnetic tunnel junctions. *Nat. Mater.* **2012**, *11*, 64–68. [CrossRef] [PubMed]

100. Kanai, S.; Nakatani, Y.; Yamanouchi, M.; Ikeda, S.; Sato, H.; Matsukura, F.; Ohno, H. Magnetization switching in a CoFeB/MgO magnetic tunnel junction by combining spin-transfer torque and electric field-effect. *Appl. Phys. Lett.* **2014**, *104*, 212406. [CrossRef]

101. Yoda, N.S.H.; Ohsawa, Y.; Shirotori, S.; Kato, Y.; Inokuchi, T.; Kamiguchi, Y.; Altansargai, B.; Saito, Y.; Koi, K.; Sugiyama, H.; et al. Voltage-Control Spintronics Memory (VoCSM) Having Potentials of Ultra-Low Energy-Consumption and High-Density. In Proceedings of the 2016 IEEE International Electron Devices Meeting (IEDM), San Francisco, CA, USA, 3–7 December 2016.

102. Shiota, Y.; Miwa, S.; Nozaki, T.; Bonell, F.; Mizuochi, N.; Shinjo, T.; Kubota, H.; Yuasa, S.; Suzuki, Y. Pulse voltage-induced dynamic magnetization switching in magnetic tunneling junctions with high resistance-area product. *Appl. Phys. Lett.* **2012**, *101*, 102406. [CrossRef]

103. Kanai, S.; Yamanouchi, M.; Ikeda, S.; Nakatani, Y.; Matsukura, F.; Ohno, H. Electric field-induced magnetization reversal in a perpendicular-anisotropy CoFeB-MgO magnetic tunnel junction. *Appl. Phys. Lett.* **2012**, *101*, 122403. [CrossRef]

104. Grezes, C.; Rozas, A.R.; Ebrahimi, F.; Alzate, J.G.; Cai, X.; Katine, J.A.; Langer, J.; Ocker, B.; Amiri, P.K.; Wang, K.L. In-plane magnetic field effect on switching voltage and thermal stability in electric-field-controlled perpendicular magnetic tunnel junctions. *AIP Adv.* **2016**, *6*, 075014. [CrossRef]

105. Shiota, Y.; Nozaki, T.; Tamaru, S.; Yakushiji, K.; Kubota, H.; Fukushima, A.; Yuasa, S.; Suzuki, Y. Evaluation of write error rate for voltage-driven dynamic magnetization switching in magnetic tunnel junctions with perpendicular magnetization. *Appl. Phys. Exp.* **2016**, *9*, 013001. [CrossRef]

106. Shiota, Y.; Nozaki, T.; Tamaru, S.; Yakushiji, K.; Kubota, H.; Fukushima, A.; Yuasa, S.; Suzuki, Y. Reduction in write error rate of voltage-driven dynamic magnetization switching by improving thermal stability factor. *Appl. Phys. Lett.* **2017**, *111*, 022408. [CrossRef]

107. Yamamoto, T.; Nozaki, T.; Shiota, Y.; Imamura, H.; Tamaru, S.; Yakushiji, K.; Kubota, H.; Fukushima, A.; Suzuki, Y.; Yuasa, S. Thermally Induced Precession-Orbit Transition of Magnetization in Voltage-Driven Magnetization Switching. *Phys. Rev. Appl.* **2018**, *10*, 024004. [CrossRef]

108. Yamamoto, T.; Nozaki, T.; Imamura, H.; Shiota, Y.; Ikeura, T.; Tamaru, S.; Yakushiji, K.; Kubota, H.; Fukushima, A.; Suzuki, Y.; et al. Write-Error Reduction of Voltage-Torque-Driven Magnetization Switching by a Controlled Voltage Pulse. *Phys. Rev. Appl.* **2019**, *11*, 014013. [CrossRef]

109. Yamamoto, T.; Nozaki, T.; Imamura, H.; Shiota, Y.; Tamaru, S.; Yakushiji, K.; Kubota, H.; Fukushima, A.; Suzuki, Y.; Yuasa, S. Improvement of write error rate in voltage-driven magnetization switching. *J. Phys. D Appl. Phys.* **2019**, *52*, 164001. [CrossRef]

110. Grezes, C.; Lee, H.; Lee, A.; Wang, S.; Ebrahimi, F.; Li, X.; Wong, K.; Katine, J.A.; Ocker, B.; Langer, J.; et al. Write Error Rate and Read Disturbance in Electric-Field-Controlled MRAM. *IEEE Magn. Lett.* **2016**, *8*, 3102705.

111. Miwa, S.; Suzuki, M.; Tsujikawa, M.; Nozaki, T.; Nakamura, T.; Shirai, M.; Yuasa, S.; Suzuki, Y. Perpendicular magnetic anisotropy and its electric-field-induced change at metal-dielectric interfaces. *J. Phys. D Appl. Phys.* **2019**, *52*, 063001. [CrossRef]

112. Bruno, P. Tight-binding approach to the orbital magnetic moment and magnetocrystalline anisotropy of transition-metal monolayers. *Phys. Rev. B* **1989**, *39*, 865–868. [CrossRef]

113. Kawabe, T.; Yoshikawa, K.; Tsujikawa, M.; Tsukahara, T.; Nawaoka, K.; Kotani, Y.; Toyoki, K.; Goto, M.; Suzuki, M.; Nakamura, T.; et al. Electric-field-induced changes of magnetic moments and magnetocrystalline anisotropy in ultrathin cobalt films. *Phys. Rev. B* **2017**, *96*, 220412(R). [CrossRef]

114. Suzuki, Y.; Miwa, S. Magnetic anisotropy of ferromagnetic metals in low-symmetry systems. *Phys. Lett. A* **2019**, *383*, 1203–1206. [CrossRef]

115. Van der Laan, G. Microscopic origin of magnetocrystalline anisotropy in transition metal thin films. *J. Phys. Condens. Mater.* **1997**, *10*, 3239–3253. [CrossRef]

116. Miwa, S.; Suzuki, M.; Tsujikawa, M.; Matsuda, K.; Nozaki, T.; Tanaka, K.; Tsukahara, T.; Nawaoka, K.; Goto, M.; Kotani, Y.; et al. Voltage controlled interfacial magnetism through platinum orbits. *Nat. Commun.* **2017**, *8*, 15848. [CrossRef] [PubMed]

117. Thole, B.T.; Carra, P.; Sette, F.; van der Laan, G. X-ray circular dichroism as a probe of orbital magnetization. *Phys. Rev. Lett.* **1992**, *68*, 1943–1946. [CrossRef] [PubMed]

118. Carra, P.; Thole, B.T.; Altarelli, M.; Wang, X. X-ray circular dichroism and local magnetic fields. *Phys. Rev. Lett.* **1993**, *70*, 694–697. [CrossRef] [PubMed]

119. Bonell, F.; Takahashi, Y.T.; Lam, D.D.; Yoshida, S.; Shiota, Y.; Miwa, S.; Nakamura, T.; Suzuki, Y. Reversible change in the oxidation state and magnetic circular dichroism of Fe driven by an electric field at the FeCo/MgO interface. *Appl. Phys. Lett.* **2013**, *102*, 152401. [CrossRef]

120. Miwa, S.; Matsuda, K.; Tanaka, K.; Kotani, Y.; Goto, M.; Nakamura, T.; Suzuki, Y. Voltage-controlled magnetic anisotropy in Fe|MgO tunnel junctions studied by x-ray absorption spectroscopy. *Appl. Phys. Lett.* **2015**, *107*, 162402. [CrossRef]

121. Suzuki, M.; Tsukahara, T.; Miyakaze, R.; Furuta, T.; Shimose, K.; Goto, M.; Nozaki, T.; Yuasa, S.; Suzuki, Y.; Miwa, S. Extended X-ray absorption fine structure analysis of voltage-induced effects in the interfacial atomic structure of Fe/Pt/MgO. *Appl. Phys. Express* **2017**, *10*, 063006. [CrossRef]

122. Bi, C.; Liu, Y.; Newhouse-Illige, T.; Xu, M.; Rosales, M.; Freeland, J.W.; Mryasov, O.; Zhang, S.; Velthuis, S.G.T.; Wang, W.G. Reversible control of Co magnetism by voltage-induced oxidation. *Phys. Rev. Lett.* **2014**, *113*, 267202. [CrossRef] [PubMed]

123. Tsukahara, T.; Kawabe, T.; Shimose, K.; Furuta, T.; Miyakaze, R.; Nawaoka, K.; Goto, M.; Nozaki, T.; Yuasa, S.; Kotani, Y.; et al. Characterization of the magnetic moments of ultrathin Fe film in an external electric field via high-precision X-ray magnetic circular dichroism spectroscopy. *Jpn. J. Appl. Phys.* **2017**, *56*, 060304. [CrossRef]

124. Yamada, K.T.; Suzuki, M.; Pradipto, A.M.; Koyama, T.; Kim, S.; Kim, K.J.; Ono, S.; Taniguchi, T.; Mizuno, H.; Ando, F.; et al. Microscopic Investigation into the Electric Field Effect on Proximity-Induced Magnetism in Pt. *Phys. Rev. Lett.* **2018**, *120*, 157203. [CrossRef] [PubMed]

125. Bauer, U.; Yao, L.; Tan, A.J.; Agrawal, P.; Emori, S.; Tuller, H.L.; van Dijken, S.; Beach, G.S. Magneto-ionic control of interfacial magnetism. *Nat. Mater.* **2015**, *14*, 174–181. [CrossRef]

126. Gilbert, D.A.; Grutter, A.J.; Arenholz, E.; Liu, K.; Kirby, B.J.; Borchers, J.A.; Maranville, B.B. Structural and magnetic depth profiles of magneto-ionic heterostructures beyond the interface limit. *Nat. Commun.* **2016**, *7*, 12264. [CrossRef] [PubMed]

127. Sakamaki, M.; Amemiya, K. Observation of an electric field-induced interface redox reaction and magnetic modification in GdOx/Co thin film by means of depth-resolved X-ray absorption spectroscopy. *Phys. Chem. Chem. Phys.* **2018**, *20*, 20004. [CrossRef] [PubMed]

128. Bauer, U.; Przybylski, M.; Kirschner, J.; Beach, G.S. Magnetoelectric charge trap memory. *Nano Lett.* **2012**, *12*, 1437–1442. [CrossRef] [PubMed]

129. Rajanikanth, A.; Hauet, T.; Montaigne, F.; Mangin, S.; Andrieu, S. Magnetic anisotropy modified by electric field in V/Fe/MgO(001)/Fe epitaxial magnetic tunnel junction. *Appl. Phys. Lett.* **2013**, *103*, 062402. [CrossRef]

130. Ong, P.V.; Kioussis, N.; Odkhuu, D.; Amiri, P.K.; Wang, K.L.; Carman, G.P. Giant voltage modulation of magnetic anisotropy in strained heavy metal/magnet/insulator heterostructures. *Phys. Rev. B* **2015**, *92*, 020407(R). [CrossRef]

131. Odkhuu, D. Giant strain control of magnetoelectric effect in Ta|Fe|MgO. *Sci. Rep.* **2016**, *6*, 32742. [CrossRef]

132. Kato, Y.; Yoda, H.; Saito, Y.; Oikawa, S.; Fujii, K.; Yoshiki, M.; Koi, K.; Sugiyama, H.; Ishikawa, M.; Inokuchi, T.; et al. Giant voltage-controlled magnetic anisotropy effect in a crystallographically strained CoFe system. *Appl. Phys. Express* **2018**, *11*, 053007. [CrossRef]

133. Nozaki, T.; Kozioł-Rachwał, A.; Skowroński, W.; Zayets, V.; Shiota, Y.; Tamaru, S.; Kubota, H.; Fukushima, A.; Yuasa, S.; Suzuki, Y. Large Voltage-Induced Changes in the Perpendicular Magnetic Anisotropy of an MgO-Based Tunnel Junction with an Ultrathin Fe Layer. *Phys. Rev. Appl.* **2016**, *5*, 044006. [CrossRef]

134. Koo, J.W.; Mitani, S.; Sasaki, T.T.; Sukegawa, H.; Wen, Z.C.; Ohkubo, T.; Niizeki, T.; Inomata, K.; Hono, K. Large perpendicular magnetic anisotropy at Fe/MgO interface. *Appl. Phys. Lett.* **2013**, *103*, 192401. [CrossRef]

135. Lambert, C.H.; Rajanikanth, A.; Hauet, T.; Mangin, S.; Fullerton, E.E.; Andrieu, S. Quantifying perpendicular magnetic anisotropy at the Fe-MgO(001) interface. *Appl. Phys. Lett.* **2013**, *102*, 122410. [CrossRef]
136. Okabayashi, J.; Koo, J.W.; Sukegawa, H.; Mitani, S.; Takagi, Y.; Yokoyama, T. Perpendicular magnetic anisotropy at the interface between ultrathin Fe film and MgO studied by angular-dependent x-ray magnetic circular dichroism. *Appl. Phys. Lett.* **2014**, *105*, 122408. [CrossRef]
137. Bonell, F.; Lam, D.D.; Yoshida, S.; Takahashi, Y.T.; Shiota, Y.; Miwa, S.; Nakamura, T.; Suzuki, Y. Investigation of Au and Ag segregation on Fe(001) with soft X-ray absorption. *Surf. Sci.* **2013**, *616*, 125–130. [CrossRef]
138. Xiang, H.S.Q.; Al-Mahdawi, M.; Belmoubaric, M.; Kasai, S.; Sakuraba, Y.; Mitani, S.; Hono, K. Atomic layer number dependence of voltage-controlled magnetic anisotropy in Cr/Fe/MgAl2O4 heterostructure. In Proceedings of the Intermag2018, Singapore, FC-03, Marina Bay Sands Convention Centere, Singapore, 23–27 April 2018.
139. Pardede, I.; Kanagawa, T.; Ikhsan, N.; Murata, I.; Yoshikawa, D.; Obata, M.; Oda, T. A Comprehensive Study of Sign Change in Electric Field Control Perpendicular Magnetic Anisotropy Energy at Fe/MgO Interface: First Principles Calculation. *IEEE Trans. Magn.* **2019**, *55*, 1700104. [CrossRef]
140. Zhang, J.; Lukashev, P.V.; Jaswal, S.S.; Tsymbal, E.Y. Model of orbital populations for voltage-controlled magnetic anisotropy in transition -metal thin films. *Phys. Rev. B* **2017**, *96*, 014435. [CrossRef]
141. Nakamura, K.; Nomura, T.; Pradipto, A.M.; Nawa, K.; Akiyama, T.; Ito, T. Effect of heavy-metal insertions at Fe/MgO interfaces on electric-field-induced modification of magnetocrystalline anisotropy. *J. Magn. Magn. Mater.* **2017**, *429*, 214–220. [CrossRef]
142. Bonaedy, T.; Choi, J.W.; Jang, C.; Min, B.-C.; Chang, J. Enhancement of electric-field-induced change of magnetic anisotropy by interface engineering of MgO magnetic tunnel junctions. *J. Phys. D Appl. Phys.* **2015**, *48*, 225002. [CrossRef]
143. Nozaki, T.; Yamamoto, T.; Tamaru, S.; Kubota, H.; Fukushima, A.; Suzuki, A.Y.; Yuasa, S. Enhancement in the interfacial perpendicular magnetic anisotropy and the voltage-controlled magnetic anisotropy by heavy metal doping at the Fe/MgO interface. *APL Mater.* **2018**, *6*, 026101. [CrossRef]
144. Ikeura, T.; Nozaki, T.; Shiota, Y.; Yamamoto, T.; Imamura, H.; Kubota, H.; Fukushima, A.; Suzuki, Y.; Yuasa, S. Reduction in the write error rate of voltage-induced dynamic magnetization switching using the reverse bias method. *Jpn. J. Appl. Phys.* **2018**, *57*, 040311. [CrossRef]
145. Tamaru, S.; Yamamoto, T.; Nozaki, T.; Yuasa, S. Accurate calculation and shaping of the voltage pulse waveform applied to a voltage-controlled magnetic random access memory cell. *Jpn. J. Appl. Phys.* **2018**, *57*, 073002. [CrossRef]
146. Matsumoto, R.; Nozaki, T.; Yuasa, S.; Imamura, H. Voltage-Induced Precessional Switching at Zero-Bias Magnetic Field in a Conically Magnetized Free Layer. *Phys. Rev. Appl.* **2018**, *9*, 014026. [CrossRef]
147. Lee, J.-W.; Jeong, J.-R.; Shin, S.-C.; Kim, J.; Kim, S.-K. Spin-reorientation transitions in ultrathin Co films on Pt(111) and Pd(111) single-crystal substrates. *Phys. Rev. B* **2002**, *66*, 172409. [CrossRef]
148. Stamps, R.L.; Louail, L.; Hehn, M.; Gester, M.; Ounadjela, K. Anisotropies, cone states, and stripe domains in Co/Pt multilayers. *J. Appl. Phys.* **1997**, *81*, 4751–4753. [CrossRef]
149. Kisielewski, M.; Maziewski, A.; Tekielak, M.; Ferré, J.; Lemerle, S.; Mathet, V.; Chappert, C. Magnetic anisotropy and magnetization reversal processes in Pt/Co/Pt films. *J. Magn. Magn. Mater.* **2003**, *260*, 231–243. [CrossRef]
150. Shaw, J.M.; Nembach, H.T.; Weiler, M.; Silva, T.J.; Schoen, M.; Sun, J.Z.; Worledge, D.C. Perpendicular Magnetic Anisotropy and Easy Cone State in Ta/Co$_{60}$Fe$_{20}$B$_{20}$/MgO. *IEEE Magn. Lett.* **2015**, *6*, 3500404. [CrossRef]
151. Fu, Y.; Barsukov, I.; Li, J.; Gonçalves, A.M.; Kuo, C.C.; Farle, M.; Krivorotov, I.N. Temperature dependence of perpendicular magnetic anisotropy in CoFeB thin films. *Appl. Phys. Lett.* **2016**, *108*, 142403. [CrossRef]
152. Park, K.-W.; Park, J.-Y.; Baek, S.-H.C.; Kim, D.-H.; Seo, S.-M.; Chung, S.-W.; Park, B.-G. Electric field control of magnetic anisotropy in the easy cone state of Ta/Pt/CoFeB/MgO structures. *Appl. Phys. Lett.* **2016**, *109*, 012405. [CrossRef]

Article

Compensating Circuit to Reduce the Impact of Wire Resistance in a Memristor Crossbar-Based Perceptron Neural Network

Son Ngoc Truong

Faculty of Electrical and Electronics Engineering, Ho Chi Minh City University of Technology and Education, Ho Chi Minh City 70000, Vietnam; sontn@hcmute.edu.vn; Tel.: +84-931-085-929

Received: 8 September 2019; Accepted: 1 October 2019; Published: 2 October 2019

Abstract: Wire resistance in metal wire is one of the factors that degrade the performance of memristor crossbar circuits. In this paper, an analysis of the impact of wire resistance in a memristor crossbar is performed and a compensating circuit is proposed to reduce the impact of wire resistance in a memristor crossbar-based perceptron neural network. The goal of the analysis is to figure out how wire resistance influences the output voltage of a memristor crossbar. It emerges that the wire resistance on horizontal lines causes the neuron's output voltage to vary more than the wire resistance on vertical lines. More interesting, the voltage variation caused by wire resistance on horizontal lines increases proportionally to the length of metal wire. The first column has small voltage variation whereas the last column has large voltage variation. In addition, two adjacent columns have almost the same amount of voltage variation. Under these observations, a memristor crossbar-based perceptron neural network with compensating circuit is proposed. The neuron's outputs of two columns are put into a subtractor circuit to eliminate the voltage variation caused by the wire resistance. The proposed memristor crossbar-based perceptron neural network is trained to recognize the 26 characters. The proposed memristor crossbar shows better recognition rate compared to the previous work when wire resistance is taken into account. The proposed memristor crossbar circuit can maintain the recognition rate as high as 100% when wire resistance is as high as 2.5 Ω. By contrast, the recognition rate of the memristor crossbar without the compensating circuit decreases by 1%, 5%, and 19% when wire resistance is set to be 1.5, 2.0, and 2.5 Ω, respectively.

Keywords: memristor; crossbar array; wire resistance; synaptic weight; character recognition

1. Introduction

Neuromorphic computing, inspired from biological perception, was introduced by C. Mead in the late 1980s [1]. It has been expected to become an alternative architecture to overcome the bottleneck of von Neumann computer architectures [1,2]. Neuromorphic computing refers to a hardware implementation of a brain-inspired system, which has the capabilities of parallel processing like a human brain. For realizing neuromorphic computing systems, various research activities, based on CPUs (Central Processing Units), GPUs (Graphics Processing Units), FPGAs (Field-Programmable Gate Arrays), analog circuits, memory circuits, etc., have been proposed in the past two decades [3–8]. These architectures are based on CMOS (Complementary-Metal-Oxide-Semiconductor) technology, which is approaching the end of their capabilities because scaling CMOS down faces several fundamental limiting factors stemming from electron thermal energy and quantum-mechanical tunneling [9,10]. The memristor crossbar array has been one of the promising candidates for realizing neuromorphic computing systems because crossbar architecture can be made with high density and low cost [11]. Memristor was postulated by Leon O. Chua in 1971 as the fourth basic circuit element and experimentally demonstrated by HP Lab in 2008 [12,13]. A memristor is a resistor with modifiable resistance, which makes it ideal for mimicking the synaptic plasticity of biological neurons [14]. The early memristor-based synaptic circuits are composed of memristors and CMOS transistors [15–17]. However, pure memristor crossbar arrays without CMOS devices seem to be more efficient in terms of their integration and power consumption [18–23]. Miao Hu et al. proposed a crossbar synaptic array that is composed of a plus and minus crossbar array representing plus- and minus-polarity connection matrices for analog neuromorphic computing [20]. Such a pure memristor crossbar array is very effective in realizing the bio-inspired systems in term of power consumption and area occupation. To reduce area and power consumption, S. N. Truong proposed a new memristor crossbar array architecture which is composed of a single memristor array and a constant-term circuit [21].

In a memristor array, some amount of voltage drop can be caused by interconnect resistance, also known as wire resistance along the row and the column lines [19,24–27]. Wire resistance degrades the performance of the circuit more seriously when the array size increases [25]. To mitigate the impact of wire resistance, several interesting schemes were proposed [24–27]. These schemes are effective when they are applied to a memristor crossbar array, in which memristors are used as binary switches between two distinct high and low resistance states (HRS and LRS, respectively). However, the impact of wire resistance in an analog memristor crossbar array for realizing the synaptic weight matrix was not fully considered. In this work, we propose a memristor crossbar array with a compensating circuit for implementing the analog synaptic array of a perceptron neural network. The impact of wire resistance is mitigated by compensating the voltage variation of two adjacent columns.

In this work, the output voltages of columns are figured out with taking the existing of wire resistance into account. The mathematical analysis and the simulation result show that the output voltage of columns increase, which is caused by the amount of voltage lost from wire resistance. The column close to the first one has a small variation of voltage, compared to the one far from the first column. From these observations, we propose a compensating circuit to mitigate the voltage variation caused by the wire resistance in a memristor crossbar array.

2. Materials and Methods

Figure 1 shows an interesting memristor array circuit for implementing the synaptic weight matrix of a perceptron neural network [21]. A single memristor array and a constant-term circuit are used for realizing the negative and positive synaptic weights, instead of using two complementary crossbar arrays [20,21].

In Figure 1, $g_{j,k}$ is the memristor's conductance at the crossing point between the jth row and the kth column. $V_{IN,j}$ is the input voltage applied to the jth row. $V_{C,k}$ is the column-line voltage on the kth column. The column line, $V_{C,F}$, is added in Figure 1 instead of using another memristor array [21]. The column line, $V_{C,F}$, is connected to the inputs, from $V_{IN,1}$ to $V_{IN,m}$. In Figure 1, $V_{C,F}$ enters G_F that constitutes an inverting OP amp with the negative feedback resistor, R_{F1}. The output voltage of G_F is V_F that is connected to all the column lines from $V_{C,1}$ to $V_{C,n}$ via R_{F2}, as shown in Figure 1. By applying Kirchhoff current law to the column line, $V_{C,F}$, we can calculate V_F and $V_{O,k}$ with Equations (1) and (2).

$$V_F = -\sum_{j=1}^{m} \frac{R_{F1}}{R_B} V_{IN,j}. \tag{1}$$

$$V_{O,k} = -\left[\sum_{j=1}^{m}\left(R_0 \cdot g_{j,k} \cdot V_{IN,j}\right) + \frac{R_0}{R_{F2}} V_F\right]. \tag{2}$$

If we choose $R_{F1} = R_{F2}$ and combining Equation (1) with Equation (2), the following Equation (3) can be obtained [21].

$$V_{O,k} = -\left[\sum_{j=1}^{m}\left(R_0 \cdot g_{j,k} - \frac{R_0}{R_B}\right)V_{IN,j}\right]. \tag{3}$$

If $-\left(R_0 \cdot g_{j,k} - \frac{R_0}{R_B}\right)$ is defined as a synaptic weight of the jth row and kth column, $w_{j,k}$, we can rewrite Equation (3) with Equation (4).

$$V_{O,k} = \sum_{j=1}^{m} w_{j,k} V_{IN,j},$$ (4)

where $w_{j,k} = R_0\left(\frac{1}{R_B} - g_{j,k}\right) = R_0\left(\frac{1}{R_B} - \frac{1}{M_{j,k}}\right)$.

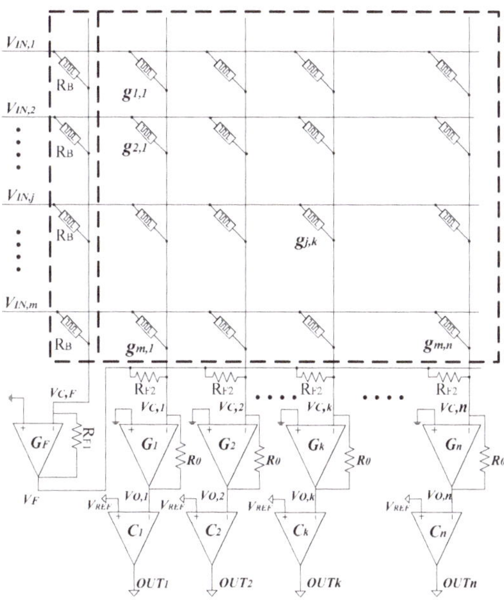

Figure 1. The memristor-based crossbar architecture with a single memristor array and a constant-term circuit for realizing the synaptic matrix of a perceptron neural network [21].

Equation (4) is used for calculating the output voltage of the kth column. The output of each column is a summation of the weighted inputs, hence each column works as a perceptron neuron. In Equation (4), $M_{j,k}$ is the memristance value of the crossing point between the jth row and kth column. R_B is a constant. The synaptic weight, $w_{j,k}$, can be decided to be either negative or positive by adjusting the memristance, $M_{j,k}$. The output of the perceptron neuron is decided by a threshold function which produces 0 or 1. By adding the comparator to the output voltage, $V_{O,k}$, we can decide if the neuron's output of the kth column, OUT_k, should be activated or not.

$$OUT_k = \begin{cases} 1, & \text{if } V_{O,k} \geq V_{REF} \\ 0, & \text{if } V_{O,k} < V_{REF} \end{cases}.$$ (5)

In previous work, the impact of wire resistance is ignored. However, in the crossbar array, the voltage drop along column and row lines cannot be omitted [19,24–27]. It becomes more serious when the array size increases [24]. The wire resistance between two adjacent junctions is modeled by a small-value resistor, r, as shown in Figure 2.

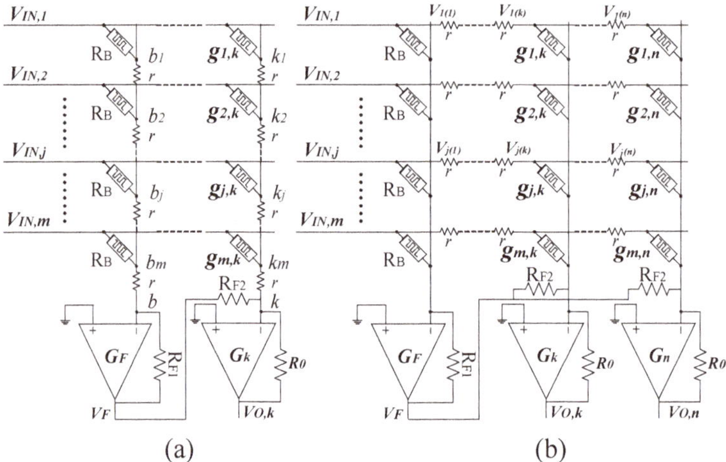

Figure 2. Wire resistance between two adjacent junctions is modeled by a small-value resistor, r, connecting between two crossing points. (**a**) Wire resistance on horizontal lines is omitted. (**b**) Wire resistance on vertical lines is omitted.

For the sake of simplicity, in this section we analyze the circuit separately with respect to the wire resistance on horizontal lines and the wire resistance on vertical lines, as shown in Figure 2a,b, respectively. We define V_{b1}, V_{b2} as the voltages of node b_1, b_2, which are on the first column. Generally, V_{bj} is the voltage of node b_j on the first column. Similarly, V_{kj} is the voltage of node k_j, which is on the jth column. Applying Kirchhoff current law for all nodes in Figure 2a, V_F and $V_{O,k}$ can be estimated as follows:

$$\frac{-V_F}{R_{F1}} = \frac{V_{IN,m}-V_{bm}}{R_B} + \ldots + \frac{V_{IN,j}-V_{bj}}{R_B} + \ldots + \frac{V_{IN,1}-V_{b1}}{R_B}$$
$$V_F = -R_{F1}\left(\sum_{j=1}^{m} \frac{V_{IN,j}}{R_B} - \sum_{j=1}^{m} \frac{V_{bj}}{R_B} \right). \tag{6}$$

$$\frac{-V_{O,k}}{R_0} = (V_{IN,m} - V_{km})g_{m,k} + \ldots + \left(V_{IN,j} - V_{kj}\right)g_{j,k} + \ldots + (V_{IN,1} - V_{k1})g_{1,k} + \frac{V_F}{R_{F2}}$$
$$V_{O,k} = -R_0\left(\sum_{j=1}^{m} V_{IN,j}g_{j,k} - \sum_{j=1}^{m} V_{kj}g_{j,k} + \frac{V_F}{R_{F2}} \right). \tag{7}$$

If we assume that $R_{F1} = R_{F2}$, Equation (7) can be simplified as follow:

$$V_{O,k} = -\left[\sum_{j=1}^{m}\left(R_0 \cdot g_{j,k} - \frac{R_0}{R_B} \right)V_{IN,j} - \sum_{j=1}^{m} R_0 V_{kj}g_{j,k} + \sum_{j=1}^{m} R_0 \frac{V_{bj}}{R_B} \right]. \tag{8}$$

By comparing Equation (8) and Equation (4), we can derive the variation of voltage, ΔV, which is caused by wire resistance on the vertical lines.

$$\Delta V = -\sum_{j=1}^{m} R_0 \frac{V_{kj}}{M_{j,k}} + \sum_{j=1}^{m} R_0 \frac{V_{bj}}{R_B}. \tag{9}$$

Here $M_{j,k}$ is the memristance of the crossing point between the jth row and the kth column. V_{bj} and V_{kj} are the voltage at nodes b_j and k_j of the first column and the kth column, respectively, as shown in Figure 2a. $M_{j,k}$ is calculated using Equation (4). It is possible to infer that the variation of voltage presented in Equation (9) can be very small because there are a negative term and a positive term in the right side of Equation (9).

In Figure 2b, wire resistance on vertical lines is omitted whereas wire resistance on horizontal lines is taken into account. The voltages applied to the columns decrease because they are lost from wire resistance. If we define $V_{j(k)}$ as the amount of voltage drop on wire resistance, which is on the jth row and between the $(k-1)$th and kth column, the voltage applied to the jth row of the kth column is calculated as Equation (10).

$$V_{IN,j(k)} = V_{IN,j} - \sum_{i=1}^{k} V_{j(i)}.$$ (10)

Here $V_{IN,j(k)}$ is the voltage applied to the jth row of the kth column. The column-line voltage on the kth column, $V_{O,k}$, can be calculated using Equation (11).

$$V_{O,k} = -\left[\sum_{j=1}^{m}\left(R_0 \cdot g_{j,k} V_{IN,j(k)} - \frac{R_0}{R_B} V_{IN,j}\right)\right].$$ (11)

By comparing Equation (11) and Equation (3), we obtain the variation of voltage, ΔV_k, of the kth column as follows.

$$\Delta V_k = \sum_{j=1}^{m} R_0 g_{j,k} V_{IN,j} - \sum_{j=1}^{m} R_0 g_{j,k} V_{IN,j(k)}.$$ (12)

Calculating $V_{IN,j(k)}$ by using Equation (10), we obtain ΔV_k as presented in Equation (13).

$$\Delta V_k = \sum_{j=1}^{m}\left(R_0 g_{j,k} \sum_{i=1}^{k} V_{j(i)}\right).$$ (13)

Here $\sum_{i=1}^{k} V_{j(i)}$ is the sum of the voltage on k resistors on the jth row. Equation (13) indicates that the output voltage of the kth column increases because of wire resistance. It is possible to infer that the column close to the first column has small voltage variation and the column far from the first column has large voltage variation. In Equation (13), the voltage variation increases proportionally to the column's index, k. Hence, it is interesting that two adjacent columns can have almost the same amount of voltage variation. Due to this reason, we propose a memristor crossbar array with compensating circuit to mitigate the voltage variation caused by wire resistance. By putting two adjacent columns into a subtraction circuit, the voltage variation can be eliminated significantly. The proposed memristor crossbar is schematically shown in Figure 3.

In Figure 3, the memristor crossbar is composed of 27 columns for recognizing 26 character images. The first column is a constant-term circuit to generate a negative voltage, as mentioned in the previous section. The remaining 26 columns represent 26 perception neurons trained to recognize the 26 characters. The differential amplifies from $G_{s,2}$ to $G_{s,26}$ are inserted into the circuit. The gain of these amplifiers is 1, so they work as the subtractors. The output voltages from $V_{O,1}$ to $V_{O,n}$ are the neuron's output of columns from Col_1 to Col_n. $V_{O,1}$ enter the comparator C_1 to decide if the neuron's output of column Col_1 should be activated or not. $V_{O,2}$ and $V_{O,1}$ go into $G_{s,2}$ that produces $V_{Os,2}$. $V_{Os,2}$ enters the comparator C_2 to decide if the neuron's output of column Col_2 should be activated or not. In general, the output voltage of the column Col_{k-1} and the column Col_k enter the subtractor $G_{s,k}$ for generating the neuron's output, $V_{Os,k}$, of the column Col_k. Using superposition theorem, $V_{Os,k}$ can be calculated with the difference of $V_{O,k-1}$ and $V_{O,k}$.

$$V_{Os,k} = -V_{O,k-1}\left(\frac{R_4}{R_3}\right) + V_{O,k}\left(\frac{R_6}{R_5 + R_6}\right)\left(\frac{R_3 + R_4}{R_3}\right).$$ (14)

If we assume that $R_3 = R_4 = R_5 = R_6$, we can obtain:

$$V_{Os,k} = V_{O,k} - V_{O,k-1}. \tag{15}$$

The differential amplifier is able to reject any signal common to both inputs. That means, if two adjacent columns have almost the same amount of voltage variation, the voltage variation is then mitigated at the output.

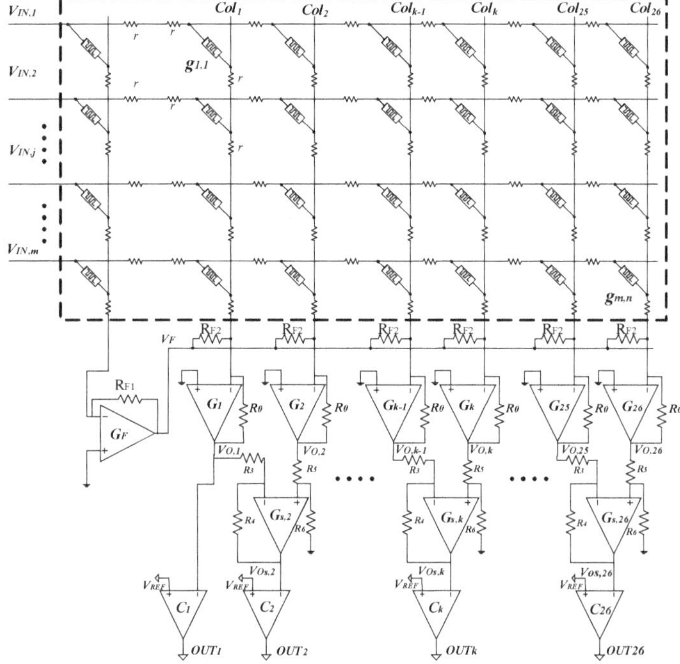

Figure 3. The proposed memristor crossbar with compensating circuit for implementing a perceptron neural network. The outputs of two adjacent columns are put into a differential amplifier working as a subtractor to eliminate the output voltage variation.

The concept of the proposed circuit is shown in Figure 4. The crossbar is trained to recognize the 26 characters from "A" to "Z". The 25th column is for recognizing the character "Y". The output of the 25th column is close to 1V when the input is "Y" and close to 0 when the other characters are applied to the input. Similarly, the neuron's output of the 26th column should be activated if the input is "Z", as indicated in Figure 4a. In Figure 4b, it is assumed that the wire resistance is present in the crossbar. The output voltage increases as reasoned in the previous section. The two last neurons recognize the input characters incorrectly, as demonstrated in Figure 4b. However, if we put the outputs of two last columns into a subtractor, the voltage variation can be mitigated significantly, as illustrated in Figure 4b. By doing this, we can maintain the recognition rate when wire resistance is present in the crossbar circuit.

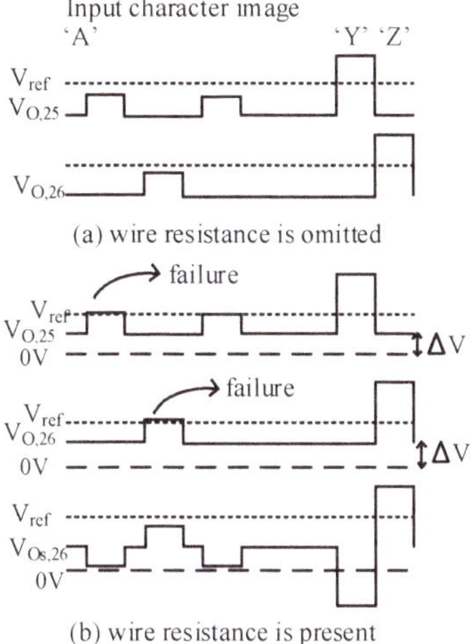

Figure 4. The concept of the proposed circuit for compensating the output voltage variation caused by wire resistance. (**a**) The ideal output of the 25th and 26th columns, which are trained to recognize character images of "Y" and "Z", respectively. (**b**) The output voltage of the 25th and 26th columns when the wire resistance is taken into account. $V_{Os,26}$ is the output of subtractor for the 26th column, as depicted in Figure 3.

3. Results

The proposed memristor crossbar circuit in Figure 3 is verified for the application of character recognition. Figure 5a shows eight × eight images of characters used in this simulation. Each character is composed of 64 black-and-white pixels. The proposed memristor crossbar is composed of 64 rows and 27 columns. The first column connects with all inputs through R_B to generate the negative voltage as mentioned in the previous section. The remaining 26 columns are for recognition of 26 characters from "A" to "Z". The 64 input voltages obtained from 64 pixels are applied to the inputs of 64 rows.

The red line in Figure 5b shows a hysteresis behavior of a real memristor based on the film structure of Pt/LaAlO$_3$/Nb-doped SrTiO$_3$ stacked layer [28]. The black line in Figure 5b represents the behavior model of the memristor used in this paper. This model can well describe various memristive behaviors that come from different kinds of memristors [29]. The circuit simulation is performed using the SPECTRE circuit simulation provided by Cadence Design Systems Inc. Memristors are modeled using Verilog-A and the CMOS technology is given by SAMSUNG 0.13 mm process technology [29,30]. The Verilog-A model parameters are presented in [28]. The wire resistance between two adjacent junctions is set to be 2.5 Ω for a 4F^2 cross-point structure [19,31]. Figure 6a shows the neuron's output of the 25th column, which is trained to be activated when character "Y" is applied to the input. Ideally, $V_{O,25}$ is close to 1V for character "Y", and close to 0V for others. However, the output voltage of the 25th column, $V_{O,25}$, is shifted up because of wire resistance, as reasoned in the previous section. Similarly, in Figure 6b, the neuron's output of the 26th column is shifted up as a result of the voltage drop along wire resistance. It can be realized that if we compare the column's output voltage, $V_{O,26}$, with the reference voltage, V_{REF}, the neuron's output of the 26th column can be activated for several input characters, which consequently degrades the recognition rate. The output voltage of the 25th column and the 26th column are put into a subtractor circuit to produce the neuron's output voltage of the 26th column, $V_{Os,26}$. By doing this, the voltage variation is mitigated significantly, as demonstrated in Figure 6c. When the character "Y" is applied to the inputs, $V_{Os,26}$ is negative, because $V_{O,25}$ is higher than $V_{O,26}$. For the character "Z", $V_{Os,26}$ is high, as indicated in Figure 6c. The simulation result shown in Figure 6c indicates that the neuron's output of the 26th column is only activated for the input character "Z", because the variation of voltage caused by wire resistance is mitigated remarkably by the subtractor circuit.

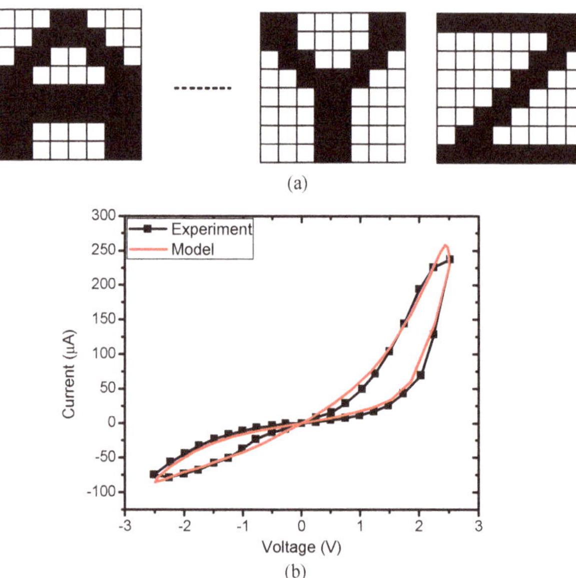

(a)

(b)

Figure 5. (**a**) The eight × eight pixels images of characters used to test the proposed memristor crossbar circuit. (**b**) The memristor's current–voltage characteristic measured from the real device and the memristor's behavior model [28,29].

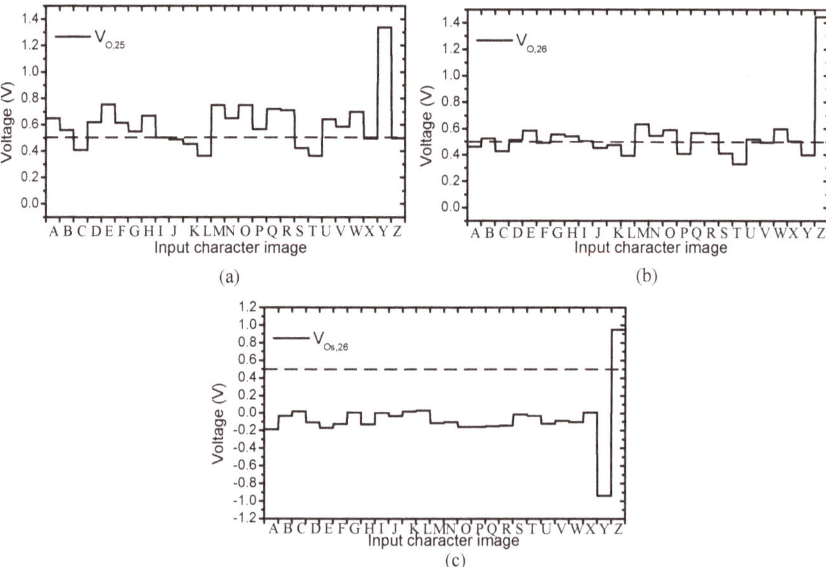

(a)

(b)

(c)

Figure 6. The simulation result of the proposed memristor crossbar array depicted in Figure 4. (**a**) The neuron's output of the 25th column without compensating circuit. (**b**) The neuron's output of the 26th column without compensating circuit. (**d**) The neuron's output of the 26th column with compensating circuit. The wire resistance between two adjacent junctions is set to be 2.5 Ω [19,28].

The proposed circuit is tested with wire resistance that is varied from 0.5 to 2.5 Ω. This range of wire resistance is commonly used and obtained from the International Technology Roadmap for Semiconductors [24,25,31–34]. Figure 7 shows the comparison of the recognition rate between the memristor crossbar without compensating circuit and the proposed memristor crossbar with compensating circuit when the wire resistance is set to be 0.5, 1.0, 1.5, 2.0, and 2.5 Ω, respectively. The recognition rate of the memristor crossbar without compensating circuit declines dramatically when wire resistance increases. In particular, the recognition rate of the memristor crossbar without compensating circuit is 99%, 95%, and 81%, when the wire resistance is set to be 1.5, 2.0, and 2.5 Ω, respectively. By contrast, the proposed memristor crossbar with compensating circuit can maintain the recognition as high as 100% when wire resistance is as high as 2.5 Ω.

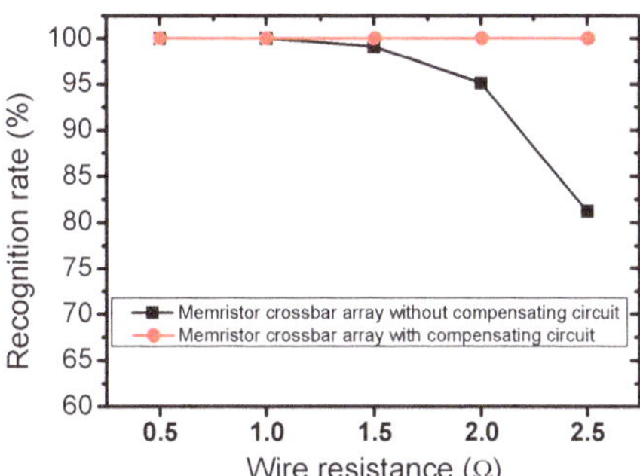

Figure 7. The comparison of the recognition rate between the memristor crossbar without compensating circuit and the proposed memristor crossbar with compensating circuit. The wire resistance is set to be 0.5, 1.0, 1.5, 2.0, and 2.5 Ω, respectively.

4. Discussion

Finally, we discuss the power and area overhead of the proposed memristor crossbar circuit. The proposed circuit uses the compensating circuit constituted by an Op-Amp and four resistors. The proposed circuit consumes more power and area, compared to the memristor crossbar without compensating circuit. However, the proposed memristor crossbar with compensating circuit shows better recognition rate by 19% than the previous memristor crossbar circuit, when wire resistance is set to be 2.5 Ω. Because wire resistance in the crossbar cannot be omitted, the proposed scheme makes the memristor crossbar-based perceptron neural network become more possible. The proposed circuit can be applied to memristor-based crossbar architectures which are used in resistive memory and artificial neural networks [34–36].

5. Conclusions

In this work, a memristor crossbar-based perceptron neural network with compensating circuit is proposed. The neuron's outputs of two columns are put into a subtractor circuit to eliminate the voltage variation caused by wire resistance. The memristor crossbar-based perceptron neural network is trained to recognize the 26 characters. The proposed memristor crossbar with compensating circuit shows better recognition rate, compared to the previous memristor crossbar without compensating circuit when wire resistance is taken into account. The simulation result shows that the proposed circuit can maintain the recognition rate as high as 100% when the wire resistance is set to be 2.5 Ω. By contrast, the recognition rate of the memristor crossbar without compensating circuit decreases by 19% when wire resistance is set to be 2.5 Ω.

Funding: This research received no external funding.

Conflicts of Interest: The author declares no conflicts of interest.

References

1. Mead, C. Neuromorphic electronic systems. *Proc. IEEE* **1990**, *78*, 1629–1636. [CrossRef]
2. Pacheco, P.S. *An Introduction to Parallel Programmin*; Elsevier: Amsterdam, The Netherlands, 2011.
3. Mirsa, J.; Saha, I. Artificial neural networks in hardware: A survey of two decades of progress. *Neurocomputing* **2010**, *74*, 239–255.
4. Himavathi, S.; Anitha, D.; Muthuramalingam, A. Feedforward neural network implementation in FPGA using layer multiplexing for effective resource utilization. *IEEE Trans. Neural Netw.* **2007**, *18*, 880–888. [CrossRef] [PubMed]
5. Du, Y.; Du, L.; Gu, X.; Du, J.; Wang, X.S.; Hu, B.; Jiang, M.; Chen, X.; Su, J.; Iye, S.S.; et al. An analog neural network computing engine using CMOS-compatible charge-trap-transistor (CTT). *IEEE Trans. Comput. Aided Des. Integr. Circuits Syst.* **2018**, *38*, 1811–1819. [CrossRef]
6. Kawaguchia, M.; Ishiib, N.; Umeno, M. Analog neural circuit and hardware design of deep learning model. *Procedia Comput. Sci.* **2015**, *60*, 976–985. [CrossRef]
7. Wang, F.; Li, Y.X. Analog Circuit Design Automation Using Neural Network-Based Two-Level Genetic Programming. In Proceedings of the 2006 International Conference on Machine Learning and Cybernetics, Dalian, China, 13–16 August 2006.
8. Shima, T.; Kimura, T.; Kamatani, Y.; Itakura, T.; Fujita, Y.; Iida, T. Neuro chips with on-chip back-propagation and/or Hebbian learning. *IEEE J. Solid-State Circuits* **1992**, *27*, 1868–1875. [CrossRef]
9. Solomon, P.M. Device innovation and material challenges at the limit of CMOS technology. *Annu. Rev. Mater. Sci.* **2000**, *30*, 681–697. [CrossRef]
10. Brđanin, T.P.; Dokić, B. Strained silicon layer in CMOS technology. *Electronics* **2014**, *18*, 63–69.
11. Kügeler, C.; Meier, M.; Rosezin, R.; Gilles, S.; Waser, R. High density 3D memory architecture based on the resistive switching effect. *Solid State Electron.* **2009**, *53*, 1287–1292. [CrossRef]
12. Chua, L.O. Memristor—The missing circuit element. *IEEE Trans. Circuit Theory* **1971**, *18*, 507–519. [CrossRef]
13. Strukov, D.B.; Sinder, G.S.; Stewart, D.R.; Williams, R.S. The missing memristor found. *Nature* **2008**, *453*, 80–83. [CrossRef] [PubMed]
14. Jo, S.H.; Chang, T.; Ebong, I.; Bhadviya, B.B.; Mazumder, P.; Lu, W. Nanoscale memristor device as synapse in neuromorphic systems. *Nano Letters* **2010**, *10*, 1297–1301. [CrossRef] [PubMed]
15. Wang, H.; Li, H.; Pino, R.E. Memristor-based synapse design and training scheme for neuromorphic computing architecture. In Proceedings of the 2012 International Joint Conference on Neural Networks (IJCNN), Brisbane, Australia, 10–15 June 2012; pp. 1–5.
16. Kim, H.; Sad, M.P.; Yang, C.; Roska, T.; Chua, L.O. Neural synapse weighting with a pulse-based memristor circuit. *IEEE Trans. Circuit Syst.* **2012**, *59*, 148–158. [CrossRef]
17. Adhikari, S.P.; Yang, C.; Kim, H.; Chua, L.O. Memristor bridge synapse-based neural network and its learning. *IEEE Trans. Neural Netw. Learn. Syst.* **2012**, *23*, 1426–1435. [CrossRef] [PubMed]
18. Chen, Y.C.; Li, H.; Zhang, W.; Pino, R.E. The 3-D stacking bipolar RRAM for high density. *IEEE Trans. Nanotechnol.* **2012**, *11*, 948–956. [CrossRef]
19. Liang, J.; Wong, H.S.P. Cross-point memristor array without cell selector—Device characteristics and data storage pattern dependencies. *IEEE Trans. Electron. Device* **2010**, *57*, 2531–2538. [CrossRef]
20. Hu, M.; Li, H.; Wu, Q.; Rose, G.S.; Chen, Y. Memristor crossbar based hardware realization of BSB recall function. In Proceedings of the 2012 International Joint Conference on Neural Networks (IJCNN), Brisbane, Australia, 10–15 June 2012; pp. 1–7.
21. Truong, S.N.; Min, K.S. New memristor-based crossbar array architecture with 50-% area reduction and 48-% power saving for matrix-vector multiplication of analog neuromorphic computing. *J. Semicond. Technol. Sci.* **2014**, *14*, 356–363. [CrossRef]
22. Soudry, D.; Castro, D.D.; Gal, A.; Kolodny, A.; Kvatinsky, S. Memristor-Based Multilayer Neural Networks with Online Gradient Descent Training. *IEEE Trans. Neural Netw. Learn. Syst.* **2015**, *36*, 2048–2421. [CrossRef]
23. Wang, L.; Shen, Y.; Yin, Q.; Zhang, G. Adaptive synchronization of memristor-based neural networks with time-varying delays. *IEEE Trans. Neural Netw. Learn Syst.* **2014**, *26*, 2033–2042. [CrossRef]
24. Linn, E.; Rosezin, R.; Kügeler, C.; Waser, R. Complementary resistive switches for passive nanocrossbar memories. *Nature Mater.* **2010**, *9*, 403–406. [CrossRef]

25. Shin, S.H.; Byeon, S.D.; Song, J.S.; Truong, S.N.; Mo, H.S.; Kim, D.J.; Min, K.S. Dynamic reference scheme with improved read voltage margin for compensating cell-position and back ground-pattern dependencies in pure memristor array. *J. Semicond. Technol. Sci.* **2015**, *15*, 685–694. [CrossRef]
26. Levisse, A.; Royer, P.; Giraud, B.; Noel, J.P.; Moreau, M.; Portal, J.M. Architecture, design and technology guidelines for crosspoint memories. In Proceedings of the 2017 IEEE/ACM International Symposium on Nanoscale Architectures (NANOARCH), Newport, RI, USA, 25–26 July 2017.
27. Giraud, B.; Makosiej, A.; Boumchedda, R.; Gupta, N.; Levisse, A.; Vianello, E.; Noel, J.-P. Advanced memory solutions for emerging circuits and systems. In Proceedings of the 2017 IEEE International Electron Devices Meeting (IEDM), San Francisco, CA, USA, 2–6 December 2017.
28. Truong, S.N.; Pham, K.V.; Yang, W.; Shin, S.; Pedrotti, K.; Min, K.S. New pulse amplitude modulation for fine tuning of memristor synapses. *Mircoelectron. J.* **2016**, *55*, 162–168. [CrossRef]
29. Yakopcic, C.; Taha, T.M.; Subramanyam, G.; Pino, R.E.; Rogers, S. A memristor device model. *IEEE Electron Device Lett.* **2011**, *32*, 1436–1438. [CrossRef]
30. Spectre®Circuit Simulator User Guide. Available online: https://www.ee.columbia.edu/~|harish/uploads/2/6/9/2/26925901/spectre_reference.pdf (accessed on 1 October 2019).
31. International Technology Roadmap for Semiconductors. 2007. Available online: https://www.semiconductors.org/wp-content/uploads/2018/08/2007Interconnect.pdf (accessed on 1 October 2019).
32. Kim, S.; Zhou, J.; Lu, W.D. Crossbar RRAM arrays: Selector device requirements during wire operation. *IEEE Trans. Electron. Devices* **2014**, *61*, 2820–2826.
33. Schindler, G.; Steinlesberger, G.; Engelhardt, M.; Steinhögl, W. Electrical characterization of copper interconnects with end-of-roadmap feature sizes. *Solid State Electron.* **2003**, *47*, 1233–1236. [CrossRef]
34. Kohonen, T. Self-organization and Associative Memory. In *Information Sciences*; Springer: Berlin/Heidelberg, Germany, 1989.
35. Li, C.; Belkin, D.; Li, Y.; Yan, P.; Hu, M.; Ge, N.; Jiang, H.; Montgomery, E.; Lin, P.; Wang, Z.; et al. Efficient and self-adaptive in-situ learning in multilayer memristor neural networks. *Nat. Commun.* **2018**, *9*, 2385. [CrossRef]
36. Caravelli, F.; Carbajal, J.P. Memristors for the curious outsider. *Technologies* **2018**, *6*, 118. [CrossRef]

 *micromachines*

Article

Comparison of the Electrical Response of Cu and Ag Ion-Conducting SDC Memristors Over the Temperature Range 6 K to 300 K

Kolton Drake, Tonglin Lu, Md. Kamrul H. Majumdar and Kristy A. Campbell *

Department of Electrical and Computer Engineering, Boise State University, Boise, ID 83725-2075, USA;
minionqb@gmail.com (K.D.); tanyalu@u.boisestate.edu (T.L.); mdkamrulhassanma@u.boisestate.edu (M.K.H.M.)
* Correspondence: kriscampbell@boisestate.edu; Tel.: +1-208-426-5968

Received: 3 September 2019; Accepted: 29 September 2019; Published: 30 September 2019

Abstract: Electrical performance of self-directed channel (SDC) ion-conducting memristors which use Ag and Cu as the mobile ion source are compared over the temperature range of 6 K to 300 K. The Cu-based SDC memristors operate at temperatures as low as 6 K, whereas Ag-based SDC memristors are damaged if operated below 125 K. It is also observed that Cu reversibly diffuses into the active Ge_2Se_3 layer during normal device shelf-life, thus changing the state of a Cu-based memristor over time. This was not observed for the Ag-based SDC devices. The response of each device type to sinusoidal excitation is provided and shows that the Cu-based devices exhibit hysteresis lobe collapse at lower frequencies than the Ag-based devices. In addition, the pulsed response of the device types is presented.

Keywords: chalcogenide; electrochemical metallization cell; electrochemical metallization (ECM); ion conduction; memristor; self-directed channel (SDC)

---

1. Introduction

Self-directed channel (SDC) memristors are a type of chalcogenide-based electrochemical metallization (ECM) device [1–7] in which it is posited that re-usable and irreversible ion-transport channels are formed within the active chalcogenide layer during the first write operation [6,7]. The persistence of these channels, even after the device is cycled between high and low resistance states, is considered the largest factor responsible for consistent SDC device state switching [6]. ECM devices of many different material types, ranging from oxides, to chalcogenides, typically using Ag or Cu as the ion source, are lauded as having the highest likelihood for success in next generation non-volatile memory, neuromorphic computing, and space applications where a robust, radiation hardened, and temperature tolerant device is desirable [1–4,6–20]. Investigation of ECM device operational theory is ongoing since device improvement and good application-based device design requires a closer understanding of how the devices work.

Recently, there has been a trend in the literature to classify all ECM device types as conductively bridged random access memory (CBRAM) devices [4,21]. This generalization is in conflict with the earlier literature where CBRAM was used to describe a specific device type in which a conductive filament is formed through a solid solution, e.g., Ge_xSe_{1-x} (or Ge_xS_{1-x}) where x < 0.33 [10–13,22–31]. These devices have also been referred to as programmable metallization cell (PCM) devices [13,26–29]. Now, the CBRAM designation is used synonymously with ECM [4] and over the years has included oxide-based materials as well as organic materials [32] and even BN films [33]. We classify the SDC memristor as an ECM device but remove it from the general classification of CBRAM for three reasons. First, the description of "GeSe-based CBRAM" is currently associated with a doped solid solution GeSe-Ag system [25,28,30]. The SDC does not contain a doped solid solution; the device does not require Ag doping, nor allow doping [7]. Unlike the GeSe-Ag CBRAM device, the SDC device structure contains material layers meant to store metal ions (the SnSe layer or other metal-chalcogenide layer—see [7] for experimental data and a discussion of the effects of the metal on SDC operation) and to facilitate fast switching; these are not present in the GeSe-Ag solid solution-based device.

Second, the fabrication methods, operation, temperature tolerance, device switching consistency, and longevity of the SDC are significantly different from the GeSe-Ag CBRAM, so separation prevents confusion between the two types of ECM devices. The SDC device can withstand higher fabrication temperatures than other typical chalcogenide-based ECM devices, which gives it more flexibility for manufacturing in a commercial facility. SDC devices have also been shown to operate at high temperatures (150 °C) for an extended time without performance degradation, and have been shown to function normally after reaching high temperatures (at least 250 °C) [6]. The SDC device fabrication is simple, requiring no photodoping or thermal annealing for incorporation of oxidizable metals. The device materials can all be sputter deposited in-situ making thin film deposition simpler as well as protecting the device material layers from oxygen and detrimental water exposure [34]. During the fabrication steps, the chalcogenide film stack is never exposed to photolithography chemicals or solvents; the final device etch step is performed by ion milling (no chemical etching), thus further preventing any water or oxygen exposure to the device active layer. The Ag-based SDC device longevity has been physically measured over a time of more than 10 years (see Supplementary Material, Figure S1).

Third, and most importantly, it is a working hypothesis that the SDC operation requires a separate metal chalcogenide layer and an amorphous active layer, such as Ge_2Se_3, which contains thermodynamically unstable homopolar bonds (such as the Ge-Ge bonds that are present in the SDC device Ge_2Se_3 active layer). Channel formation then occurs through an irreversible chemical reaction between the device material layers upon the first programming event [7,35]; the combination of these layers will reactively generate permanent channels, i.e., Ag or Cu ion transport routes, through the SDC device Ge_2Se_3 active layer via a chemical reaction preferentially with the Ge-Ge bond sites. Once the channel is formed, it is permanent under similar operating conditions, with the device state change depending on Ag or Cu ion movement within the established channel. There is no "dissolution" of randomly ordered conductive filaments into the material matrix film which is the hallmark of CBRAM [36,37]; there is simply movement of metal ions within a well-defined transport route. Ag or Cu ions can move into or out of the channels, corresponding to a write or erase for modification to a lower or higher resistance. The channels enable more consistent and predictable switching within a device as well as between different devices compared with the other chalcogenide-based ECM device types. After channel formation, the SnSe layer can be considered an intermediate layer, or 'stepping stone' for oxidation of Ag or Cu, and storage of metal ions. The formed channels assist the device in fast and consistent switching since it allows formation and storage of oxidizable metal ions instead of overcrowding and saturating the active glass layer. The desired morphology of the SnSe layer is thus one that is disordered, with a large surface area for Ag or Cu (and their ions) to react with SnSe [7].

A similar approach has recently been used in amorphous carbon (a-C) ECM devices which use Ag as the oxidizable metal [38]. In this case, a layer of AgInSbTe was used to buffer the a-C film from oversaturation of Ag, as well as to provide a location for Ag-ion storage. Similarly to the SDC device compared with and without the SnSe layer [7], these a-C devices exhibited highly uniform switching, high cycling endurance, and fast switching times only in the presence of the storage layer.

Even though there are ECM devices with materials systems that appear similar to the SDC due to the chemical elements present in the device [22], the extreme differences in device operation, fabrication and stability justify placing the SDC memristors in their own ECM subcategory as a "self-directed channel".

In this work, we compare the direct current (DC) (quasi-static) and pulse electrical response of Ag and Cu SDC memristors as a function of temperature from 6 K to 300 K, and discuss the device stability under various programming conditions.

2. Materials and Methods

2.1. Device Structure and Fabrication

Devices were fabricated in the Idaho Microfabrication Laboratory at Boise State University on 100 mm p-type wafers in a stacked layer structure (Figure 1). The device size is defined by the bottom electrode contact area and is 2 μm in diameter. The devices were fabricated with either a Ag or Cu layer as the mobile ion source layer. The active layer, responsible for device resistance switching is the bottom Ge_2Se_3 layer in contact with the bottom W electrode, within the nitride opening. The details of the purpose of each thin film layer have been described previously [6] and a full discussion can be found there. In brief, the SnSe layer assists in the formation of the self-directed channel within the active layer and acts as a cation storage layer. The two Ge_2Se_3 layers surrounding the Ag or Cu layer enable thin film adhesion and photolithography. The active switching layer is the bottom Ge_2Se_3 layer.

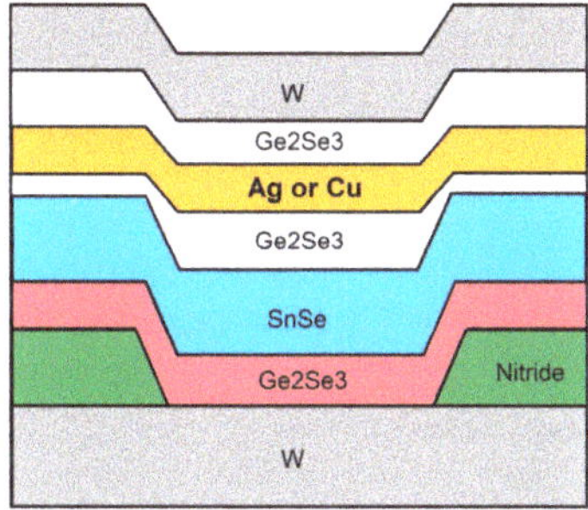

Figure 1. Self-directed channel (SDC) device structure described in [6]. The target layer thicknesses were (from bottom to top): Ge_2Se_3 (300 Å)/SnSe (800 Å)/Ge_2Se_3 (150 Å)/Ag (500 Å)/Ge_2Se_3 (100 Å)/W (400 Å). The top three layers below the W top electrode, corresponding to Ge_2Se_3/Ag/Ge_2Se_3, mix during fabrication, becoming one conductive layer.

Prior to thin film deposition using an AJA International ATC Orion 5 UHV Magnetron sputtering system, the wafers were sputtered with Ar^+ to prepare the bottom electrode surface. This was followed by *in-situ* sputter deposition of all of the remaining device layers, including a top W electrode capping layer. *In-situ* deposition of layers was performed to minimize the potential for detrimental water vapor on the device [34]. Final device etching was performed with a Veeco ME1001 ion-mill (Veeco, Plainview, NY, USA).

2.2. Electrical Measurements

Temperature control and sample probing was performed using a Lake Shore CRX-4K probe station, two Lake Shore Model 340 temperature controllers (Lake Shore Cryotronics, Inc., Westerville, OH, USA), a SHI RDK-408D2 closed-cycle refrigerator (Sumitomo (SHI) Cryogenics of America, Inc., Allentown, PA, USA) with a controlled temperature range of 5 K to 400 K, and an RC-EM10-208230-60 CE liquid helium recirculating chiller. Lake Shore ZN50R alumina ceramic probe cards with 25 μm Tungsten tips were used for measurements. Probe cards were anchored to the sample stage with copper braiding to ensure temperature equilibration between stage and probe. Vacuum was maintained and monitored with a Varian V-81 turbo pump.

DC (quasi-static), sinusoidal excitation, and pulsed measurements were made using a Keysight B1500A Semiconductor Parameter Analyzer equipped with two Waveform Generator/Fast Measurement Units (WGFMUs) (Keysight, Inc., Santa Rosa, CA, USA). The WGFMUs allowed direct measurement of the current through the device during testing without external circuits or current limiting series resistors. The sweep rate for a DC measurement depends on the voltage and current ranges used, which are varied depending upon the sample measurement temperature and write/erase measurements;

however, for all measurements the sweep rate is in the range of 0.14 to 0.2 V/s (switching voltage vs sweep rate is shown in the Supplementary Material, Figure S2).

At least 10 unique devices were measured at every temperature, for both the Ag and Cu-based devices. Three trials of temperature measurements were performed over a two year period after wafer fabrication. During the storage periods, the samples were maintained in the dark, at ambient temperature. Since the effect of cold temperature on the devices was unknown at the start of the experiment, wafer pieces were measured in the order of decreasing and increasing temperature. It was determined that the temperature order of measurement did not influence the measurement outcome. Therefore, for the experimental data provided in this work, the samples were brought to a base temperature of 6 K and equilibrated for 30 min prior to commencing measurements. The temperature was raised for each subsequent temperature measurement, with an equilibration of at least 30 min at each temperature prior to the measurement.

All DC sweep measurements consisted of the sequence: Write 1-Erase-Write 2-Read. The Write sweeps applied a positive potential to the device top electrode and used a 10×10^{-6} A compliance current. The Erase sweep applied a negative potential to the top electrode; a 10×10^{-3} A compliance current was applied. A +20 mV Read sweep was applied to the top electrode to read the final written resistance state after the Write 2 step (Read). In all measurements, the bottom electrode was maintained at ground and the top electrode potential was varied. For the Ag devices, the Write 1 and Write 2 sweeps were performed over the range of 0 to 3 V for T $\geq$ 150 K, and 0 to 5 V for T < 150 K. For the Cu devices, the Write 1 range was 0 to 3 V for T $\geq$ 140 K and 0 to 5 V for T < 140 K. The Erase voltage ranges for each type of device were the same, except in the negative potential direction.

Prior to the first (Write 1) sweep, all devices tested were in a pristine (never previously tested) state. Table 1 summarizes the sweep measurement and Read voltage for each resistance type. The conductance of a device was calculated from 1/R, where R is the measured resistance.

Table 1. I–V sweep resistance measurement descriptions.

Resistance	Resistance Measurement Sweep
Initial, R_i	+20 mV on Write 1
First Write, R_{W1}	−20 mV on Erase
Erased, R_E	+20 mV on Write 2
Second Write, R_{W2}	+20 mV on Read

3. Results

3.1. DC (Quasi-Static) Measurements

Representative DC I–V measurement curves for each measurement sequence at each temperature are shown in Figures 2–4. Write 1 and Write 2 curves are shown in Figure 2a,b for Ag, and Figure 2c,d for Cu. The Erase sweeps are shown in Figure 3a,b for Ag devices and Figure 4a–d for Cu devices.

The Write 1 sweep is the first time voltage is applied to a pristine device. This measurement can therefore provide the initial device resistance when measured at +20 mV during the Write 1 sweep. The Write I–V curves in Figure 2 are typical for an SDC device. The Write sweep starts at 0 V and the potential is increased until the device transitions to a low-resistance state, at which point the current reaches the compliance current. In the pristine state, Ag-based SDC devices initially have a very high resistance (GΩ range), and exhibit either an instantaneous increase in current to the compliance value during the Write 1 sweep, or an exponential rise in current with applied voltage, depending upon temperature. The exponential increase in current is present in the low temperature Write 1 sweeps for Ag and Cu (Figure 2a,c), and Cu Write 2 sweeps (Figure 2), but is absent in the Write 2 sweep for Ag (Figure 2b).

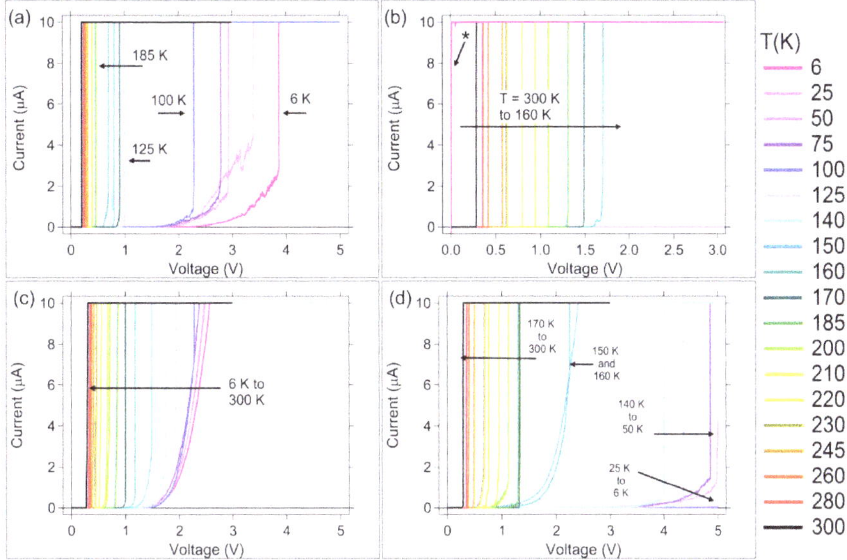

Figure 2. Representative Write I–V curves as a function of temperature. Ag devices: (**a**) Write 1; (**b**) Write 2. Cu devices: (**c**) Write 1; (**d**) Write 2. The * in (**b**) corresponds to I–V curves of broken (shorted) devices. Note: the Write 1 sweep voltage maximum for measurements below 150 K was 5 V, compared to 3 V used for T ≥ 150 K.

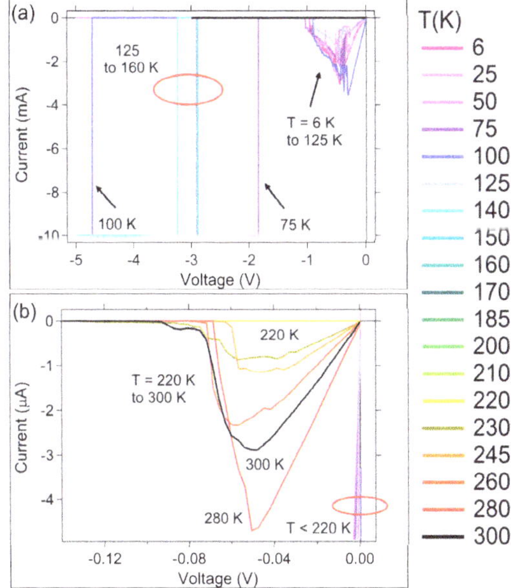

Figure 3. Ag device representative Erase I–V curves for all temperatures (see legend). (**a**) Full scale view; (**b**) expanded low I–V region.

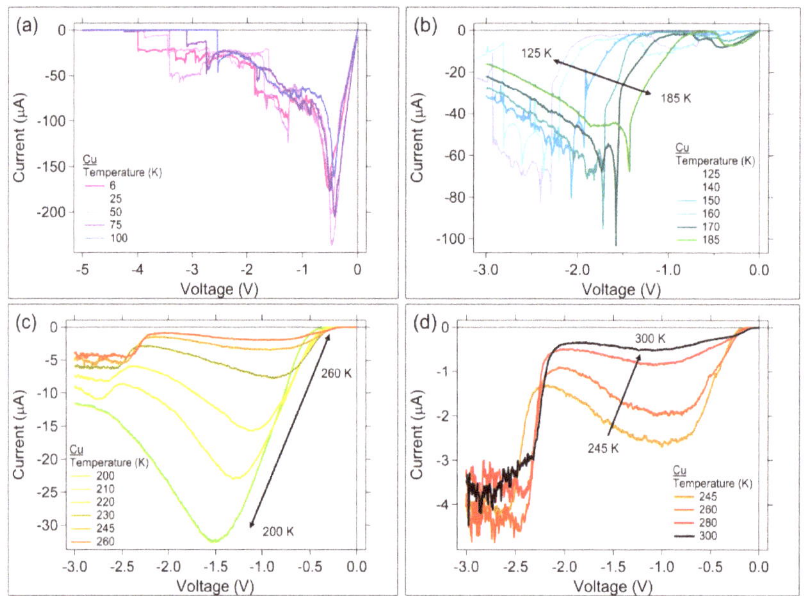

Figure 4. Cu device erase regions for (**a**) 6 K ≤ T ≤ 100 K; (**b**) 125 K ≤ T ≤ 185 K; (**c**) 260 K ≤ T ≤ 200 K; and (**d**) 245 K ≤ T ≤ 300 K.

The Write 1 I–V sweeps, Figure 2a,c for both device types, show that devices switch with an increasing switching voltage as the temperature is reduced. At 6 K, the Ag devices switch at approximately 4 V. However, as can be seen in the erase sweeps (Figure 3a,b), not all of the Ag devices that switched at low temperature can erase. These devices appear as 'shorts' on the Write 2 I–V sweeps (denoted by * in Figure 2b). The devices that were switched at low temperature could not be cycled back, despite the large compliance current on the Erase sweep of 10×10^{-3} A, compared to 10×10^{-6} A on the Write sweep. Measurements upon return to room temperature indicated the devices were destroyed.

An Erase sweep corresponds to application of a reverse polarity potential to the top electrode. Since the erase occurs after the SDC device was written, the low potential region of the Erase I–V curve for the Ag device (Figure 3a,b) shows a mostly linear behavior up to a peak current value, as expected for a device programmed into a resistive state. Beyond the peak current voltage, the device experiences negative resistance as it transitions towards a high resistance state. Erase sweeps are shown in an expanded view in Figure 3b for Ag devices, in which only the data for T ≥ 220 K is in view since the peak current is 1000 times lower than for T < 220 K (low temperature, higher currents, are dominant in the full-scale view in Figure 3a). The current required to erase the Ag devices that switched at temperatures below 150 K is in the mA range, which is three orders of magnitude higher than under room temperature conditions (Figure 3b). For the temperatures between 140 K ≤ T ≤ 220 K, the devices that were switched during the Write 1 sweep (Figure 2b) did not latch the low resistance state, so no Erase current is measured (it is within the noise of the instrument and too low to observe in Figure 3a or Figure 3b).

It is reasonable to expect that not all Cu or Ag ions generated during the Write sweep get reduced within the duration of the Write sweep. It is anticipated that when the voltage is removed, there is a concentration of ions still within the active layer channels that are oxidized. In other words, not all ions generated during the Write sweep will reach a conductive contact point, directly or indirectly, with the bottom electrode during the measurement. These ions can remain within the channel, diffusing towards a more energetically stable location in the channel or the SnSe layer. As temperature is reduced, the possibility of the ions diffusing is reduced, making it more likely that the low temperatures yield more excess ions within the devices.

The low temperature Ag device Erase I–V curves (Figure 3a) are much different than the curves for higher temperatures (Figure 3b). For the 300 K to 230 K Erase sweep data (Figure 3b) there is a maximum current peak at approximately −0.05 V, and a linear slope leading up to this peak from 0 V. These data are consistent with the rupture, or loss of contact, of a conduction path between the electrodes. In contrast, a broad peak in the Ag device Erase sweeps (Figure 3a) at temperatures below 150 K occurs between −0.3 to −0.5 V. This voltage range is 10 times greater than the potential needed to break conductive contact between electrodes for 230 K ≤ T ≤ 300 K. Cyclic voltammograms for Sn-Ag systems [39] can show broad peaks between −0.2 V and −0.8 V, depending upon the concentrations of Ag and Sn in the system, and formation of a Sn-Ag alloy. The observed broad peak for T < 150 K is within the range observed in cyclic voltammetry of systems comprising the formation of Sn-Ag alloys [39]. Since Sn can participate in redox reactions during the channel formation and subsequent switching cycles, it is possible that the maximum sweep voltage (5 V) during the low temperature Write sweep is high enough to oxidize Sn. Any excess Ag^+ that remained upon removal of the Write sweep potential would still be present within the device, with ion diffusion occurring much slower at the low temperatures. Reduction of the excess Ag^+ is therefore possible during the Erase sweep, and would appear more prominently at low temperatures where higher ion concentration is expected. Interestingly, the low temperature Write 1 I–V curves exhibit an exponential increase in current, not observed in the higher temperature curves (Figure 2a). This, in combination with the Erase peak occurring within a Sn-Ag alloy redox potential range, could indicate Sn involvement in switching. With the Ag consumed in an alloying reaction with Sn, Ag would no longer be available for ion movement during programming. A Sn-Ag alloy could produce a permanently conductive pathway within the channel. The Ag devices in the low temperature range were readily damaged by application of potentials higher in magnitude than −1 V during the erase, as seen by the sharp transitions to compliance current on the Erase sweep (Figure 3a). In order to erase the devices that did write at low temperature, currents as high as almost 4 mA were needed. The formation of a Sn-Ag alloy may be why the devices that switch at low temperature are 'shorted' and no longer function.

Similarly to the Ag case, at low temperatures, the Cu devices required higher applied potentials to switch; the Cu device threshold voltage increased as the temperature decreased (Figure 2c). If the only redox consideration were Cu, it could be concluded that the oxidation of Cu during the Write will occur first through $Cu \rightarrow Cu^{2+}$, then as voltage is further increased, it would go directly from $Cu \rightarrow Cu^+$. Therefore, it would be expected that for the cold temperature Write sweeps, the generation of Cu^+ would be possible due to the increased Write Sweep potential. It is further expected that there would be excess Cu^+ and Cu^{2+} in the device upon removal of the Write potential; at cold temperatures, the diffusion of the ions would be significantly reduced, thus keeping a higher concentrations of ions within the channel. Upon application of an Erase sweep, these excess ions could be reduced. Reduction of the Cu ions would appear in the Erase sweep as a reduction of $Cu^+ \rightarrow Cu$ at a potential near −0.5 V. A peak around −0.45 V is observed for the Erase sweeps between 5 to 100 K, Figure 4a. This is not observed in the other Erase temperature ranges, where the Write 1 sweep voltage maximum was 3 V, instead of 5 V, and therefore was likely not high enough to achieve the $Cu \rightarrow Cu^+$ oxidation. However, as in the case for Ag, a contribution due to Sn redox reactions cannot be ruled out.

The Erase sweeps for the Cu sample have at least four temperature regions with differing I–V curve characteristics (Figure 4a–d). Since Cu can oxidize during the Write sweep to both Cu^+ and Cu^{2+} depending upon the magnitude of the applied potential, it is expected that the Cu device Erase I–V curves could be more complicated than the Ag device Erase curves, and potentially have between two and three peaks corresponding to different Cu ion reduction potentials. Based on cyclic voltammograms [34] it is expected that the peak near the lowest magnitude Erase potential corresponds to a $Cu^+ \rightarrow Cu$ reduction. The next highest potential would correspond to a $Cu^{2+} \rightarrow Cu$ reduction, and the highest potential to the $Cu^{2+} \rightarrow Cu^{+}$ reduction. However, Cu can also form an alloy with Sn [40]. Therefore, the observed peaks may be complicated by multiple redox reactions of Cu and Sn.

The Erase peak potentials for the Cu-based devices have a significant temperature dependence above 125 K. Figure 4b–d show the Erase sweeps for temperatures from 125 K to 300 K. For 185 K ≥ T ≥ 125 K (Figure 4b), there are multiple small peaks on I-V curves between −3 V to −1.4 V. There are also low amplitude broad peaks between −0.3 V to −1.25 V. In all cases, the I-V curves exhibit a temperature dependence, with peak shifting to lower voltage as the temperature is increased. The higher voltage region peaks exhibit an increasing number of sharp peaks as the temperature is reduced. Similar sharp peaks have been observed in a Cu-Sn alloy reaction [40]. Within the temperature range 260 K ≥ T ≥ 200 K (Figure 4c), the largest Erase peak voltage (at approximately −1.5 V for the 200 K trace) and current at the peak, has a significant dependence on temperature, with the peak voltage and current decreasing with an increase in temperature. This is also the temperature range where the Cu device exhibits a negative slope in the $\ln(1/R_{W1})$ vs 1000/T plot (Figure 5b). The peaks that occur between −2 and −3 V in Figure 4 could correspond to a $Cu^{2+} \rightarrow Cu^{+}$ reduction due to excess Cu^{2+} present in the channel following the Write sweep [34].

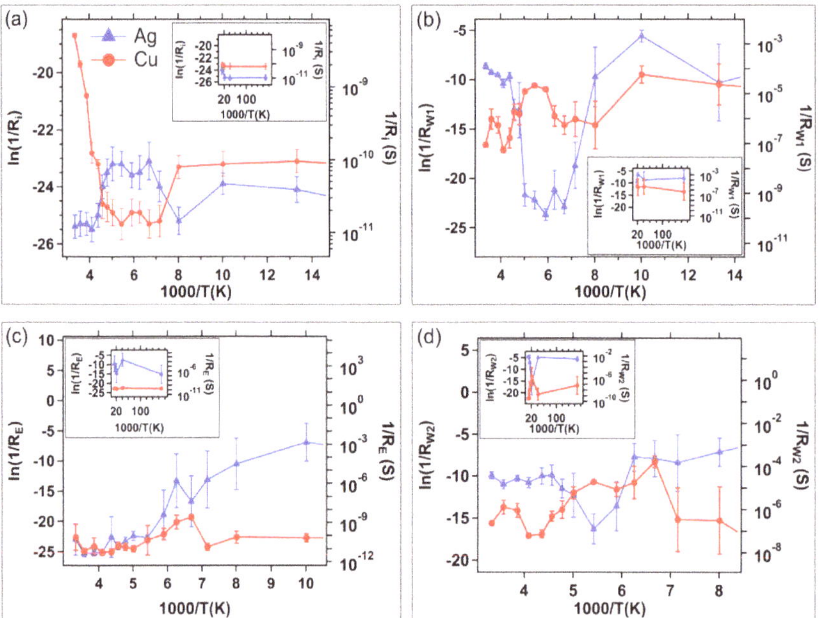

Figure 5. Average conductance versus inverse temperature for (**a**) initial resistance, R_i; (**b**) written resistance after the first write sweep, R_{W1}; (**c**) erased resistance, R_E; and (**d**) written resistance after the second write sweep, R_{W2}. In each graph, the Cu device data is represented by circles; Ag devices as triangles. The inset of each graph is the extension of the data into the coldest temperature region. Error bars are one standard deviation.

The average initial, written, and erased resistances as a function of temperature are provided in Figure 5, plotted as $\ln(1/R)$ vs 1000/T. Error bars correspond to 1 standard deviation. The lowest temperature region is displayed in the inset of each plot.

As discussed, and apparent in the Write 2 sweep, Figure 2, and the Erase sweep in Figure 3, Ag devices did not always switch at low temperatures. The inset of Figure 5c shows this clearly: instead of erasing to high resistance, R_E is a similar magnitude to R_{W1} and R_{W2} at temperatures below 125 K.

This indicates that the Ag devices were damaged or permanently altered during the Write 1 sweep. The Cu devices clearly erased to high resistances over the entire temperature range.

The switching (threshold) voltage, V_{th}, for a Write voltage sweep is identified as the potential at which a large current jump is initiated towards the compliance current value. These switching voltages were determined for each I-V trace of the Write 1 and Write 2 sweeps and have been plotted as a function of temperature in Figure 6 as V_{th1} and V_{th2}, respectively. There is an exponential relationship between V_{th} and T for both threshold voltages above 150 K. This is clear in the inset graph which plots $Ln(V_{th})$ vs 1000/T and the corresponding linear fit. No data is available for V_{th2} for the Ag devices operated below 150 K due to low temperature operational damage (Figure 6b).

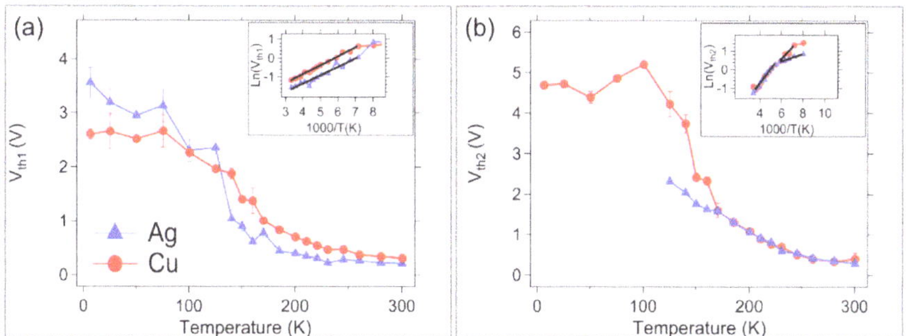

Figure 6. Average write threshold voltages as a function of temperature. (**a**) First Write, V_{th1}; (**b**) Second Write, V_{th2}. The inset in (a) and (b) is the $Ln(V_{th})$ vs 1000/T plot showing the Arrhenius behavior for the threshold voltage of both device types. Error bars represent one standard deviation.

3.2. Sinusoidal Excitation and Pulsed Response

A sinusoidal input signal was applied to each device type, and the device response as a function of frequency of the input signal was measured (Figure 7). Both device types exhibit the characteristic fingerprint of memristors, a pinched hysteresis loop, under sinusoidal excitation (Figure 7) [41]. In both cases, the device response is pinched at the origin, and the hysteresis lobe area is decreased to zero as the input signal frequency is increased. Cu devices (Figure 7a) display flattened hysteresis lobes at a low frequency of 100 Hz, whereas for the Ag devices (Figure 7b) this occurs at 10 kHz.

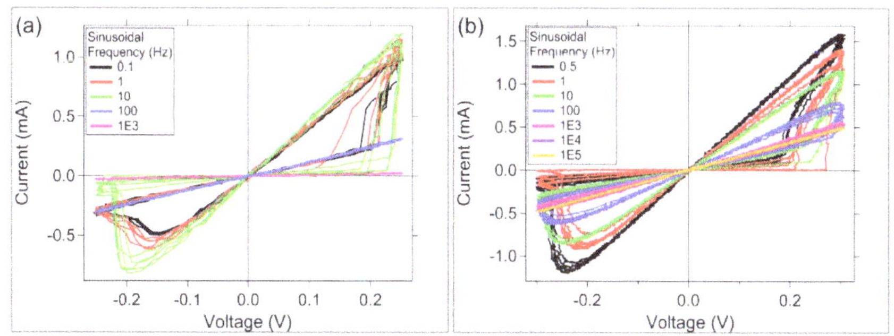

Figure 7. Cu and Ag device response to a sinusoidal input with varying frequency. (**a**) Cu device; (**b**) Ag device. T = 300 K. Six cycles at each frequency are shown.

The pulse response is measured by applying a programming voltage pulse sequence (as labeled in Figure 8) to the memristor. The response of the memristor is determined by the current measured through it during application of the voltage pulse. The current measurement is opposite polarity from the voltage pulse sequence due to the instrument set up; a negative current is measured when voltage is positive. An adjustment of the data to the correct sign of current is not made, since this allows current and voltage data to be displayed on the same graph (right and left axes, respectively) with minimal interference.

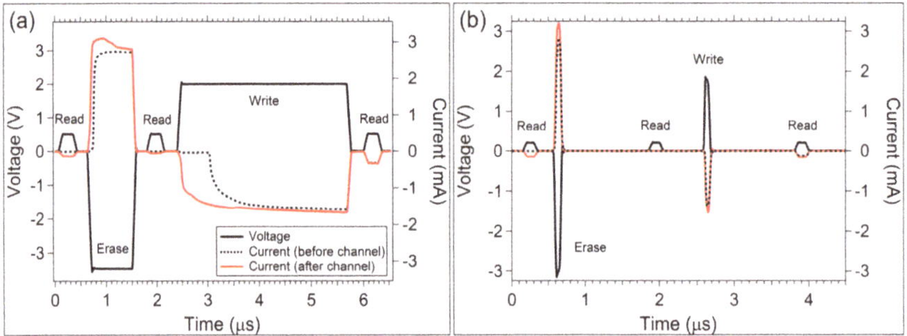

Figure 8. Cu and Ag device response to a programming pulse sequence. The solid black trace corresponds to the applied voltage pulse sequence. The dashed lines correspond to the current measured during the voltage pulse sequence for a pristine (before channel formation) device. The solid red line corresponds to the current measured during the next applied pulse sequence after channel formation. (**a**) Cu device; (**b**) Ag device. T = 300 K.

The response of Cu devices to the programming pulse sequence is provided in Figure 8a. A pristine Cu device (never switched previously, but from the R_i data in Figure 5a, does appear to have some Cu diffused into the active layer) was tested. The current through the pristine device during the voltage pulse sequence is given by the dashed line trace in Figure 8a. A Read pulse was applied first; the current response during the Read pulse is too low to observe on the mA scale, indicating that R_I was higher than 10 MΩ. The second Read pulse also shows no measureable current, indicating the device is still in a high resistance state following the Erase pulse (as anticipated). Given that the Cu device was pristine, the self-directed channel has not yet been formed in the active layer at this point in the pulse sequence (i.e., prior to the Write pulse). Channel formation happens during the first Write pulse. Note that there is approximately 500 ns delay from the initiation of the Write pulse and the current response. This delay is likely due to oxidation of Cu, and the chemical reaction taking place within the SnSe layer and active layer to form the channel. The measured current through the device during the final Read pulse indicates that the device was written to a low resistance state during the Write pulse.

A second pulse sequence was applied to the same Cu device two minutes after the previous measurement. The current response to this second sequence is given by the red trace in Figure 8a. The device was still in a low resistance state from the previous measurement (as indicated by the current through the device during the first Read pulse of the second applied pulse sequence, red trace Figure 8a). However, the amplitude of the current response during the Read pulse is lower than the Read current measurement at the end of the first pulse sequence, indicating that the Cu device exhibits a drift in the programmed resistance state. Following the Erase pulse of the second pulse sequence, the Read pulse indicates the device resistance was increased successfully. Application of the Write pulse on this second pulse sequence does not have the delay in device response that was observed in the first Write pulse, as expected since the channel was formed on the prior pulse sequence, and storage of Cu is presumed to be present in the SnSe layer.

The Ag devices, Figure 8b, did not exhibit the large delay in initial switching during channel formation. In this case, if there is a delay during channel formation, it is beyond the resolution of the pulse timing.

4. Discussion

Even though Figure 5 provides the measured conductance as a function of temperature, it must be noted that while the conductance plots in Figure 5 are plotted in an Arrhenius form ($\ln(1/R)$ vs $1/T$), these data are not typical conductance vs temperature measurements where one could determine conduction activation energies accurately, or reliably investigate conduction mechanisms. The conductance value at each temperature is determined from the resistance that the device achieved upon switching *at a particular temperature*. There are many factors that go into device switching at each temperature for the SDC device. Some examples include the temperature dependence of the chemical reaction between Ag or Cu and the SnSe layer and induced reactions in the active layer; movement of mobile ions through a variable-stiffness glass network; constricted channel for ion motion (e.g., due to cold temperature volume contractions); and the typical DC conductivity mechanistic concerns (e.g., Fermi energy level and dominant electron conduction mechanism at each temperature [42–44]). It could also be reasoned that even programming a set of devices to a state value, and then subjecting the devices to a set of varying temperatures and measuring conductivity could also confound the mechanism analysis. The amorphous chalcogenide materials tend to be flexible and can move (constrict volume, expand volume, pull away from interfaces) which could have a ripple effect around any ions within the material or provide alternative electron conduction pathways as a function of temperature. In this work, the switching properties at a given temperature were studied, not how a pre-programmed property changed as a function of temperature.

Despite the stated concerns, several observations can be made from the DC switching data as a function of temperature. The lower initial resistance of the Cu devices at 300 K in Figure 5a (especially compared to the R_E values, Figure 5c) indicate that the Cu devices have experienced Cu diffusion into the active layer over time while stored at ambient room temperature. This limits the Cu device data retention. This is not the case for the Ag devices. This conclusion can be reached for three reasons: (1) the lower initial resistance of the Cu device; (2) since both the Ag and Cu devices are pristine in the initial resistance measurement, the active layers should be the same and give the same Ri throughout the temperature measurement range; and (3) the erased resistance of the Cu device is the same as the Ag device when they are at higher temperatures (Figure 5c), as expected if excess Ag and Cu have been removed from the channel during the reverse potential sweep.

The Cu migration may be responsible for the well-behaved switching of the Cu devices at low temperatures during the Write 1 sweep (Figure 2c). Since this diffused Cu may be removed during an Erase sweep, it could account for the worse switching observed in the I–V curves for the Write 2 sweeps (Figure 2d) compared to the Write 1 I–V curves.

The Cu devices survive switching at temperatures down to 6 K. Figure 4 shows the Cu device Erase I-V sweeps and it is clear that devices erase at all temperatures. This is supported by the R_E data shown in Figure 5c. In addition to the robustness of the Cu-based SDC device, a Cu-silica memristor was also shown to survive operation at 4 K [11].

The Ag device I-V curves have three distinct temperature transitions at 230 K, 210 K, and 140 K in which the ability of the device to write varies. The R_{W2} and R_E data for the Ag devices indicate that below 150 K these devices are damaged if they are operated. This is not the case for the Cu devices. However, it is notable that the Cu devices have higher Write resistances when operated at T > 200 K (Figure 5b,d). It is interesting to note that the effect of temperature on R_{W1} and R_{W2} for Ag and Cu devices is opposite. Taking R_{W1} as an example, above 200 K, the slope of the $\ln(1/R_{W1})$ vs 1000/T plot (Figure 5b) is negative for the Ag devices, but positive for the Cu devices. Between temperatures of 200 K and 150 K, the Ag devices exhibit high resistance trough with little resistance change; the Cu devices exhibit a hill-like peak of decreasing resistance. A similar, but less pronounced, effect is seen for R_{W2} in Figure 5d.

The Ag device R_{W1} resistances are higher than those for the Cu devices at temperatures between 200K and 150 K (Figure 5b) since within this range, the Ag devices do not switch out of the high resistance state. When the temperature is further reduced to a range where T < 150 K, the devices 'break' (the exact 'breaking' temperature is unknown). As previously mentioned, Ag devices switched at these temperatures are damaged. The formation of a Sn-Ag alloy at the lower temperatures might be the cause of the inability to erase devices that have been written at those temperatures.

The write threshold voltages for each device type as a function of temperature are in Figure 6. The Write 2 threshold voltage is the same between both device types (Figure 6b). This seems logical if a channel is formed on the first write and used for small movement of mobile ions within the channel during subsequent programming events. Note again that there is a divergence in device response between Ag and Cu near 150 K. This is likely due to differences in Ag or Cu participation in the chemical reaction of channel formation and Ag or Cu storage in the SnSe layer that become relevant when higher voltages are applied. In addition, the first and second write thresholds exhibit an exponential dependence on temperature between 300 K and 150 K. This exponential behavior has been attributed to the collective motion of carriers in metal-insulator transition studies, for example two-terminal VO_2 devices [45]. In the SDC device case, this could correspond to the collective motion of mobile ions, or to the formation of a Sn-Ag alloy (or Sn-Cu alloy) [39,40].

Sinusoidal excitation (Figure 7) and pulse studies (Figure 8) can offer insights into ion movement and channel formation for each device type. During sinusoidal excitation, the Cu devices show flattened hysteresis lobes at an input signal frequency of 100 Hz, whereas Ag devices achieve flattened lobes at approximately 10 kHz. The significance of this is still not understood, however, the two device types demonstrate differences in switching speed. The possibility of using the frequency at which lobes flatten as a predictor of device switching speed would offer a simple way to predict device speed prior to more complicated pulsed measurements.

5. Conclusions

The electrical behavior of the Ag and Cu-based memristors over a large temperature range is complex. The factors that contribute to device operation are varied and include the effects of temperature on the active Ge_2Se_3 material layer's flexibility, the chemical reaction involved in formation of the self-directed channels, and the redox reactions of Ag, Cu, and Sn from the SnSe layer. The Ag-based devices appear to be damaged when operated at low temperatures. However, it is possible this is due only to the increased potential applied during the Write sweeps at lower temperatures and a resultant alloy formation with Sn. Further work is underway to quantify the effects of the interaction between the SnSe layer and Ag and Cu during device operation and to understand any potential alloy formation between Sn and Ag or Cu.

Interestingly, the Cu-based devices showed a migration of Cu through the active layer over time. This migration is detrimental for long term data storage since the device will lose any programmed data state. The Ag-based devices did not exhibit this response.

The Ag-based devices appear to exhibit faster pulsed programming switching during the first programming Write cycle. Faster response time of the Ag devices was also observed in the sinusoidal excitation measurements where the Ag-devices exhibited a flattening of the characteristic memristor hysteresis loop at 10 kHz, whereas the Cu-based devices exhibited flattening at 100 Hz.

The formation of the self-directed channels as a function of the SnSe layer should be studied through the replacement of Sn within that layer, with different metals. It is possible that any alloy formation between the mobile ion and the metal from the metal chalcogenide layer could have a significant impact on device performance, and be a method of selected device performance tuning. If Sn-Ag alloy is responsible for the device damage when higher voltages at low temperatures, it may be possible to change the metal in the metal-chalcogenide layer to one less likely to alloy with Ag.

Supplementary Materials: The following are available online at http://www.mdpi.com/2072-666X/10/10/663/s1, Figure S1, a data retention plot for Ag-based SDC devices. Figure S2, Ag-based device switching voltage vs sweep rate.

Author Contributions: Conceptualization, K.A.C.; Data curation, K.D. and K.A.C.; Formal analysis, T.L., M.K.H.M. and K.A.C.; Funding acquisition, K.A.C.; Investigation, K.D. and K.A.C.; Methodology, K.A.C.; Supervision, K.A.C.; Writing—original draft, K.A.C.

Funding: This research was partially funded by a grant from the United States Air Force Office of Scientific Research, DEPSCoR Grant No. FA9550-07-1-0546, and by the United States Air Force Research Laboratory, Grant No. FA9453-08-2-0252.

Acknowledgments: K.A.C. would like to thank Prabesh Subedi and Wesley Butler for assisting with data analysis.

Conflicts of Interest: The authors declare no conflict of interest.

References

1. Valov, I.; Waser, R.; Jameson, J.R.; Kozicki, M.N. Electrochemical metallization memories-fundamentals, applications, prospects. *Nanotechnology* **2011**, *22*, 254003. [CrossRef] [PubMed]
2. Waser, R.; Ielmini, D.; Akinaga, H.; Shima, H.; Wong, H.-S.P.; Yang, J.J.; Yu, S. Introduction to nanoionics elements for information technology. In *Resistive Switching*; Ielmini, D., Waser, R., Eds.; Wiley-VCH: Weinheim, Germany, 2016; pp. 1–29.
3. Zidan, M.A.; Chen, A.; Indiveri, G.; Lu, W.D. Memristive computing devices and applications. *J. Electroceram.* **2017**. [CrossRef]
4. Waser, R.; Dittmann, R.; Menzel, S.; Noll, T. Introduction to new memory paradigms: Memristive phenomena and neuromorphic applications. *Faraday Discuss.* **2019**, *213*, 11–27. [CrossRef] [PubMed]
5. Menzel, S.; Tappertzhofen, S.; Waser, R.; Valov, I. Switching Kinetics of Electrochemical Metallization Memory Cells. *Phys. Chem. Chem. Phys.* **2013**, *15*, 6945–6952. [CrossRef] [PubMed]
6. Campbell, K.A. Self-Directed Channel Memristor for High Temperature Operation. *Microelectron. J.* **2017**, *59*, 10–14. [CrossRef]
7. Campbell, K.A. Self-Directed Channel Memristor: Operational Dependence on the Metal-Chalcogenide Layer. In *Handbook of Memristor Networks*; Springer: New York, NY, USA, 2019; p. 38.
8. Rajendran, B.; Alibart, F. Neuromorphic computing based on emerging memory technologies. *IEEE J. Emerg. Sel. Top. Circuits Syst.* **2016**, *6*, 198–211. [CrossRef]
9. Kumar, S.; Strachan, J.P.; Williams, R.S. Chaotic dynamics in nanoscale NbO_2 Mott memristors for analogue computing. *Nature* **2017**, *548*, 318–321. [CrossRef] [PubMed]
10. Van den Hurk, J.; Havel, V.; Linn, E.; Waser, R.; Valov, I. Ag/GeSx/Pt-based complementary resistive switches for hybrid CMOS/ nanoelectronic logic and memory architectures. *Sci. Rep.* **2013**, *3*, 2856. [CrossRef] [PubMed]
11. Chen, W.; Chamele, N.; Gonzalez-Velo, Y.; Barnaby, H.J.; Kozicki, M.N. Low-temperature characterization of Cu-Cu: Silica-based programmable metallization cell. *IEEE Electron Device Lett.* **2017**, *38*, 1244–1247. [CrossRef]
12. Kozicki, M.N.; Barnaby, H.J. Conductive bridging random access memory-materials, devices and applications. *Semicond. Sci. Technol.* **2016**, *31*, 113001. [CrossRef]
13. Mahalanabis, D.; Barnaby, H.J.; Gonzalez-Velo, Y.; Kozicki, M.N.; Vrudhula, S.; Dandamudi, P. Incremental resistance programming of programmable metallization cells for use as electronic synapses. *Solid-State Electron.* **2014**, *100*, 39–44. [CrossRef]
14. Campbell, K.A.; Drake, K.T.; Barney Smith, E.H. Pulse shape and timing dependence on the spike-timing dependent plasticity response of ion-conducting memristors as synapses. *Front. Bioeng. Biotechnol.* **2016**, *4*, 97. [CrossRef] [PubMed]
15. Wang, Z.; Minghui, Y.; Zhang, T.; Cai, Y.; Wang, Y.B.; Yang, Y.; Huang, R. Engineering incremental resistive switching in TaO_x based memristors for brain-inspired computing. *Nanoscale* **2016**, *8*, 14015–14022. [CrossRef] [PubMed]
16. Wang, Z.Q.; Xu, H.Y.; Li, X.H.; Yu, H.; Liu, Y.C.; Zhu, X.J. Synaptic learning and memory functions achieved using oxygen ion migration/diffusion in an amorphous InGaZnO memristor. *Adv. Funct. Mater.* **2012**, *22*, 2759–2765. [CrossRef]
17. Pickett, M.D.; Medeiros-Ribeiro, G.; Williams, R.S. A scalable neuristor built with Mott memristors. *Nat. Mater.* **2013**, *12*, 114–117. [CrossRef] [PubMed]

18. Mahalanabis, D.; Sivaraj, M.; Chen, W.; Shah, S.; Barnaby, H. Demonstration of spike timing dependent plasticity in CBRAM devices with silicon neurons. In Proceedings of the IEEE International Symposium on Circuits and Systems (ISCAS), Montreal, QC, Canada, 22–25 May 2016. [CrossRef]

19. Gaba, S.; Sheridan, P.; Zhou, J.; Choi, S.; Lu, W. Stochastic memristive devices for computing and neuromorphic applications. *Nanoscale* **2013**, *5*, 5872–5878. [CrossRef] [PubMed]

20. Edwards, A.H.; Barnaby, H.J.; Campbell, K.A.; Kozicki, M.N.; Liu, W.; Marinella, M.J. Reconfigurable memristive device technologies. *Proc. IEEE* **2015**, *103*, 1004–1033. [CrossRef]

21. Arita, M.; Ohno, Y.; Murakami, Y.; Takamizawa, K.; Tsurumaki-Fukuchi, A.; Takahashi, Y. Microstructural transitions in resistive random access memory composed of molybdenum oxide with copper during switching cycles. *Nanoscale* **2016**, *8*, 14754–14766. [CrossRef]

22. Schindler, C.; Valov, I.; Waser, R. Faradaic currents during electroforming of resistively switching Ag-Ge-Se type electrochemical metallization memory cells. *Phys. Chem. Chem. Phys.* **2009**, *11*, 5974–5979. [CrossRef]

23. Jameson, J.R.; Gilbert, N.; Koushan, F.; Saenz, J.; Wang, J.; Hollmer, S.; Kozicki, M. Effects of cooperative ionic motion on programming kinetics of conductive-bridge memory cells. *Appl. Phys. Lett.* **2012**, *100*, 023505. [CrossRef]

24. Jameson, J.R.; Gilbert, N.; Koushan, F.; Saenz, J.; Wang, J.; Hollmer, S.; Kozicki, M.; Derhacobian, N. Quantized conductance in $Ag/GeS_2/W$ conductive-bridge memory cells. *IEEE Electron Device Lett.* **2012**, *33*, 257–259. [CrossRef]

25. Kozicki, M.N.; Mitkova, M. Mass transport in chalcogenide electrolyte films–materials and applications. *J. Non-Cryst. Solids* **2006**, *352*, 567–577. [CrossRef]

26. Wang, F.; Dunn, W.P.; Jain, M.; De Leo, C.; Vickers, N. The effects of active layer thickness on programmable metallization cell based on Ag-Ge-S. *Solid-State Electron.* **2011**, *61*, 33–37. [CrossRef]

27. Kamalanathan, D.; Russo, U.; Ielmini, D.; Kozicki, M.N. Voltage-driven on-off transition and tradeoff with program and erase current in programmable metallization cell (PMC) memory. *IEEE Electron Device Lett.* **2009**, *30*, 553–555. [CrossRef]

28. Mitkova, M.; Kozicki, M.N. Silver incorporation in Ge-Se glasses used in programmable metallization cell devices. *J. Non-Cryst. Solids* **2002**, *299*, 1023–1027. [CrossRef]

29. Kamalanathan, D.; Akhavan, A.; Kozicki, M.N. Low voltage cycling of programmable metallization cell memory devices. *Nanotechnology* **2011**, *22*, 254017. [CrossRef] [PubMed]

30. Wang, Y.; Mitkova, M.; Georgiev, D.G.; Mamedov, S.; Boolchand, P. Macroscopic phase separation of Se-rich $(x<1/3)$ ternary $Ag_y(Ge_xSe_{1-x})_{1-y}$ glasses. *J. Phys. Condens. Matter* **2003**, *15*, S1573–S1584.

31. Lee, D.; Oukassi, S.; Molas, G.; Carabasse, C.; Salot, R.; Perniola, L. Memory and energy storage dual operation in chalcogenide-based CBRAM. *IEEE Electron Dev. Soc.* **2017**, *5*, 283–287. [CrossRef]

32. Song, M.-J.; Kwon, K.-H.; Park, J.-G. Electro-forming and electro-breaking of nanoscale Ag filaments for conductive-bridging random-access memory cell using Ag-doped polymer-electrolyte between Pt electrodes. *Sci. Rep.* **2017**, *7*, 3065. [CrossRef]

33. Jeon, Y.-R.; Abbas, Y.; Sokolov, A.S.; Kim, S.; Ku, B.; Choi, C. Study of in situ silver migration in amorphous boron nitride CBRAM device. *ACS Appl. Mater. Interfaces* **2019**, *11*, 23320–23336. [CrossRef]

34. Valov, I.; Tsuruoka, T. Effects of moisture and redox reactions in VCM and ECM resistive switching memories. *J. Phys. D Appl. Phys.* **2018**, *51*, 413001. [CrossRef]

35. Li, P.; Wang, Q.; Deng, G.; Guo, X.; Jiang, W.; Liu, H.; Li, F.; Thanh, N.T.K. A new insight into the thermodynamical criterion for the preparation of semiconductor and metal nanocrystals using a polymerized complexing method. *Phys. Chem. Chem. Phys.* **2017**, *19*, 24742–24751. [CrossRef] [PubMed]

36. Yalon, E.; Kalaev, D.; Gavrilov, A.; Cohen, S.; Riess, I.; Ritter, D. Detection of the conductive filament growth direction in resistive memories. In Proceedings of the 72nd Device Research Conference, Santa Barbara, CA, USA, 22–25 June 2014; pp. 299–300.

37. Celano, U.; Goux, L.; Belmonte, A.; Opsomer, K.; Franquet, A.; Schulze, A.; Detavernier, C.; Richard, O.; Bender, H.; Jurczak, M.; et al. Three-dimensional observation of the conductive filament in nanoscaled resistive memory devices. *Nano Lett.* **2014**, *14*, 2401–2406. [CrossRef] [PubMed]

38. Tao, Y.; Li, X.; Xu, H.; Wang, Z.; Ding, W.; Liu, W.; Ma, J.; Liu, Y. Improved uniformity and endurance through suppression of filament overgrowth in electrochemical metallization memory with AgInSbTe buffer layer. *Electron Dev. Soc.* **2018**, *6*, 714–720. [CrossRef]

39. Shaban, M.; Kholidy, I.; Ahmed, G.M.; Negem, M.; El-Salam, H.M.A. Cyclic voltammetry growth and characterization of Sn-Ag alloys of different nanomorphologies and compositions for efficient hydrogen evolution in alkaline solutions. *RSC Adv.* **2019**, *9*, 22389–22400. [CrossRef]
40. Walsh, F.C.; Low, C.T.J. Composite, multi-layer and three-dimensional substrate supported tin based electrodeposits from methanesulfonic acid. *Trans. Inst. Met. Finish.* **2016**, *94*, 152–158. [CrossRef]
41. Chua, L. Everything you wish to know about memristors but are afraid to ask. *Radioengineering* **2015**, *24*, 331–368. [CrossRef]
42. Chiang, T.-H.; Wager, J.F. Electronic conduction mechanisms in insulators. *IEEE Trans. Electron Dev.* **2018**, *64*, 223–230. [CrossRef]
43. Bychkov, E. Superionic and ion-conducting chalcogenide glasses: Transport regimes and structural features. *Sol. Stat. Ion.* **2009**, *180*, 510–516. [CrossRef]
44. Chiu, F.-C. A review on conduction mechanisms in dielectric films. *Adv. Mater. Sci. Eng.* **2014**, 578168. [CrossRef]
45. Jo, H.; Kim, M.-W.; Hong, W.-K. Voltage sweep direction-dependent metal-insulator transition in a single-crystalline VO_2 nanobeam embedded in a insulating layer. *J. Alloy. Compd.* **2017**, *720*, 445–450. [CrossRef]

Article

A Floating Gate Memory with U-Shape Recessed Channel for Neuromorphic Computing and MCU Applications

Lu-Rong Gan †, Ya-Rong Wang †, Lin Chen *, Hao Zhu and Qing-Qing Sun

State Key Lab. of ASIC and System, School of Microelectronics, Fudan University, Shanghai 200433, China
* Correspondence: linchen@fudan.edu.cn; Tel.: +86-21-5566-4324
† These authors contributed equally to this work.

Received: 17 July 2019; Accepted: 22 August 2019; Published: 23 August 2019

Abstract: We have simulated a U-shape recessed channel floating gate memory by Sentaurus TCAD tools. Since the floating gate (FG) is vertically placed between source (S) and drain (D), and control gate (CG) and HfO_2 high-k dielectric extend above source and drain, the integrated density can be well improved, while the erasing and programming speed of the device are respectively decreased to 75 ns and 50 ns. In addition, comprehensive synaptic abilities including long-term potentiation (LTP) and long-term depression (LTD) are demonstrated in our U-shape recessed channel FG memory, highly resembling the biological synapses. These simulation results show that our device has the potential to be well used as embedded memory in neuromorphic computing and MCU (Micro Controller Unit) applications.

Keywords: U-shape recessed channel; floating gate; neuromorphic computing; MCU (microprogrammed control unit)

1. Introduction

With the popularity of intellectualization in medical devices, automotive electronics, smart grid, green energy, wearing equipment, smartcards, and the rise of the Internet of things, Microprogrammed Control Unit (MCU) has been widely used in industrial control and consumer electronics markets and has shown tremendous growth potential in the next few years. To reduce peripheral discrete devices and increase applicability, MCU tends to store programs and small amounts of data through embedded non-volatile memory (NVM). Therefore, with the expanding scale of semiconductor devices and the increasing density of transistors, embedded flash memory, as an important branch of flash products, is more and more widely used in the booming MCU market, and its requirement of integration density is higher and higher [1–4]. With the development of Moore's law, the traditional horizontal channel embedded flash memory has limited miniaturization capability and encountered the small size effect. The leakage caused by this effect will affect the memory's judgment of 0/1 state, which is a serious problem to be avoided, especially in the development of multi-value storage of floating gate (FG) memory.

Today digital computers are based on von Neumann architecture where the memory and processor are physically separated. This fundamentally limits the development of modern computers [5]. Envisioned by Carver Mead in 1990, neuromorphic computing seeks inspirations from the massive parallelism, robust computation, and high energy efficiency of the human brain and can potentially give rise to a revolutionary computing technology that fundamentally overcomes the von Neumann bottleneck in conventional digital computers [6–10]. Synapse is the basic unit in biological nervous system, which connects between two neurons and response differently to incident signals [11]. The change of the strength of synaptic weights caused by memorization events is in charge of encoding and storing memory. Mimicking the physiological synaptic behaviors by using electronic devices is the most important step for neuromorphic systems [12]. The embedded flash memory can emulate the synaptic behaviors such as long-term potentiation (LTP) and long-term depression (LTD), and a high accuracy of more than 1% can be obtained in the application of neuromorphic computing [13]. However, the slow operation speed of traditional embedded floating-gate memory and its limited miniaturization ability hinder its further development in neuromorphic computing [14].

For the first time, this paper proposes a new FG memory structure (UFGM) based on NAND flash programming method and U-shape recessed channel for the applications of neuromorphic computing and MCU. Since the floating gate is vertically placed between source and drain, and control gate and HfO_2 high-k dielectric extend above source and drain, the integrated density can be well improved. The enlarged tunneling area and enhanced tunneling rate dramatically increase the tunneling current when the device is turned on, and the erasing and programming speed of the device are respectively decreased to 75 ns and 50 ns. Therefore, UFGM can quickly adjust synaptic weights during long-term potentiation (LTP) and long-term depression (LTD) operation. In addition, the off-leakage current of UFGM is suppressed because of the extended physical channel length [15–18], which is conducive to reducing the power consumption whether it is used as a synaptic device in the application of neuromorphic computing or MCU. Furthermore, for UFGM, because FG is U-shape embedded, there is no FG capacitive coupling crosstalk between cell and cell in the storage matrix.

2. Device Structure

We have simulated two devices with Sentaurus TCAD tools. Their difference is the doping type of FG. The first device structure is shown in Figure 1a and its FG is p^+-doped. The second device structure is shown in Figure 1b and its FG is n^+-doped.

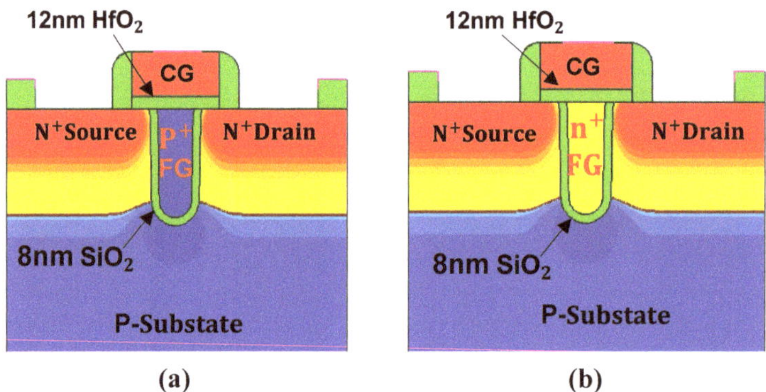

Figure 1. The device structure of (**a**) a new FG memory structure UFGM with p^+ floating gate (FG) and (**b**) UFGM with n^+ FG.

Take the first device as an example. The p^+-doped FG is buried vertically between source (S) and drain (D), and S and D are cut off, and the channel becomes U-shape recessed. This can save area to increase device density, reduce short-channel effects, reduce cell-to-cell coupling, and suppress the off-leakage current. These features will facilitate the applications of UFGM in neuromorphic computing and MCU.

The traditional SiO_2 blocking layer between the polysilicon control gate (CG) and p^+-doped FG is replaced with 12 nm HfO_2 high-k dielectric, and CG and HfO_2 high-k dielectric extend above S and D. The advantage of this is that the inversion and accumulation of electrons and holes on both sides of S and D can be directly controlled by CG through HfO_2 high-k dielectric, which will greatly enhance Fowler–Nordheim (F-N) tunneling rate. Another advantage is that FG is coupled to CG directly through HfO_2 high-k dielectric, and the coupling capacitance is increased, so the CG potential can be dropped to FG more effectively, thus enhancing FN tunneling rate. In terms of the tunneling area, UFGM also shows its advantage. Compared with the horizontal channel, the U-shape recessed channel can increase the effective tunneling area approximately twice under the same feature size. The enlarged tunneling area and enhanced tunneling rate can dramatically increase the tunneling current when the device is turned on.

We also simulated two devices with original SiO_2 based FG for comparison. The first device structure is shown in Figure 2a and its FG is p^+-doped. The second device structure is shown in Figure 2b and its FG is n^+-doped. The fabrication process of the device is similar to that of the UFGM with HfO_2 based FG, except that the 12 nm HfO_2 high-k dielectric material is replaced by 12 nm SiO_2.

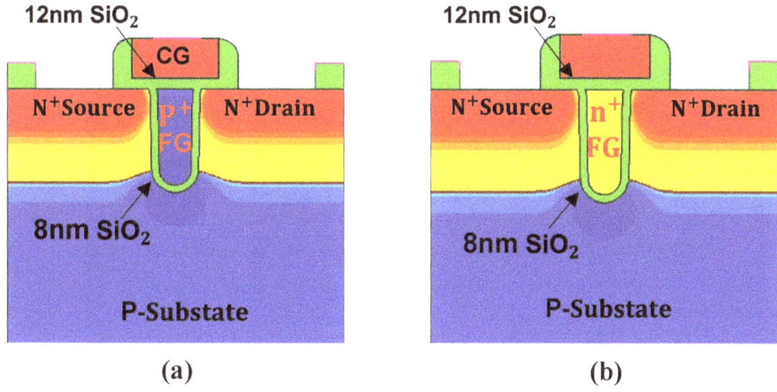

Figure 2. The device structure of UFGM with SiO_2 based (**a**) p^+ FG and (**b**) n^+ FG.

3. Electrical Characteristics

Table 1 contains the main physical models used in electrical simulation. The non-local tunneling model is powerful. It can deal with any shape of barrier and take into account the carrier heating. It allows users to describe tunneling between valence band and conduction band, and approximates several different tunneling probabilities. Non-local tunneling includes FN tunneling.

Table 1. Main physical models selection.

Interface	Physical Mechanism	Model Selection
Oxide/FG poly	Nonlocal tunneling	eBarrierTunneling hBarrierTunneling
Oxide/silicon	Nonlocal tunneling	eBarrierTunneling hBarrierTunneling

We studied the change in the FG potential during one operation period. There are similar trends in the two kinds of devices. As described in Figure 3, under the same conditions, the amount of change in the FG potential gradually increases as V_{cg} increases. Due to the capacitive coupling, a change in the FG potential will cause a drift in the device threshold voltage, which will be used to distinguish between state 0 and state 1 during the reading operation. In the erasing/programming operation, there is a balance between the voltage magnitude and the time setting. Take the UFGM with p^+ FG as an example, at V_{CG} = 10 V and time = 50 ns, the FG potential drops by 0.0528 V, while at V_{CG} = 13 V and time = 50 ns, the FG potential drops by 1.8527 V. However, by extending the bias time at V_{CG} = 10 V, we can get the same FG potential change as at V_{CG} = 13 V and time = 50 ns. The erasing and writing speed can be manually adjusted with different voltage and the time of reading and writing sequence. Therefore, the specific setting of working voltage and time should be carried out under the specific requirements of high speed or low power design.

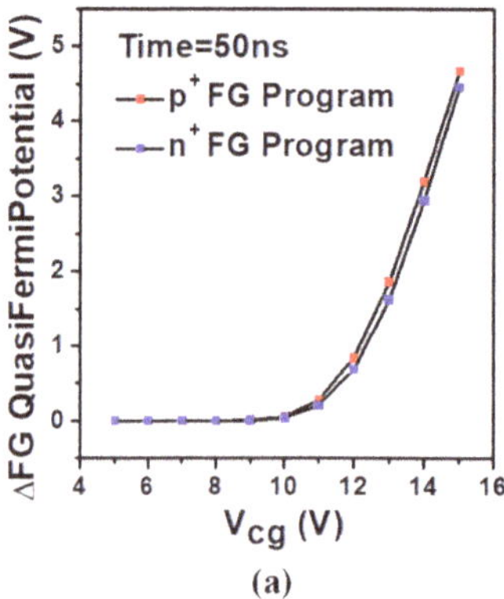

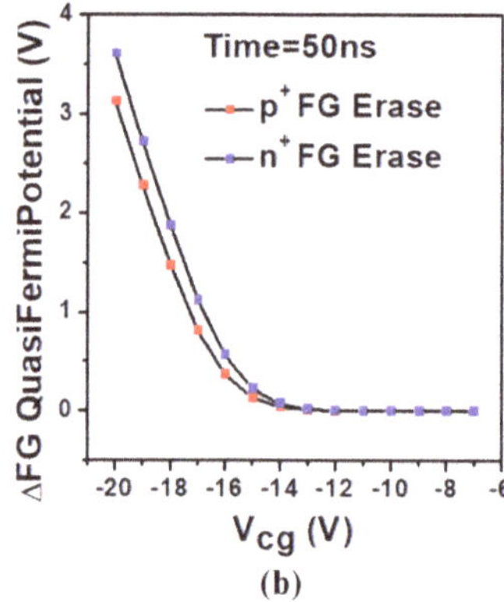

Figure 3. FG potential shift in UFGM as a function of V_{CG} after (**a**) 50 ns programming operation and (**b**) 50 ns erasing operation. The other contacts are set to 0 V.

There are also some slight gaps between two kinds of devices. In the programming operation, the amount of change in the FG potential of the UFGM with p^+ FG is much larger than the UFGM with n^+ FG, which means the UFGM with p^+ FG responds faster. For example, at $V_{CG} = 15$ V and time = 50 ns, the FG potential of the UFGM with p^+ FG drops by 4.6726 V and the FG potential of the UFGM with n^+ FG drops by 4.4548 V. In the erasing operation, the amount of change in the FG potential of the UFGM with n^+ FG is much larger than the UFGM with p^+ FG, which means the UFGM with n^+ FG responds faster. For example, at $V_{CG} = -15$ V and time = 50 ns, the FG potential of the UFGM with p^+ FG increases by 0.1327 V and the FG potential of the UFGM with n^+ FG increases by 0.2259 V. As a conclusion, these two devices have their own advantages. In the application of neuromorphic computing and MCU, we can choose the suitable device according to actual needs.

We also studied the change of FG potential of UFGM based on SiO_2 under different operating voltage. There is a similar trend in these two devices. As shown in Figure 4a, the variation of FG potential increases with the increase of V_{cg} in programming operation. At $V_{cg} = 15$ V, the potential change of p^+ UFGM based on SiO_2 is 0.2637 V, but at the same voltage, the potential change of p^+ UFGM based on HfO_2 can reach 4.6726 V. When $V_{cg} = 20$ V, the potential change of p^+. UFGM based on SiO_2 can reach 4.1173 V, which is still lower than that of UFGM based on p^+ HfO_2 when $V_{cg} = 15$ V. As shown in Figure 4b, in the erasing operation, the change in FG potential gradually decreases as V_{cg} increases. At $V_{cg} = -20$ V, the potential change of n^+ UFGM based on SiO_2 is 3.2647 V while the potential change of n^+ UFGM based on HfO_2 can reach 3.6059 V at the same operation voltage. By comparing the potential changes, we can find that UFGM based on HfO_2 has obvious speed advantages over UFGM based on SiO_2 in both programming and erasing operations.

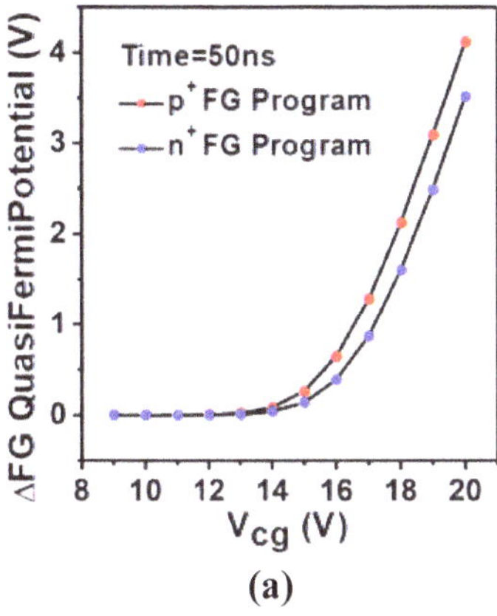

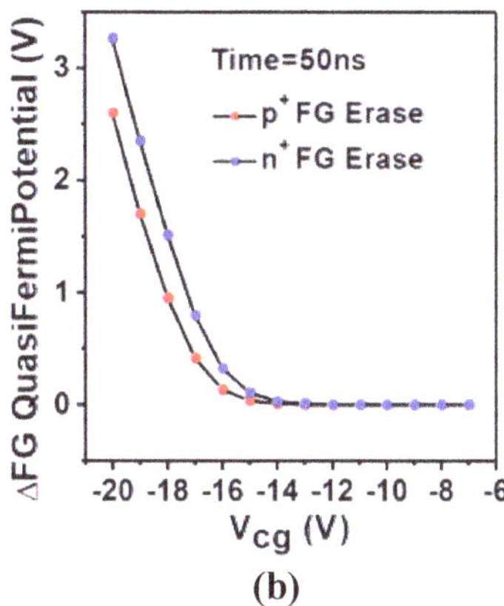

(a)　　　　(b)

Figure 4. FG potential shift in UFGM with SiO_2 based FG as a function of V_{CG} after (**a**) 50 ns programming operation and (**b**) 50 ns erasing operation. The other contacts are set to 0 V.

Figure 5a,b describes the change of the FG potential with time, and the operating voltage scheme as shown in Table 2. During the programming operation, as described in Figure 5a, potential gradually decreases as time increases. The potential decreases approximately linearly in the first 1 µs, and with the increase of time, the potential decreases slowly and finally tends to saturation state. However, the time of linear change of potential is close to 1 µs, and the change of FG potential is about 2.0212 V, which is already enough to distinguish state 0 and state 1. For example, in this paper, we only need 50 ns of operation time. In LTP/LTD operations, there is also sufficient time for weights to approximate linear variations. Similarly, during the erasing operation, as described in Figure 5b, FG potential gradually increases as time increases and the potential increases approximately linearly in the first 1 µs. With the increase of time, the potential increases slowly and finally tends to saturation state. The time of linear change of potential during erasing operation is close to 1.6 µs, and the change of FG potential is about 1.4957 V, which is also enough to distinguish 0/1 state.

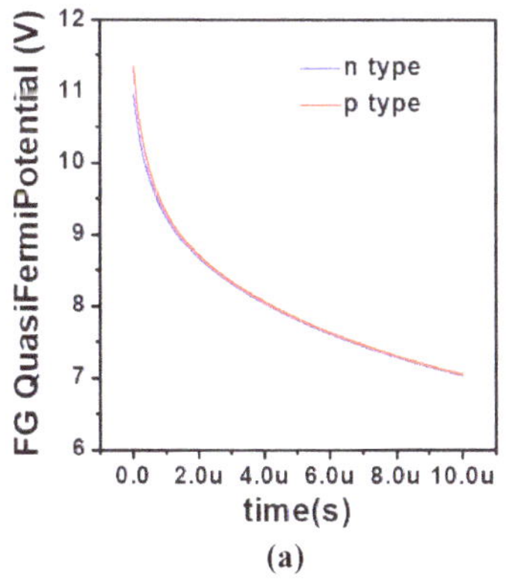

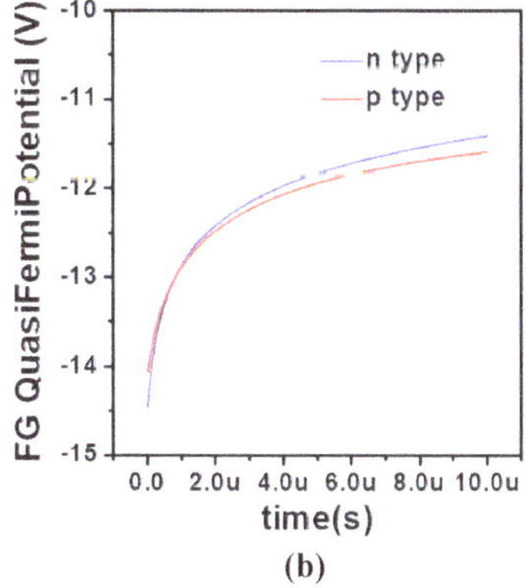

(a)　　　　(b)

Figure 5. FG potential in UFGM as a function of time after (**a**) programming operation and (**b**) erasing operation using the operation voltage scheme in Table 2.

Table 2. Operation voltage and time of UFGM with p^+ FG.

Voltage or Time	Program	Erase	Read	Standby
V_{CG} (V)	11	−15	1.5	0
V_D (V)	0	0	2	0
V_S (V)	0	0	0	0
V_{Sub} (V)	0	0	0	0
Figure 6 Time (ns)	50	75	50	50
Figure 7 Time (ns)	1	1.5	1	2

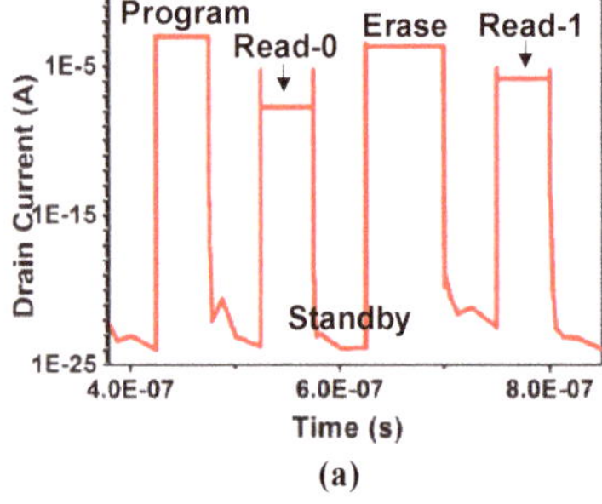

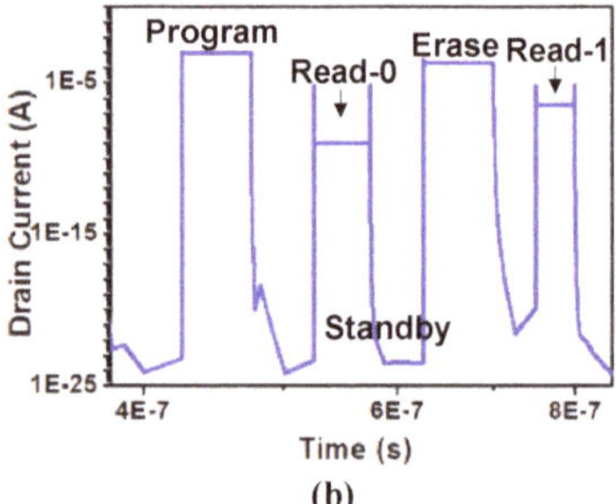

Figure 6. The I_d change curve of (**a**) UFGM with p^+ FG and (**b**) UFGM with n^+ FG with time in a transient simulation using the operation voltage scheme in Table 2.

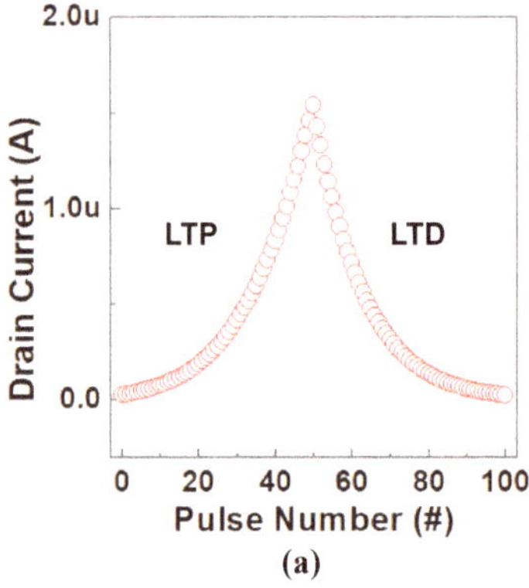

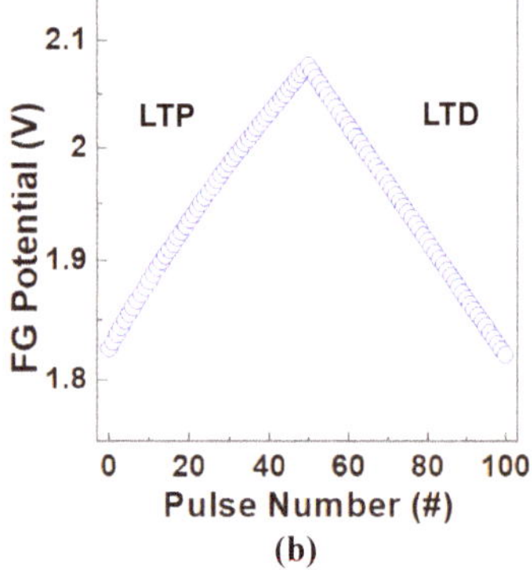

Figure 7. Long-term potentiation (LTP)/long-term depression (LTD) characteristics of UFGM with p^+ FG: (**a**) Drain current and (**b**) FG potential vary with the number of pulses in a transient simulation using the operation voltage scheme in Table 2.

Figure 6 is the drain current (I_d) curve of UFGM cell extracted in the second cycle. The operation voltage and time settings of UFGM with p^+ FG are given in Table 2. According to the simulation experience, the current is more stable and reproducible from the second cycle. The drain current curve of UFGM with p^+ FG and UFGM with n^+ FG cells are shown in Figure 6a,b, respectively. There are also similar trends in the two kinds of devices. As can be seen from Figure 6a, after 50 ns programming operation, a small I_d of about 1.84×10^{-8} A can be read and state 0 is successfully written. After 75 ns erasing operation, a large current of about 1.42×10^{-6} A can be read under the same reading voltage, and state 1 is successfully written. The I_{ON}/I_{OFF} ratio is over 77. As can be seen from Figure 6b, after 50 ns programming operation, a small I_d of about 1.01×10^{-9} A can be read and state 0 is successfully written. After 75 ns erasing operation, a large current of about 3.81×10^{-7} A can be read under the same reading voltage, and state 1 is successfully written. The I_{ON}/I_{OFF} ratio is over 376. In the application of MCU, we need to distinguish the state "0" and the state "1" as clearly as possible, so the difference value between I_{ON} and I_{OFF} should be as large as possible to achieve this distinction, so it is more appropriate to use the UFGM with p^+ FG at this time. In the application of neuromorphic computing, for example, we build a neural network to do weight updates, the I_{ON}/I_{OFF} ratio should be as large as possible to get as many adjustable current states as possible. Here, the UFGM with n^+ FG is more suitable. From the simulation results, we can see that a high-speed embedded FG memory with good characteristics of scaling down is realized, which has the potential to be well applied to neuromorphic computing and MCU.

In the biological brain, the energy efficiency of synaptic transmission is not fixed, which changes with the change of synaptic activity pattern. In many synapses, repeated stimuli can produce an increase or decrease in synaptic weights up to hours or even days. Synaptic weights refer to the strength or magnitude of synaptic weights between the presynaptic and postsynaptic nodes. The enhancement of synaptic weight is called long-term potentiation (LTP), and the reduction of synaptic weight is called long-term depression (LTD). LTP and LTD are the material basis for learning and memory formation [19]. We use the UFGM with p^+ FG as an example to simulate the LTP and LTD characteristics of synapses.

Figure 7 shows the LTP and LTD characteristics of UFGM with p^+ FG. The operation voltage and time settings of pulses applied to UFGM with p^+ FG are given in Table 2. Each erasing/programming pulse is followed by a 1 ns read pulse to monitor the erasing/programming effect. As shown in Figure 7a, the current flowing through the device increases with the increase of the number of pulses, which means the UFGM with p^+ FG exhibits obvious LTP characteristics under a series of pulses with a width of 1.5 ns width and an amplitude of −15 V. Changing the direction of the programmable pulse, setting the pulse width to 1 ns, the amplitude to 11 V, the current flowing through the device decreases gradually with the increase of the number of pulses, and the device shows obvious LTD characteristics. As shown in Figure 7b, when the pulse width is 1.5 ns, the amplitude is −15 V, the potential of the FG increases with the increase of the number of pulses, and the threshold voltage of the device reduces gradually. At a constant reading voltage, the device shows obvious LTP characteristics. Similarly, by changing the direction of the programmed pulse, the device is stimulated by a pulse with a pulse width of 1 ns, an amplitude of 11 V. The potential of the device decreases gradually with the increase of the number of pulses, thus the threshold voltage of the device increases gradually. At the same constant reading voltage, the distinct LTD characteristics can be displayed.

The linearity in weight update refers to the linearity of the curve between the device conductance and the number of identical programming pulses. Ideally, this should be a linear and symmetrical relationship that maps the weight of the algorithm directly to the conductance of the device [14]. This nonlinearity/asymmetry is undesirable because the weight changes depend on the current weight, or in other words, the weight updates are historically relevant [20–22]. As can be seen from Figure 7, the drain current and potential curves of UFGM with p^+ FG have good linearity and symmetry, which means the weight update of UFGM with p^+ FG has excellent linearity and symmetry. This can avoid the loss of learning accuracy of neural networks due to nonlinearity/asymmetry.

4. Conclusions

In this research, we designed and simulated two new structures of U-shape recessed channel FG memory using Sentaurus TCAD tools. After 50 ns programming operation and 75 ns erasing operation, the I_{ON}/I_{OFF} ratio of the UFGM with p^+ FG is over 77, while the I_{ON}/I_{OFF} ratio of the UFGM with n^+ FG is over 376. When a series of continuous pulse operations are applied, the UFGM shows obvious LTP and LTD characteristics. The increase in operating speed, the decrease in short-channel effects and cell-to-cell coupling of FG, the enhanced tunneling rate, the excellent LTP and LTD characteristics, and the increased scaling down ability of the device due to structural changes, make it suitable for the use as an embedded FG memory in neuromorphic computing and MCU.

Author Contributions: Q.-Q.S. and L.-R.G. conceived and designed the experiments; L.-R.G. performed the experiments; Y.-R.W., L.-R.G., L.C. and H.Z. contributed to the data analysis; Y.-R.W. and L.-R.G. completed the manuscript preparation.

Funding: The authors would like to acknowledge the financial support in part by the NSFC (61704030 and 61522404), Shanghai Rising-Star Program (19QA1400600), the Program of Shanghai Subject Chief Scientist (18XD1402800), and the Support Plans for the Youth Top-Notch Talents of China.

Conflicts of Interest: The authors declare no conflict of interest.

References

1. Hidaka, H. Evolution of embedded flash memory technology for MCU. In Proceedings of the IEEE International Conference on IC Design & Technology, Kaohsiung, Taiwan, 2–4 May 2011; pp. 1–4.

2. Shum, D.; Luo, L.Q.; Kong, Y.J.; Deng, F.X.; Qu, X.; Teo, Z.Q.; Liu, J.Q.; Zhang, F.; Cai, X.S.; Tan, K.M.; et al. 40nm embedded self-aligned split-gate flash technology for high-density automotive microcontrollers. In Proceedings of the IEEE International Memory Workshop (IMW), Monterey, CA, USA, 14–17 May 2017; pp. 1–4.

3. Hatanaka, M.; Hidaka, H. Value creation in SOC/MCU applications by embedded non-volatile memory evolutions. In Proceedings of the IEEE Asian Solid-State Circuits Conference, Jeju, South Korea, 12–14 November 2007; pp. 38–42.

4. Hidaka, H. Applications and Technology Trend in Embedded Flash Memory. In *Embedded Flash Memory for Embedded Systems: Technology, Design for Sub-systems, and Innovations*; Springer: Cham, Switzerland, 2018; pp. 7–27.

5. Jiang, J.; Guo, J.; Wan, X.; Yang, Y.; Xie, H.; Niu, D.; Yang, J.; He, J.; Gao, Y.; Wan, Q. 2D MoS$_2$ Neuromorphic Devices for Brain-Like Computational Systems. *Small* **2017**, *13*, 1700933. [CrossRef] [PubMed]

6. Mead, C. Neuromorphic electronic systems. *Proc. IEEE* **1990**, *78*, 1629–1636. [CrossRef]

7. Indiveri, G.; Liu, S.C. Memory and information processing in neuromorphic systems. *Proc. IEEE* **2015**, *103*, 1379–1397. [CrossRef]

8. Kuzum, D.; Yu, S.; Wong, H.S.P. Synaptic electronics: Materials, devices and applications. *Nanotechnology* **2013**, *24*, 382001. [CrossRef] [PubMed]

9. Merolla, P.A.; Arthur, J.V.; Alvarez-Icaza, R.; Cassidy, A.S.; Sawada, J.; Akopyan, F.; Jackson, B.L.; Imam, N.; Guo, C.; Nakamura, Y.; et al. A million spiking-neuron integrated circuit with a scalable communication network and interface. *Science* **2014**, *345*, 668–673. [CrossRef] [PubMed]

10. Zhu, J.D.; Yang, Y.C.; Jia, R.D.; Liang, Z.; Zhu, W.; Ur Rehman, Z.; Bao, L.; Zhang, X.; Cai, Y.; Song, L.; et al. Ion gated synaptic transistors based on 2D van der Waals crystals with tunable diffusive dynamics. *Adv. Mater.* **2018**, *30*, 1800195. [CrossRef]

11. Wang, Z.Q.; Xu, H.Y.; Li, X.H.; Yu, H.; Liu, Y.C.; Zhu, X.J. Synaptic learning and memory functions achieved using oxygen ion migration/diffusion in an amorphous InGaZnO memristor. *Adv. Funct. Mater.* **2012**, *22*, 2759–2765. [CrossRef]

12. Kong, L.-A.; Sun, J.; Qian, C.; Gou, G.Y.; He, Y.K.; Yang, J.L.; Gao, L.Y. Ion-gel gated field-effect transistors with solution-processed oxide semiconductors for bioinspired artificial synapses. *Org. Electron.* **2016**, *39*, 64–70. [CrossRef]

13. Guo, X.; Bayat, F.M.; Prezioso, M.; Chen, Y.; Nguyen, B.; Do, N.; Strukov, D.B. Temperature-insensitive analog vector-by-matrix multiplier based on 55 nm NOR flash memory cells. In Proceedings of the Custom Integrated Circuits Conference (CICC), Austin, TX, USA, 30 April–3 May 2017; pp. 1–4.

14. Yu, S.M. Neuro-inspired computing with emerging nonvolatile memorys. *Proc. IEEE* **2018**, *106*, 260–285. [CrossRef]

15. Heinrich, A.; Loth, S.A. A Logical Use for Atoms. *Science* **2011**, *332*, 1039–1040. [CrossRef] [PubMed]

16. Hamamoto, T.; Ohsawa, T. Overview and Future Challenges of Floating Body RAM (FBRAM) Technology for 32nm Technology Node and Beyond. *Solid State Electron.* **2009**, *53*, 676–683. [CrossRef]

17. Jiang, S.Y.; Yuan, Y.; Wang, X.; Chen, L.; Zhu, H.; Sun, Q.Q.; Zhang, D.W. A Semi-Floating Gate Transistor with Enhanced Embedded Tunneling Field Effect Transistor. *IEEE Electron Device Lett.* **2018**, *39*, 1497–1499.

18. Wang, W.; Wang, P.F.; Zhang, C.M.; Lin, X.; Liu, X.Y.; Sun, Q.Q.; Zhou, P.; Zhang, D.W. Design of U-shape channel tunnel FETs with SiGe source regions. *IEEE Trans. Electron Devices* **2013**, *61*, 193–197. [CrossRef]

19. Nicholls, J.G.; Martin, A.R.; Brown, D.A.; Diamond, M.E.; Weisblat, D.A.; Fuchs, P.A. *From Neuron to Brain*; Sinauer Associates, Inc.: Sunderland, MA, USA, 2001; pp. 317–332.

20. Chen, P.Y.; Lin, B.; Wang, I.; Hou, T.-H.; Ye, J.; Vrudhula, S.; Seo, J.-S.; Cao, Y.; Yu, S. Mitigating effects of non-ideal synaptic device characteristics for on-chip learning. In Proceedings of the IEEE/ACM International Conference on Computer-Aided Design, Austin, TX, USA, 2–6 November 2015; pp. 194–199.

21. Wang, I.-T.; Chang, C.-C.; Chiu, L.-W.; Chou, T.; Hou, T.-H. 3D Ta/TaOx/TiO$_2$/Ti synaptic array and linearity tuning of weight update for hardware neural network applications. *Nanotechnology* **2016**, *27*, 365204. [CrossRef] [PubMed]

22. Burr, G.W.; Shelby, R.M.; Sidler, S.; di Nolfo, C.; Jang, J.; Boybat, I.; Shenoy, R.S.; Narayanan, P.; Virwani, K.; Giacometti, E.U.; et al. Kurdi; Experimental demonstration and tolerancing of a large-scale neural network (165000 synapses) using phase-change memory as the synaptic weight element. *IEEE Trans. Electron Devices* **2015**, *62*, 3498–3507. [CrossRef]

Article

A RISC-V Processor with Area-Efficient Memristor-Based In-Memory Computing for Hash Algorithm in Blockchain Applications

Xiaoyong Xue [1], Chenzedai Wang [1], Wenjun Liu [1,*], Hangbing Lv [2,*], Mingyu Wang [1,*] and Xiaoyang Zeng [1]

1   State Key Laboratory of ASIC and System, Fudan University, Shanghai 201203, China
2   Key Laboratory of Microelectronics Devices and Integrated Technology, Institute of Microelectronics of the Chinese Academy of Sciences, Beijing 100029, China
*   Correspondence: wjliu@fudan.edu.cn (W.L.); lvhangbing@ime.ac.cn (H.L.); mywang@fudan.edu.cn (M.W.); Tel.: +86-021-51355200-987 (W.L.)

Received: 31 May 2019; Accepted: 16 August 2019; Published: 16 August 2019

Abstract: Blockchain technology is increasingly being used in Internet of things (IoT) devices for information security and data integrity. However, it is challenging to implement complex hash algorithms with limited resources in IoT devices owing to large energy consumption and a long processing time. This paper proposes a RISC-V processor with memristor-based in-memory computing (IMC) for blockchain technology in IoT applications. The IMC-adapted instructions were designed for the Keccak hash algorithm by virtue of the extendibility of the RISC-V instruction set architecture (ISA). Then, a RISC-V processor with area-efficient memristor-based IMC was developed based on an open-source core for IoT applications, Hummingbird E200. The general compiling policy with the data allocation method is also disclosed for the IMC implementation of the Keccak hash algorithm. An evaluation shows that >70% improvements in both performance and energy saving were achieved with limited area overhead after introducing IMC in the RISC-V processor.

Keywords: in-memory computing; memristor; RISC-V; Internet of things; blockchain

---

1. Introduction

Internet of things (IoT) refers to the network of different physical devices, which enables them to collect and exchange data [1,2]. With the development of telecommunication, computers, and integrated circuits, IoT is being increasingly applied in commercial fields such as modern agriculture, driverless vehicles, smart cities, etc., which promise to become vital parts of global economics [3]. However, as billions of IoT devices are connected to the continuously growing networks, security appears to be a major concern. IoT devices collect a great amount of private information, which is vulnerable to attacks if not well protected. Moreover, most of the devices are resource-constrained and, thus, heavy cryptographic approaches are difficult to implement.

Recently, a trend emerged to exploit the blockchain technology in IoT devices for information security and data integrity [4]. The blockchain is a peer-to-peer (P2P) ledger which was first used in the Bitcoin cryptocurrency for economic transactions [5]. Bitcoin users that are known by a changeable public key generate and broadcast transactions to the network to transfer money. These transactions are pushed into a block by users. Once a block is full, the block is appended to the blockchain by performing a mining process. To mine a block, some specific nodes known as miners try to solve a cryptographic puzzle named proof of work (POW), and the node that solves the puzzle first mines the new block to the blockchain, as shown in Figure 1. Because of its distributed, secure, and private nature, the blockchain can enable secure messaging between devices in an IoT network. In this approach, the blockchain treats message exchanges between devices similar to financial transactions in a Bitcoin network. To enable message exchanges, devices leverage smart contracts which model the agreement between two parties. The distributed datasets maintained by blockchain technology also allow the data to be safely stored by different peers, and people are not required to entrust IoT data produced by their devices to centralized companies [6]. Moreover, the blockchain technology lowers the cost of the deployment of the IoT devices and makes it safe and easy for users to pay for the data on IoT devices [7].

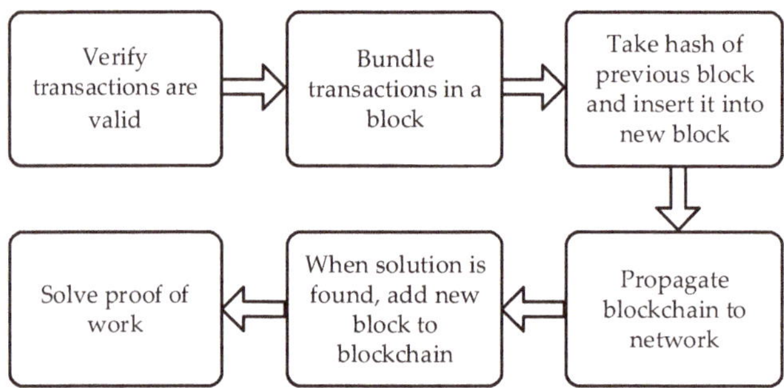

Figure 1. Bitcoin mining process using blockchain technology.

However, the hardware for mining in IoT devices has to be lightweight, low-cost, and energy-efficient to adapt the blockchain technology. IoT devices are often deployed in nonhuman conditions to a great extent and are powered through batteries, calling for extremely low cost and low energy consumption [8]. However, when using a general processor, i.e., central processing unit (CPU) or graphics processing unit (GPU), to implement the blockchain, it is likely to consume too much energy, resulting in frequent recharging or a short battery lifetime. Resorting to a conventional application-specific integrated circuit (ASIC) or coprocessor can help to reduce energy consumption and improve speed, but will induce considerable area cost [9].

In-memory computing (IMC) provides a promising alternative. In a general processor, the data transfer on the bus between the central processing unit (CPU) and the memory leads to large power consumption and limited performance, i.e., memory bottleneck. To address this issue, IMC modifies the memory to be able to perform some regular logic operations such as AND, OR, and exclusive or (XOR) [9]. Especially for data vectors with large bit width, IMC can accomplish the AND/OR/XOR operation in one read access, saving both execution time and power consumption. Static random-access memory (SRAM) can be employed in IMC, but its cell size is too large with 6–10 transistors and it also needs constant power to hold the data, incurring considerable area cost and standby power [10].

Emerging memory technologies, especially memristors, feature a simple cell structure, high density, three-dimensional (3D) stackability, good compatibility with complementary metal–oxide–semiconductor (CMOS) processes, and non-volatility [11]. Recently, memristors were investigated to realize IMC using a one-transistor-one-memristor (1T1R) array accompanied by modified peripheral circuits [12]. However, it is still difficult to rely on memristor-based IMC alone to implement the hash algorithm in blockchain technology. A processor is still required to perform the data allocation, as well as other complexed logic operations. For resource-limited IoT devices, the processor should be flexible to support memory computation instructions while incurring small power consumption and area cost. Thanks to its simplicity, scalability, fast speed, and low power, the RISC-V processor is believed to be competent for the abovementioned requirements [13,14]. The instruction set architecture (ISA) of the RISC-V is designed to avoid over-architecting, while supporting command extension to achieve high flexibility [13]. Nevertheless, for the practical integration of IMC in RISC-V, the corresponding compiling policy and data allocation method still need specific consideration.

This paper proposes a RISC-V processor with memristor-based IMC for blockchain technology in IoT applications. The IMC-adapted instructions are designed for the Keccak hash algorithm by virtue of the extendibility of the RISC-V ISA. Then, a RISC-V processor with area-efficient memristor-based IMC is developed based on the open-source core, Hummingbird E200. The general compiling policy with data allocation method is also disclosed for the Keccak hash algorithm. An evaluation shows that remarkable improvements in performance and energy consumption are achieved with limited area overhead after introducing IMC.

The rest of the paper is organized as follows: Section 2 gives the IMC-adapted ISA design for the hash algorithm. Section 3 describes the RISC-V processor architecture with IMC and the implementation of IMC. Section 4 provides the policy for compiling and data allocation. Section 5 presents the evaluation, and Section 6 concludes this paper.

2. IMC-Adapted ISA Design for Hash Algorithm

2.1. Hash Algorithm in Blockchain Technology

A blockchain is literally a chain of blocks, each of which has a block header containing the hash value of its parent block to ensure the integrity of the chain [5]. With the rapid development of both computer hardware and software, traditional hash algorithms like Message-Digest algorithm 4 (MD4), Message-Digest algorithm 5 (MD5), and Secure Hash Algorithm 1 (SHA-1) were cracked. Therefore, the United States (US) National Institute of Standards and Technology (NIST) selected the Keccak sponge function family as the third-generation secure hash algorithm (SHA-3) to ensure the security of hash algorithms [15,16].

Keccak or SHA-3 shares a structure involving sponge functions with different parameters. The default Keccak sponge function works on a 1600-bit state array, which is logically a three-dimensional array with a row and column width of five and a lane width of 64. The array is often denoted as [x][y][z] in GF(2), where $0 \leq x \leq 4$, $0 \leq y \leq 4$, and $0 \leq z \leq 63$.

The process of the Keccak sponge function consists of two phases, i.e., the absorbing phase and the squeezing phase. In the absorbing phase, the r-bit input blocks are XORed into the first r bits of the state, interleaved with a permutation called Keccak-f permutation; when all input blocks are processed, the sponge construction switches to the squeezing phase. In the squeezing phase, the first r bits of the state are returned as output blocks, interleaved with Keccak-f permutation; the number of output blocks is chosen at will by the user. Here, the value r is the bit rate. The process of the Keccak sponge function is actually an iteratively executed Keccak-f permutation, which takes most of the executing time. By default, 24 Keccak-f permutations take place for one permutation of sponge function.

Keccak-f permutation consists of five steps, which are the θ step, ρ step, π step, χ step, and ι step. The corresponding calculations of the five steps are shown in Equations (1)–(7). More detailed information for the algorithm can be found in Reference [17]. Table 1 summarizes the main processes performed in the five steps where the calculations of large vectors are hopefully implemented by the IMC.

$$a[x][y][z] \leftarrow a[x][y][z] + \sum_{y'=0}^{4} a[x-1][y'][z] + \sum_{y'=0}^{4} a[x+1][y'][z-1]. \tag{1}$$

$$a[x][y][z] \leftarrow a[x][y]\left[z - \frac{1}{2}(t+1)(t+2)\right]. \tag{2}$$

$$\begin{pmatrix} 0 & 1 \\ 2 & 3 \end{pmatrix}^{t} \begin{pmatrix} 1 \\ 0 \end{pmatrix} = \begin{pmatrix} x \\ y \end{pmatrix}, \ 0 \leq t \leq 24 \text{ or } x = y = 0, \ t = -1. \tag{3}$$

$$a[x][y] \leftarrow a[x'][y'], \ \begin{pmatrix} x \\ y \end{pmatrix} = \begin{pmatrix} 0 & 1 \\ 2 & 3 \end{pmatrix}\begin{pmatrix} x' \\ y' \end{pmatrix}. \tag{4}$$

$$a[x] \leftarrow a[x] + (a[x+1]+1)a[x+2]. \tag{5}$$

$$a[0][0] \leftarrow a[0][0] + R_i. \tag{6}$$

Table 1. Five steps in Keccak-*f* permutation.

Step	Equations	Main Process
θ	(1)	Massive 64-bit and 320-bit bitwise XOR operations, a few 64-bit shift operations
ρ	(2), (3)	Massive 64-bit shift operations and data copying
π	(4)	Massive 64-bit data copying
χ	(5)	Massive bitwise 320-bit logic operations (XOR, OR and AND)
ι	(6)	Massive operations on one 64-bit binary string

2.2. IMC-Adapted ISA Design

Before proposing the RISC-V processor with IMC for SHA-3, the characteristics hidden in Keccak calculations and how to adapt the ISA to support the IMC are investigated. Many operations in SHA-3, especially the sheet and plane logic operations, require frequent memory access and can be greatly optimized by adopting IMC, since they are 320 bits long while a processor is often 32-bit or 64-bit. RISC-V ISA is highly extendable and provides the users with four custom operations in its base instruction set and long custom instruction sets to be defined in the future [12]. To improve SHA-3 performance, only a few IMC instructions are needed; thus, this work employs the four custom operations to adapt IMC. The long custom instruction sets are reserved for more IMC operations as needed.

The operations in Keccak-*f* permutation can be classified into four different types, which are (1) long bitwise logic operations (both 64 bits and 320 bits), (2) 64-bit shift operations on a 320-bit binary string, (3) 64-bit data copying, and (4) operations on one 64-bit binary string. For these four types of operations, the first three can be easily implemented by IMC technology. Based on the above analysis, three kinds of IMC operations are adopted, including 320-bit bitwise logic operations (XOR, OR, and AND), 64-bit shift operation (SHIFT), and 64-bit data copying operation (CP). In addition, an operation that copies 64-bit data to all columns in another row address (copy to all columns, CPA) is needed for data allocation purposes (see Section 4). CPA operations are also needed in the θ step and χ step for data allocation purposes. Table 2 shows the IMC operations involved in different steps of Keccak-*f* permutation.

Table 2. In-memory computing (IMC) applications in Keccak-*f* permutation. XOR—exclusive or; SHIFT—64-bit shift operation; CPA—copy to all columns; CP—64-bit data copying operation.

Step	IMC Involved
θ	XOR, SHIFT, CPA
ρ	SHIFT, CP
π	CP
χ	XOR, OR, AND, CPA
ι	None

Table 3 shows the detailed IMC instruction definition. The IMC logic instructions including XOR, OR, and AND perform the 320-bit logic operation with operands from addresses (BA + A1) and (BA + A2), and store the results in (BA + A0). A0, A1, and A2 are addresses either from immediate or registers, depending on 3-bit I/R, and BA is an address from a register. SHIFT instruction performs the 64-bit circular right shift on (BA + A1) by SA[6:0] amount and stores the result in (BA + A0). A0 and A1 are addresses either from immediate or from registers, depending on 2-bit I/R. The addresses used in 320-bit operations are all 9-bit row addresses; thus, only 9 bits in the address are valid. The normal read loads the 32-bit data from memory address (rs + Imm[11:0]) to register rd. The normal write stores the 32-bit word data in register rs2 to memory address (rs1 + Imm[11:0]). For CP and CPA instructions, when Flag = 0, the data in memory address (BA + A1 + Col[2:0]) are copied to address (BA + A2 + Col[5:3]) for CP; when Flag = 1, the data in memory row address (BA + A1 + Col[3:0]) are copied to all the columns in row address (BA + A2) for CPA. The reserved bits in the IMC-adapted ISA can be used for more functions if necessary.

Table 3. IMC-adapted instruction definition list.

Bit	31–30	29–25	24–20	19–15	15–13	12	11–7	6–0
XOR	00	A1	A2	BA	I/R		A0	Custom0
OR	01	A1	A2	BA	I/R		A0	Custom0
AND	10	A1	A2	BA	I/R		A0	Custom0
SHIFT	11	A1	SA[5:0]	BA	I/R	SA[6:0]	A0	Custom0
Normal read	Imm[11:0]			rs	Reserved		rd	Custom1
Normal write	Imm[11:5]		rs2	rs1	Reserved		Imm [4:0]	Custom2
CP and CPA	0 Flag	A1	A2	BA	I/R		Col[5:0]	Custom3

3. RISV Processor with IMC

3.1. Processor Architecture

RISC-V foundations introduced a few open-source RISC-V processor cores. This work chose Hummingbird E200 as the original processor because it was designed for IoT applications and optimized for low power and area costs [18].

The original Hummingbird E200 processor employs two static random-access memories (SRAMs) as working memories, one for instructions and the other for data. This work adds an additional memory module, i.e., the IMC module, which includes an IMC core based on a memristor and a customized IMC controller to interact with the control and operation module (COM) in the CPU core, as shown in Figure 2. Some modifications are also made inside the processor without changing the original functions; thus, the generality is not destroyed after adding IMC functions. The memory controller is not reused for the IMC module because it has more functions than a traditional SRAM. Therefore, a separate controller is designed inside the IMC module (as discussed in Section 3.2).

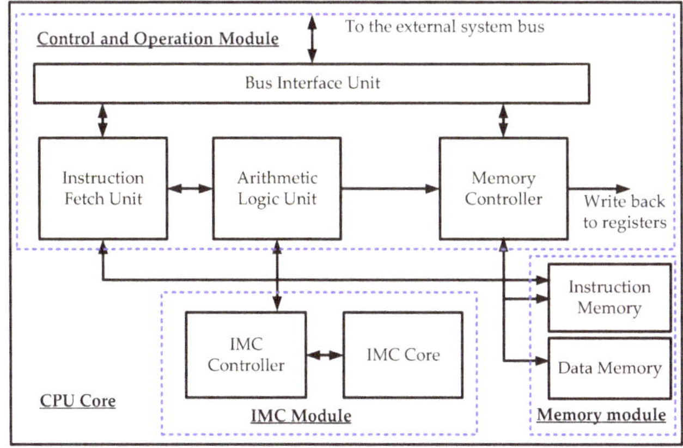

Figure 2. Modified RISC-V processor core with in-memory computing (IMC).

3.2. IMC Implementation

3.2.1. IMC Core Architecture and Assistant Logic

The IMC core is designed to implement the IMC instructions. It consists of an advanced row decoder, a write buffer, a memristor array, an IMC read-out circuit, a 64-bit shifter, and a mode selector, as shown in Figure 3.

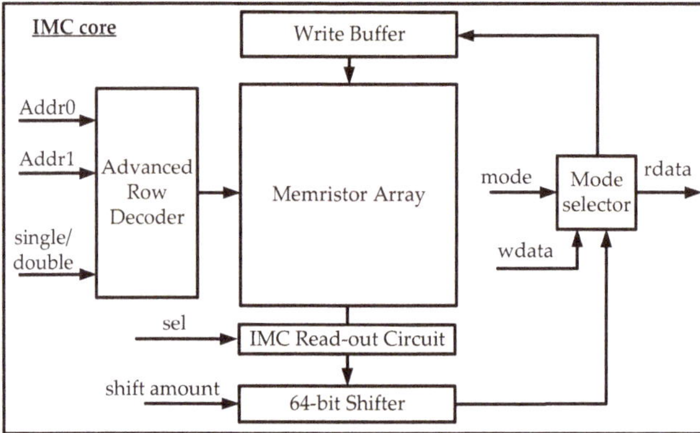

Figure 3. IMC core architecture.

The read-out circuit is specially designed to implement the IMC logic instructions. The memristor array stores the data which participate in the IMC computations. These two modules are indispensable for IMC and are described in Section 3.2.2. The rest of the IMC core includes assistant circuits, which help to implement the IMC instructions and the control of the IMC core.

The advanced row decoder can either activate two row addresses simultaneously to execute IMC logic instructions or only one address to execute read/write instructions. The 64-bit shifter implements the 64-bit circular shift operation, and is disabled when other operations are performed. The mode selector decides whether the data are loaded out to the registers or written to the memory (either 64-bit data or 320-bit data) for CP and CPA. The write buffer is used when the data are written to the memristor array. A selection signal is sent to the Bitline (BL) calculator inside the IMC read-out circuit to determinate the IMC logic type. It should be noted that some control circuits are not shown in Figure 3 for conciseness.

3.2.2. IMC Memristor Array and Read-Out Circuit

In-memory computing implements all the 320-bit bitwise logic operations including AND, OR, and XOR operations in the hash algorithm using memristor-based IMC technology. As shown in Figure 4, a one-diode-one-memristor (1D1R) crossbar array is proposed with the IMC read-out circuit to realize the logic operations. The diode helps to restrain the disturbance of sneaking current to write/read, and logic operations; the memristor features unipolar set and reset operations. Using the diode as the selector, the 1D1R cell achieves higher density than the 1T1R cell [19]. Moreover, the diode selector and the memristor can both be integrated in the back end of line (BEOL) of the standard CMOS process. Therefore, the IMC core can be physically stacked by placing peripheral circuits on the substrate and lower interconnect metals, and the 1D1R crossbar array on the middle or upper interconnect metals. This can save area further, in accordance with the low-cost requirement of IoT devices.

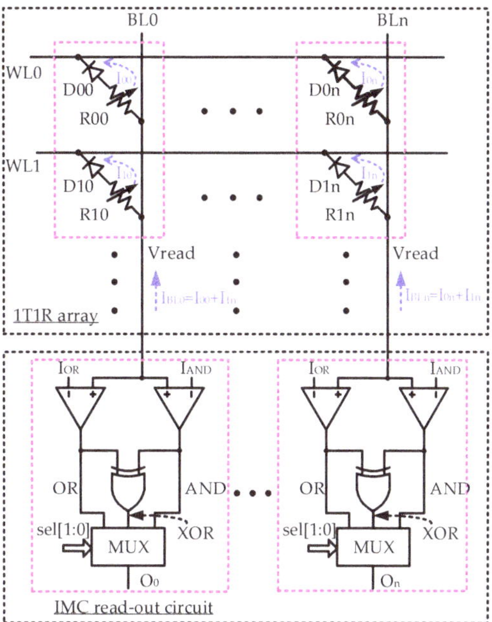

Figure 4. Memristor array with IMC read-out circuit.

The data are written into the 1D1R array by the processor in advance. The operation table of the 1D1R memristor array is shown in Table 4. Here, Vset/Vreset/Vread stand for the set/reset/read voltage of the memristor, and Vt is the threshold voltage of diode selector. The low-resistance state (LRS) of the memristor stands for logic "1", while the high-resistance state (HRS) stands for logic "0". To perform IMC, two selected wordlines (WLs), e.g., WL0 and WL1, are activated while applying proper read voltage (Vread) on the bitlines, e.g., BL0–BLn. The sum of currents along the same bitline (BL), e.g., I_{BL0} and I_{BLn}, are compared with two reference currents, I_{OR} and I_{AND}. The HRS is usually 10 times larger than the LRS [10], meaning that $I_{LRS} \gg I_{HRS}$, where I_{HRS} and I_{LRS} stand for the typical read currents for HRS and LRS, respectively. Therefore, the typical values of I_{OR} and I_{AND} can be set as $0.5 \times I_{LRS}$ and $1.5 \times I_{LRS}$. When I_{SUM} is larger than I_{OR}, the signal OR becomes logic "1", implying that at least one of the two activated memristors along the same bitline is the LRS. When I_{SUM} is larger than I_{AND}, the signal AND becomes logic "1", implying that both activated memristors along the same bitline are the LRS. By sending the results of OR and AND to an XOR gate, the XOR result is obtained at the output O_0–O_n. According to the control signal sel[1:0] from the assistant logic circuit, the corresponding result is written back to the 1D1R array in the next clock cycle. To perform 320-bit operations, this work adopts a 20-kb memristor array with 64 rows and 320 columns.

Table 4. Operation table of one-diode-one-memristor (1D1R) memristor array for IMC. HRS—high-resistance state; LRS—low-resistance state.

Operation Mode	Wordline (WL)		Bitline (BL)	
	Selected	Un-Sel	Selected	Un-Sel
Set (HRS→LRS)	0	Vset	Vset + Vt	0
Reset (LRS→HRS)	0	Vreset	Vreset + Vt	0
Logic (Read)	0	Vread	Vread + Vt	0

4. IMC Compiling Policy and Data Allocation Method

4.1. IMC Compiling Policy

In a traditional general processor, it is up to the software programmers to decide how to store the data needed, and the compiler to decide where to store them [20]. However, when it comes to IMC instructions, the programmer also has to decide whether to perform the computation with Arithmetic Logic Unit (ALU) or with IMC, requiring a special compiling policy. In addition, as mentioned in Section 3, only data in the same column and different rows can perform IMC logic operations; thus, IMC requires a different data allocation policy.

When a 32-bit vector is to be calculated with another 32-bit vector, ALU can finish this process in one clock cycle if the data are already cached in the registers, but IMC needs two clock cycles. This indicates that IMC consumes more processing time than ALU when performing simple logics. However, if both vectors are originally in the memory, ALU needs two additional clock cycles to load them out, and another clock cycle to store them into the memory if needed. This consumes more time than IMC. More generally, for a certain part of the algorithm with A 32-bit inputs, N steps of basic 32-bit operations, and Y 32-bit outputs (including long-lifetime intermediate results that cannot be cached in general registers), ALU takes $(A + N + Y)$ clock cycles to process, whereas IMC needs $2N$. Therefore, ALU should be used to perform calculations when

$$A + N + Y < 2N,\tag{7}$$

i.e.,

$$N > A + Y.\tag{8}$$

Similarly, if the vectors are 64-bit long, ALU needs at least 2–6 clock cycles to finish this operation, but IMC needs only two clock cycles anyway; thus, IMC should be used to perform the calculations. This works better for vectors with widths larger than 64 bits. To sum up, for 32-bit vectors, ALU performs better when Equation (8) is satisfied, and, for 64-bit or longer vectors, IMC is always better.

4.2. Data Allocation Method for SHA-3

In terms of data allocation, IMC logics require any data processed to be in the same columns and different rows, and then data in the same row are handled simultaneously. Therefore, it is required that data placed in the same columns should frequently be operands of IMC operations, and data in the same row should share the same IMC operations frequently.

Considering the regular features in Keccak-*f* permutation and the general compiling policy, we decided to adopt the data allocation method as shown in Figure 5. The 1600-bit state array is placed in row addresses R0–R4, and five 64-bit words are located in each row address with column address C0–C4, denoted as A(x,y). The five permutation steps are processed as below.

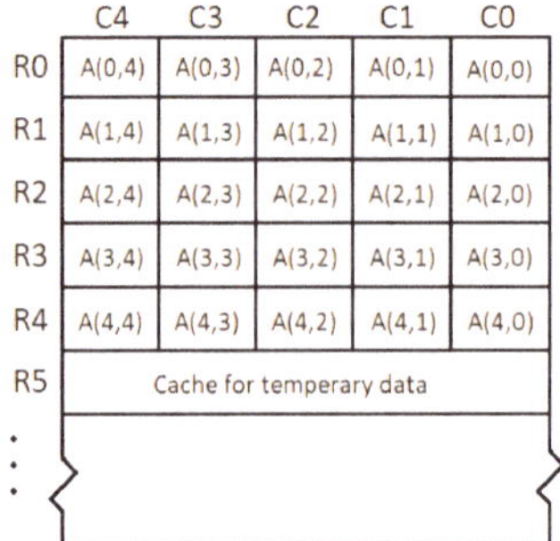

Figure 5. Data mapping of the 1600-bit state array.

a.   θ step

Perform XOR operations and get the XOR result of R0–R4, and then put the result in R5. Copy the result in C0–C4 of R5 to all columns in R6–R10 by performing CPA operations. Perform SHIFT operations on R6–R10 with the result placed in R11–R15. Then, XOR operations with the result placed in R0–R4 can be performed to finish the θ step.

b.   ρ and π step

The ρ step and π step can be processed in a mixed way. Copy all data from R0–R4 to R5–R9; then, perform SHIFT operation to get the rotated value (stored temporarily in R10) and CP operations to update the data in R0–R4.

c.   χ step

Copy data in C0–C4 of R0 to R5–R9 by CPA operations, and perform NOT, AND, and XOR operations in succession and update R0. Repeat this process five times so that all R0–R4 rows are updated. Note that the NOT operation can be performed by XOR with an all-1 vector.

d.   ι step

In the ι step, there are lots of frequently used data and few long vectors; thus, ALU is used to perform this operation, and the instructions can be given by a C compiler.

5. Evaluation

5.1. Evaluation Methods

The proposed RISC-V processor with IMC for the Keccak algorithm was evaluated against the baseline one without IMC in terms of area, processing time, and energy consumption. The evaluation was carried out using the 28-nm process parameters.

For area evaluation, the control and operation module in Verilog hardware description language (HDL) format was firstly compiled by a Synopsys design compiler to acquire the equivalent gate count, which was then multiplied by the size of two-input NAND gate, i.e., NAND2, in the 28-nm process to get the corresponding area. The total area of the processor was calculated by summing the area of the control and operation module, the area of two working SRAM memories, and the area of the 20-kb IMC module.

Figure 6 gives the evaluation method for processing time and energy consumption. To evaluate the processing time, the Keccak process was simulated in a Synopsis VCS Verilog simulator [21]. A 7-byte binary string was adopted as the test input. By simply compiling the C source code of the Keccak algorithm, the baseline processor could give the SHA-3 value through a non-IMC method. Then, by adding IMC instructions into the compiled machine codes of Keccak algorithm, the IMC-extended processor could give the SHA-3 value through an IMC method. The processing time can be acquired from the simulation log files. The energy evaluation was based on the simulation results of processing time. Firstly, the executed instructions in both cases were counted from the simulation log files separately. Then, based on the average energy consumption of individual instructions, the total energy consumption could be obtained by weighted summation.

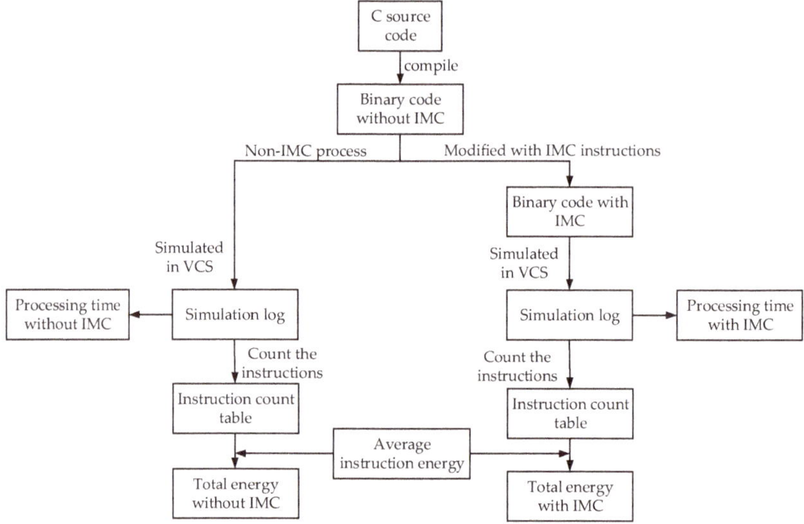

Figure 6. Evaluation method for processing time and energy consumption.

5.2. Area Overhead

The equivalent gate count of the control and operation module, i.e., COM, was compiled to be about 110 K. Given the size of NAND2 to be 0.9 μm × 0.56 μm, the area of COM was about 0.006 mm^2. The two working SRAM memories both had a capacity of 64 kb. The SRAM cell size was 0.12 μm^2 and the total area of two working SRAMs was 0.028 mm^2 [22]. For the IMC module, the count of IMC read-out circuits was required to be as many as 320 to support 320-bit bitwise logic operations. Assuming that each IMC read-out circuit had a size of 2 μm × 4 μm, the total area of IMC read-out circuits was 0.0026 mm^2. The area of the advanced row decoder was estimated to be 0.001 mm^2, i.e., 50 μm × 20 μm. The areas of the other circuits in the IMC module were relatively small and were estimated to be 0.0005 mm^2. By 3D stacking, the 20-kb memristor array of the 1D1R cell would not bring additional area cost. To sum up, the area of the IMC module was about 0.004 mm^2. Figure 7 shows the area comparison of the baseline and the RISC-V processor with IMC. The IMC module brings an area overhead of about 12%. However, the memristor array in the IMC module also plays the part of data cache; thus, the capacity of SRAM memory for data can be reduced, alleviating the area overhead. When the capacity of SRAM memory for data is reduced by 20 kb, the total area is reduced by about 0.003 mm^2, and the area overhead is decreased to only 3%.

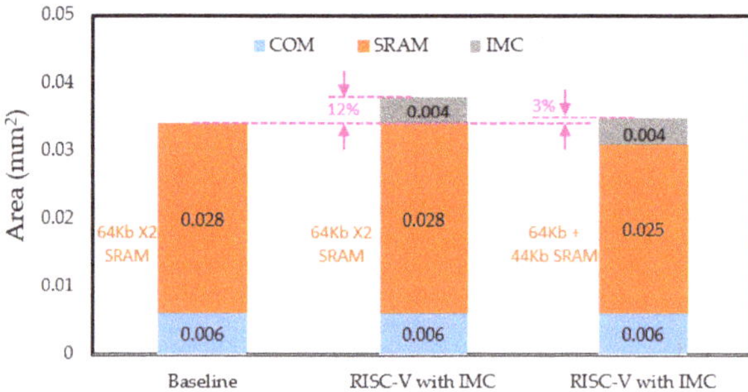

Figure 7. Area comparison between the baseline and the proposed RISC-V with IMC.

5.3. Performance Improvement

The processing time of the baseline RISC-V processor and the proposed one with IMC can be easily given by the simulator. The simulation was performed at a clock frequency of 62.5 MHz. Since our IMC technology accelerates each round in the Keccak-*f* permutation, both the processing time in one round and the overall process were considered, as shown in Figure 8. The processor can achieve over 70% improvement in terms of processing time for both one round and the overall process.

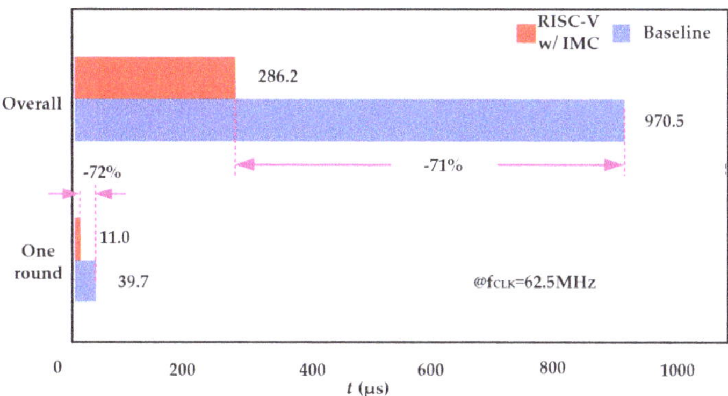

Figure 8. Comparison of processing time for the Keccak algorithm.

5.4. Energy Reduction

The average energy consumption for different operations was firstly characterized in the 28-nm process, as shown in Table 5. The energy consumed by SRAM read or write was similar. The 1D1R memristor cell consumed more energy than SRAM for read and write due to large active currents. Furthermore, the write operation of the memristor was even more energy-consuming than the read. Since the IMC logic was performed mainly by the read operation, the IMC readout circuits and other peripheral circuits still brought additional energy consumption by about 50%. All the parameters were closely relevant to the circuit design techniques and can be further optimized.

Table 5. The average energy consumption for different operations in the 28-nm process.

Operation	Energy (pJ)
ALU	70
SRAM read/write	0.1/bit
memristor read	0.3/bit
memristor write	0.6/bit
memristor logic	0.45/bit

The average energy for each instruction is described in Table 6. The energy of ALU instruction refers to the energy consumed by the control and operation module to fetch an instruction from instruction SRAM, and then to decode and execute the instruction. The energy of SRAM read/write refers to the energy consumed by the normal ALU instruction and the energy to read/write 32-bit data from/to the data SRAM. The energy of IMC read/write refers to the energy consumed by the normal ALU instruction and the energy to read/write 32-bit data from/to the memristor array. The energy of IMC CP refers to the energy consumed by the normal ALU instruction and the energy to read 64-bit data from the memristor array and then write it to another address in the memristor array. The energy of IMC CPA refers to the energy consumed by the normal ALU instruction and the energy to read 64-bit data from the memristor array and then write it to five addresses in the same row of the memristor array. The energy of IMC logic refers to the energy consumed by the normal ALU instruction and the energy to perform 320-bit IMC logic and then write the 320-bit result back to the memristor array. The energy of 320-bit IMC SHIFT refers to the energy consumed by the normal ALU instruction and the energy to read 320-bit data from the memristor array and then write it back to the memristor array after shifting. It should be noted that IMC instructions usually consume more energy than ALU and SRAM read/write (R/W) instructions. With the development of memristor technology, the power consumption can be expected to decrease.

Table 6. The average energy consumption for each instruction in the 28-nm process.

Instruction	Main Actions	Energy (pJ)
ALU	Fetch, decode and execute the instruction	70
SRAM read/write	ALU, 32-bit SRAM read/write	73.2
IMC read	ALU, 32-bit memristor read	82.8
IMC write	ALU, 32-bit memristor write	89.2
IMC CP	ALU, 64-bit memristor read and write	134
IMC CPA	ALU, 64-bit memristor read and 320-bit memristor write	287.6
IMC Logic (AND, OR, and XOR)	ALU, 320-bit memristor logic and write	406
IMC SHIFT	ALU, 320-bit memristor read and write	390

Like the processing time, both the energy consumption in one round of Keccak-*f* permutation and the overall process were considered. Figure 9 gives the comparison of instruction count of the baseline RISC-V processor and the one with IMC. In one round of Keccak-*f* permutation, the instruction counts of ALU and SRAM R/W were greatly reduced and the total instruction count was reduced by 83% after introducing IMC. The reduced instructions mean less data transferred between the memory and the ALU and also less workload for the ALU. As a result, the energy consumption in one round of Keccak-*f* permutation was reduced by 72%, as shown in Figure 10. Among the IMC instructions, the IMC logic brought the most energy consumption, accounting for more than 60%. Although the IMC instructions are generally energy-consuming, remarkable energy reduction was still achieved owing to the sharp reduction in instruction count. The reductions in instruction count and energy consumption for the overall Keccak process show similar trends to one round of Keccak-*f* permutation, achieving reductions of 81% and 70% after introducing IMC, respectively, as shown in Figures 11 and 12. It should be noted that our simulation adopted a 7-byte binary string as the Keccak input, and, if the input data were infinitely long, the energy improvement tended to approximate to the extent of one round of Keccak-*f* permutation.

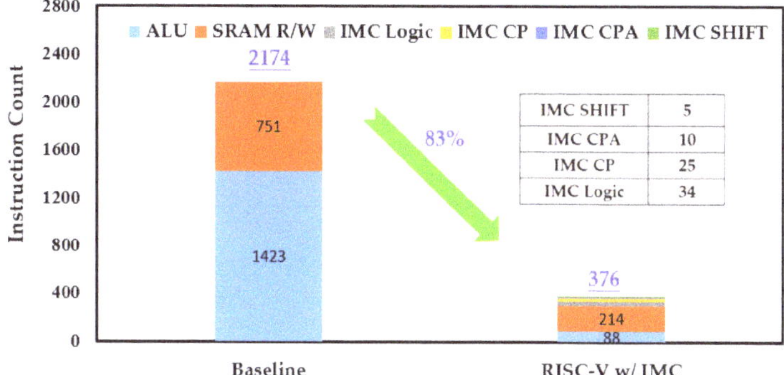

Figure 9. Comparison of instruction count in one round of Keccak-*f* permutation.

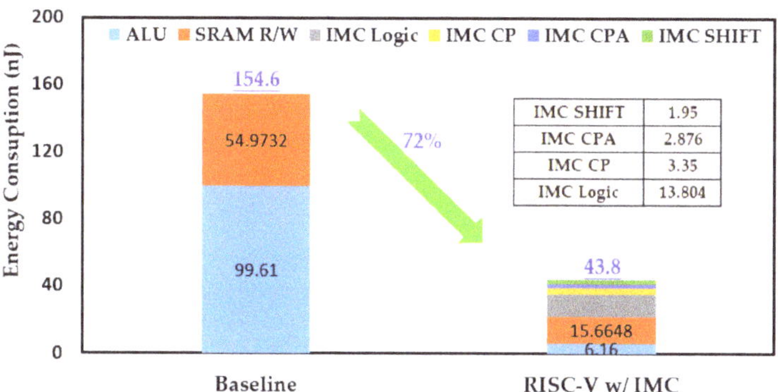

Figure 10. Comparison of energy consumption in one round of Keccak-*f* permutation.

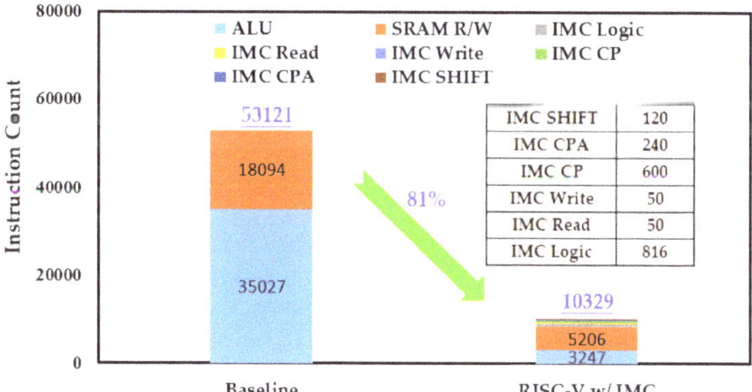

Figure 11. Comparison of instruction count in overall Keccak process.

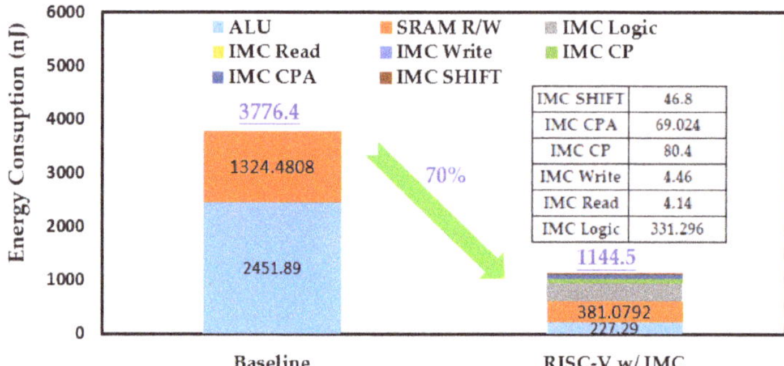

Figure 12. Comparison of energy consumption in overall Keccak process.

5.5. Comparison with Mainstream Mining Platforms and SRAM-Based IMC

Table 7 gives a comparison of the proposed memristor-based IMC with mainstream mining platforms and SRAM-based IMC. For the CPU, GPU, and ASIC, we selected i5 2500k from Intel, Tesla S1070 from Nvidia, and Antminer S4 from Bitmain, all of which were implemented in the 28/32-nm technology node. The performance was measured by the hash rate, i.e., hash operations performed in one second (H/s). The SRAM-based IMC was evaluated using the same method with the memristor-based one in this work.

Table 7. Comparison of memristor-based IMC with central processing unit (CPU), graphics processing unit (GPU), application-specific integrated circuit (ASIC), and SRAM-based IMC.

Mining Platform	Performance (H/s)	Active Power (Watts)	Energy Efficiency (J/H)	Area (mm^2)
CPU (i5 2500K) [23]	4.80×10^4	90	1.88×10^{-3}	large scale
GPU (Tesla S1070) [23]	1.55×10^8	8.00×10^2	5.16×10^{-6}	large scale
ASIC (Antminer S4) [23]	2.00×10^9	1.40×10^3	7.00×10^{-7}	large scale
SRAM-based IMC	1.03×10^3	8.80×10^{-4}	8.50×10^{-7}	3.50×10^{-2}
Memristor-based IMC	1.03×10^3	1.17×10^{-3}	1.14×10^{-6}	5.50×10^{-2}

From the comparison, two key points are worth mentioning. Firstly, although the CPU/GPU/ASIC provided higher performance than the SRAM/memristor-based IMC RISC-V, the latter consumed much less power (more than four orders of magnitude), which satisfies the low-power requirement of IoT applications. The large power of the CPU/GPU/ASIC also brings the need for cooling facilities. Moreover, a great number of IoT devices can also be coordinated to acquire high performance [4]. Secondly, memristor-based IMC brings less area cost (>50%) than SRAM-based IMC. Currently, SRAM-based IMC is more energy-efficient (~30%) than memristor-based IMC. The reason is that the write and read operations of the memristor consume much larger current. However, with the development of memristor technology, it is believed that the power consumption of the memristor will decrease. Moreover, the nonvolatility of the memristor enables the IMC module to fully power-off without data loss during standby mode, helping to reduce the total standby power.

6. Conclusions

Security for private information on IoT devices is becoming increasingly important. The hash function used in blockchain helps to ensure information security, as well as data integrity. However, the corresponding hardware in IoT devices is challenging when realizing the complex hash algorithm owing to a low energy budget. This paper proposes combining in-memory computing with the highly extensible RISC-V for lower-power hash algorithm implementation. Remarkable improvements in both performance and energy consumption were achieved with limited area overhead. Further work may involve general compiling techniques to help the processor with IMC to realize diverse functions.

Author Contributions: X.X. conceived the idea. C.W. and X.X. designed the architecture, and performed the simulation. H.L., M.W., W.L., and X.Z. took part in the discussion and provided expertise. X.X. and W.L. supervised the work.

Funding: This research was funded by the National Natural Science Foundation of China (61874028, 61704029, 61834009, 61774041), Science and Technology Commission of Shanghai Municipality (17ZR1546800, 19520711500), the Huawei Innovation Research Program (HIRP), Guangdong Province Key Technologies Research and Development Program (2019B010128001), the MOST of China under Grant 2016YFA0203800, the Strategic Priority Research Program of the Chinese Academy of Sciences under Grant No. XDPB12, and National Key Technologies Research and Development Program of China (2017YFB0405600).

Conflicts of Interest: The authors declare no conflicts of interest.

References

1. Ashton, K. That 'internet of things' thing. *RFID J.* **2009**, *22*, 97–115.
2. Pujolle, G. An autonomic-oriented architecture for the internet of things. In Proceedings of the IEEE John Vincent Atanasoff 2006 International Symposium on Modern Computing, Sofia, Bulgaria, 3–6 October 2006.
3. Madakam, S.; Ramaswamy, R.; Tripathi, S. Internet of Things (IoT): A literature review. *J. Comput. Commun.* **2015**, *3*, 164. [CrossRef]
4. Dorri, A.; Kanhere, S.S.; Jurdak, R.; Gauravaram, P. Blockchain for IoT security and privacy: The case study of a smart home. In Proceedings of the IEEE International Conference on Pervasive Computing and Communications Workshops, Kona, HI, USA, 13–17 March 2017.
5. Nakamoto, S. Bitcoin: A Peer-to-Peer Electronic Cash System. Available online: http://citeseerx.ist.psu.edu/viewdoc/summary?doi=10.1.1.221.9986 (accessed on 20 May 2019).
6. Zhang, Y.; Wen, J. The IoT electric business model: Using blockchain technology for the internet of things. *Peer Peer Netw. Appl.* **2017**, *10*, 983–994. [CrossRef]
7. Banerjee, M.; Lee, J.; Choo, K.K.R. A blockchain future for internet of things security: A position paper. *Digit. Commun. Netw.* **2018**, *4*, 159–160. [CrossRef]
8. Gubbi, J.; Buyya, R.; Marusic, S.; Palaniswami, M. Internet of Things (IoT): A vision, architectural elements, and future directions. *Future Gener. Comput. Syst.* **2013**, *29*, 1645–1660. [CrossRef]
9. Zhang, Y.; Yang, K.; Saligane, M.; Blaauw, D.; Sylvester, D. A compact 446 Gbps/W AES accelerator for mobile SoC and IoT in 40 nm. In Proceedings of the IEEE Symposium on VLSI Circuits (VLSI-Circuits), Honolulu, HI, USA, 15–17 June 2016.
10. Zhang, Y.; Xu, L.; Yang, K.; Dong, Q.; Jeloka, S.; Blaauw, D.; Sylvester, D. Recryptor: A reconfigurable in-memory cryptographic Cortex-M0 processor for IoT. In Proceedings of the IEEE Symposium on VLSI Circuits (VLSI-Circuits), Kyoto, Japan, 5–8 June 2017.
11. Xue, X.; Jian, W.; Yang, J.; Xiao, F. A 0.13 μm 8 Mb Logic-Based CuxSiyO ReRAM with Self-Adaptive Operation for Yield Enhancement and Power Reduction. *IEEE J. Solid State Circuits* **2013**, *48*, 1315–1322. [CrossRef]
12. Chen, W.; Lin, W.; Lai, L.; Li, S.; Hsu, C.-H.; Lin, H.-T.; Lee, H.-Y.; Su, J.-W.; Xie, Y.; Sheu, S.-S.; et al. A 16Mb dual-mode ReRAM macro with sub-14ns computing-in-memory and memory functions enabled by self-write termination scheme. In Proceedings of the IEEE International Electron Devices Meeting (IEDM), San Francisco, CA, USA, 2–6 December 2017.
13. Lee, Y.; Waterman, A.; Avizienis, R.; Cook, H.; Sun, C.; Stojanović, V.; Asanović, K. A 45 nm 1.3 GHz 16.7 double-precision GFLOPS/W RISC-V processor with vector accelerators. In Proceedings of the European Solid-State Circuits Conference (ESSCIRC), Venice Lido, Italy, 22–26 September 2015.
14. The RISC-V Instruction Set Manual, Volume I: User-Level ISA, Document Version 2.2. University of California: Berkeley, CA, USA, 2017. Available online: https://riscv.org/specifications/ (accessed on 3 January 2019).
15. Huang, S.; Wang, X.; Xu, G.; Wang, M.; Zhao, J. Conditional cube attack on reduced-round Keccak sponge function. In *Annual International Conference on the Theory and Applications of Cryptographic Techniques*; Springer: Cham, Switzerland, 2017.

16. Dinur, I.; Morawiecki, P.; Pieprzyk, J.; Srebrny, M.; Straus, M. Cube attacks and cube-attack-like cryptanalysis on the round-reduced Keccak sponge function. In *Annual International Conference on the Theory and Applications of Cryptographic Techniques*; Springer: Berlin, Germany, 2015.

17. Keccak Specifications, Version 2, Team Keccak. 2009. Available online: https://keccak.team/obsolete/Keccak-specifications-2.pdf (accessed on 3 January 2019).

18. The Ultra-Low Power RISC Core. Available online: https://github.com/SI-RISCV/e200_opensource (accessed on 18 May 2019).

19. Zhou, K.; Xue, X.; Yang, J.; Xu, X.; Lv, H.; Wang, M.; Jing, M.; Liu, W.; Zeng, X.; Chung, S.S.; et al. Nonvolatile Crossbar 2D2R TCAM with Cell Size of 16.3 F^2 and K-means Clustering for Power Reduction. In Proceedings of the IEEE Asian Solid-State Circuits Conference (A-SSCC), Tainan, Taiwan, 5–7 November 2018.

20. Udayakumaran, S.; Barua, R. Compiler-decided dynamic memory allocation for scratch-pad based embedded systems. In Proceedings of the 2003 International Conference on Compilers, Architecture and Synthesis for Embedded Systems, San Jose, CA, USA, 30 October–1 November 2003.

21. Synopsys VCS Verilog Simulator. Available online: http://www.synopsys.com/products/simulation/simulation.html (accessed on 3 January 2019).

22. Sinangil, M.E.; Mair, H.; Chandrakasan, A.P. A 28 nm high-density 6T SRAM with optimized peripheral-assist circuits for operation down to 0.6 V. In Proceedings of the IEEE International Solid-State Circuits Conference (ISSCC), San Francisco, CA, USA, 20–24 February 2011.

23. Difference between ASIC, GPU, and CPU Mining. Available online: https://cointopper.com (accessed on 15 July 2019).

Article

An Ultra-Area-Efficient 1024-Point In-Memory FFT Processor

Hasan Erdem Yantir [1,*], **Wenzhe Guo** [1], **Ahmed M. Eltawil** [2], **Fadi J. Kurdahi** [2] **and Khaled Nabil Salama** [1,*]

[1]   Sensors Lab, Advanced Membranes & Porous Materials Center (AMPMC), Computer, Electrical and Mathematical Sciences and Engineering Division, King Abdullah University of Science and Technology (KAUST), Thuwal 23955-6900, Saudi Arabia

[2]   Center for Embedded and Cyber-physical Systems, University of California, Irvine, CA 92697, USA

*   Correspondence: hasan.yantir@kaust.edu.sa (H.E.Y.); khaled.salama@kaust.edu.sa (K.N.S.)

Received: 30 June 2019; Accepted: 30 July 2019; Published: 31 July 2019

Abstract: Current computation architectures rely on more processor-centric design principles. On the other hand, the inevitable increase in the amount of data that applications need forces researchers to design novel processor architectures that are more data-centric. By following this principle, this study proposes an area-efficient Fast Fourier Transform (FFT) processor through in-memory computing. The proposed architecture occupies the smallest footprint of around 0.1 mm^2 inside its class together with acceptable power efficiency. According to the results, the processor exhibits the highest area efficiency (FFT/s/area) among the existing FFT processors in the current literature.

Keywords: Fast Fourier Transform; in-memory computing; associative processor; non-von neumann architecture

---

1. Introduction

Today's processor-centric design principle of computer architectures causes a great deal of energy waste. This is mainly because processing on the data is performed far away from the data [1]. Moreover, even though the processor systems are highly optimized, the data units are not considered much. On the other hand, computer applications are becoming increasingly data hungry. This became an indispensable fact especially after the rise of artificial intelligence (AI) and deep-learning domains for which big data is necessary [2]. Therefore, data movement energy dominates to compute energy in a traditional computer architecture serving today's computational needs. For example, memory access nearly consumes 1000× the energy of a complex addition operation [3]. Since the amount of data required increases, this adversely affects the efficiency of computers. Not only for AI but also in all domains ranging from signal processing to robotics, an efficient and memory-optimized computation is desired for the sake of specific advantages. Therefore, this fact forces researchers to find alternative computation methodologies. A paradigm shift to perform the computation with minimal data movement is needed by computer scientists. The most reasonable way to achieve this is by making the computation more data-centric than at present processor-centric. This research goal is investigated by many different methodologies. In the ideal case, the most advantageous computing methodology is in-memory computing means that data is processed where it resides.

In-memory computing can be achieved through different methodologies [4]. The most straightforward method is placing memory and processor inside the same chip to facilitate ultra-fast data processing instead of moving the data through the slow buses between the different chips [5]. Even though the idea seems as simple, this combination requires special fabrication in-chip manufacturing. The architecture targets to combine the processor logic with a stack of through-silicon-via (TSV) bonded memory die [6,7]. The logical core of the memory system is a kind of single instruction multiple data

(SIMD) processor where the different memory portions are directly connected to different cores, thus increasing the overall system bandwidth. As another methodology, some researchers focus to insert extra abilities to existing memory chips with minimal modifications. As a basic motivation example, 5% of the overall cycles in Google's data centers are spent on *memcopy* and *memmove* operations [8]. If a dynamic random-access memory (DRAM) has the capability to exchange data between its rows without processor intervention, then these operations would not have to be carried over the central processing unit (CPU). The study in [9] modifies the DRAM chip to perform this operation directly inside the DRAM without moving the data. The modification increases the DRAM area only 0.01%. Emergence of the new nonvolatile memory (NVM) technologies such as resistive RAM (ReRAM) and phase change memory (PCM) created a widespread adaptation for in-memory processing due to their inherently analog processing capability, high density, and scalability [10,11]. There are many studies that aim to perform in-memory computation by using NVMs, but with different methodologies [12,13]. An example of this kind in-memory computation methods is using memristor crossbars where the crossbar is configured in a way to perform corresponding specific operations. When an input is applied to the programmed crossbar, its corresponding output becomes the result of the programmed operation [14]. The study in [15] exploits the memristor crossbar for approximate addition and multiplication operations. The prime architecture proposed in [16] uses memristor crossbars to create a neural network realization which is the fundamental operation in deep learning. Another approach of in-memory computing is integrating simple logic structures in each memory cell [17]. The study in [18] proposes an architecture in which the memory cells can both store the data and perform simple computations on it. Furthermore, two or more cells can be combined to perform more complex operations. Another study in [19] proposes a systolic three-layer memory structure consists of memory, routing, and logic planes.

As another methodology, associative in-memory processing performs the in-memory computation by using look-up tables of the arithmetic and logical operations. Unlike the traditional von Neuman or near-memory computation in which the data sent to a processor for computation, associative processors (AP) sent the functionality (i.e., operation) over the data without moving it. In other words, functionality is performed directly inside the memory. Table 1 summarizes the comparison between these methodologies. According to the specifications, in-memory processing provides the broadest constraint in terms of bandwidth. With the invention of resistive memory devices such as ReRAM [20], STT-RAM (spin-transfer torque random-access memory) [21] this convention has started to gain popularity recently. Since there are numerous studies performing in-memory computation through different approaches, the study in [18] puts an extra effort for the taxonomy and proposes a classification into four groups; computation-near-memory (processor and memory in the same chip), computation-in-memory (computation is performed in the peripheral circuitry of the memory), computation-with-memory (LUTs are used for computation), and logic-in-memory (the memory cells have the computation ability). Regarding this classification, associative processing can be considered to be a computation-with-memory approach.

Table 1. Computation types with respect to memory.

Computation Type	Data Location	Functionality Location	Bandwidth Constraint
Traditional	Separate IC	Processor	Inter-chip Bus
Near-memory	Same IC	Processor	In-chip Bus
In-memory	Same IC	Memory	Memory Capacity

In this study, a fast and efficient in-memory accelerator/processor is proposed for the Fast Fourier Transform (FFT) which is the most important and extensively used algorithm in signal processing. Since the computation domain has already reached to big data era, signal processing architectures should be reconsidered from the data perspective. As a supportive case from health industry, magnetic resonance imaging (MRI) requires huge data sampling and processing for better patient diagnosis. Fast Fourier Transform is heavily used during the processing [22]. If the computation is inadequate in performing the FFT at enough speed, the patient must stay longer inside the MRI [23,24], therefore be exposed to more stress. On the contrary, if the sampled signal size is decreased, the accuracy is affected negatively which is not acceptable in the health industry. Therefore, a fast FFT processor is required to acquire both enough accuracy and processing speed. The proposed architecture exploits the different FFT computation methodologies which have a coherence inherently for in-memory computing to come up with the efficient architectures. The study also proposes the overall integration solution in which accelerator can be used as a standalone processor on its own.

The rest of the paper is organized as follows: In the following section, the fundamental idea of associative computing together with the architecture is presented. Section 3 introduces the proposed two architectures of in-memory FFT processor that are throughput-optimized and area-optimized, respectively. Experimentation and evaluation results are discussed in Section 4. The final section concludes the work.

2. In-memory Associative Processing

Associative in-memory processing is a computing paradigm aims to perform the operations on the data by using associativity principles [25]. The proposed FFT processor in this study bases the associative in-memory processing. All primitive FFT operations are performed on the input data placed inside the memory without moving it. The following two subsections form a background on the AP architecture as well as how associative computing is performed.

2.1. Associative Computing

Figure 1 shows the overall architecture of an AP in detail. The key component of an AP is a content addressable memory (CAM) [26,27]. A CAM is used to access the data by its content on the contrary to the traditional memory where the data is accessed by its address. The CAM stores the data on which the operations are performed. The figure shows the SRAM-based CAM cell structure. In this cell, the one-bit data is stored by a coupled inverter where each inverter supports to the other to keep its logical value. Associative processing exploits the associativity feature of the CAMs hence the name comes from. The basic operation on a CAM is done through the *key*, *mask*, and *tag* registers which are managed by the *controller*. A search operation inside the CAM can be performed as follows; First, the content which searched for inside the CAM is written to the key. The mask register identifies the columns on which the search is performed. If the content is found in a row, the corresponding tag register of this row becomes logic-1.

In addition to CAM, AP needs an address decoder for the communication with the outer system. This outer system can directly be a data source or a processor. Depending on the usage, AP can function as either a standalone processor or an accelerator. The computation inside the CAM is performed in a SIMD fashion. On the other hand, the traditional processors or outer systems (e.g., sensors) provide the data as sequential. To interact between these two different systems, an address decoder is used to feed or output data as sequential by activating the specific rows of the CAM.

As detailed in the next subsection, APs are very powerful for performing parallel operations when the provided data is on the same row. On the other hand, if the benchmark requires computation not only as pairwise (e.g., vector dot product) but also between the different pairs (e.g., matrix multiplication), it needs data exchange between the rows. For this purpose, a *switching matrix* is used to move the data as column-wise between the APs or to the same AP. This communication must be configurable if the processor supports different kinds of tasks with different communication patterns. On the other hand, if the processors is an application specific, it can be fixed. Figure 1 shows these two kinds of approaches in the interconnection matrix.

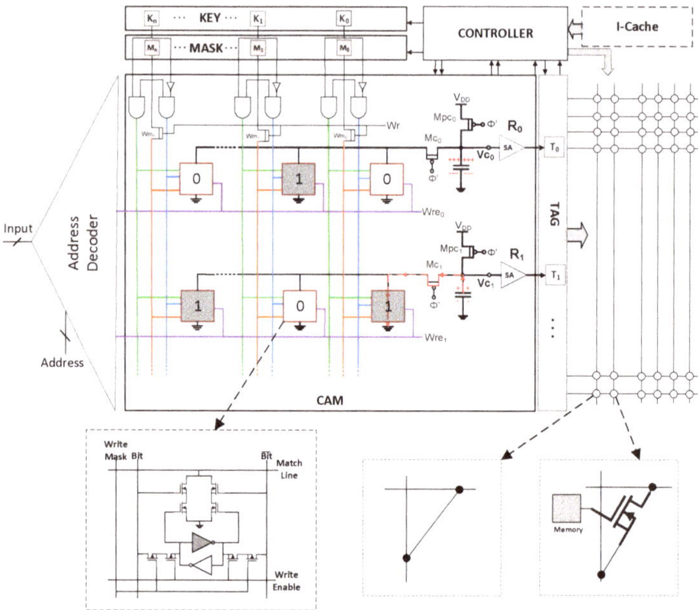

Figure 1. Associative processor architecture.

2.2. Operation

The main idea of associative in-memory processing is performing the function/operation on the data without moving it. In traditional processors, to perform an operation on a set of data, the data is moved to the processor through the special high-speed buses and brought to the processor. Inside the processor, the data passes through the functionality (e.g., a full adder or multiplier) and the computed results are written back to the memory. Unlike this approach of sending data over functionality, in in-memory associative processing, the functionality is sent over the data (see Table 1). Even though this approach seems unconventional, the CAM structure inside the AP makes it feasible.

The operations on the AP are performed through the compare and write cycles. During the compare cycle, a specific key (data) is searched for inside the CAM and in the write cycle, the specific data can be written to the columns which have the searched content (i.e., matched as a result of compare operation). Since a specific content can be selected in the CAM through the compare cycles, the corresponding function on this specific content can be applied to data inside the CAM. As an example, to perform the logical NOT operation (i.e., $B = {\sim}A$ where column B is initialized with logic-0), the CAM is searched for logic-0 on the input column (i.e., column A) and a logic-1 is written to the column B of the matched rows. Therefore, the logical not operation can be applied to the data which is logic-0. In the end, the rows with logic-1 in Column A have logic-0 in Column B and vice versa. Therefore, by applying the special functionality with respect to the searched content, the intended function can be performed. The functionalities of the AP operations are defined by look-up tables (LUTs). Depending on the LUT, the corresponding functionality is applied to the rows of the CAM separately. Table 2 shows two example LUTs for in-place addition (i.e., $B \leftarrow B + A$) and subtraction (i.e., $B \leftarrow B - A$) operations where Cr and Br are stand for carry and borrow respectively. The operations are performed as bitwise, starting from the least significant bit (LSB) of the operand towards the most significant bit (MSB). On each bit, the LUT passes are applied through the compare and write cycles. As an example of addition operation, in the first LUT pass, "011" is searched for in the CAM array for Cr, B, and A bits respectively during the compare cycle and then "10" is written to the Cr and B columns of the matched rows. The entries of LUT are iteratively applied to all bits of B and A in sequence by following the order. The comment column indicates the order that LUT entries are applied required to perform the operation correctly. Some LUT entries are unnecessary and do not participate in the result; therefore, they are indicated as NC (no change) in the comment column. The studies in [27–29], show the detailed examples of some arithmetic and logical AP operations in detail together with the step-by-step illustrations.

Table 2. LUTs for addition and subtraction.

			Addition			Subtraction		
Compare			*Write*			*Write*		
Cr/Br	*B*	*A*	*Cr*	*B*	*Comment*	*Br*	*B*	*Comment*
0	0	0	0	0	*NC*	0	0	*NC*
0	0	1	0	1	*2nd Pass*	1	1	*1st Pass*
0	1	0	0	1	*NC*	0	1	*NC*
0	1	1	1	0	*1st Pass*	0	0	*2nd Pass*
1	0	0	0	1	*3rd Pass*	1	1	*4th Pass*
1	0	1	1	0	*NC*	1	0	*NC*
1	1	0	1	0	*4th Pass*	0	0	*3rd Pass*
1	1	1	1	1	*NC*	1	1	*NC*

3. FFT Processor Architecture

The Fourier transform is a function used to decompose the given signals into its sinusoidal components [30]. It is used in nearly all scientific domains ranging from signal processing to artificial intelligence. In 1965, Cooley and Tukey proposed a faster algorithm named FFT to compute the Discrete Fourier Transform (DFT) [31] where the complexity of the transform decreased to $\mathcal{O}(n \log_2 n)$ from $\mathcal{O}(n^2)$. The proposed faster methodology consists of the interleaved computation stages where each stage composes of basic butterfly operations performed on data pairs. Since the algorithm is highly parallel, it inherently provides a widespread adaptation for in-memory associative processing both has a computation structure in an SIMD fashion [32]. On the other hand, the architecture requires some modifications to fulfill the requirements of an efficient processing platform. The following subsections detail the proposed implementations of FFT on in-memory AP in a hierarchical manner.

3.1. Butterfly Operation

The butterfly operation is the fundamental building block of an FFT stage. Figure 2 shows the simplest butterfly diagram consisting of two inputs, two outputs and one exponential coefficient (twiddle factor) where all numbers are complex (i.e., $X_0, X_1 = \text{butterfly}(e_0, x_0, x_1)$).

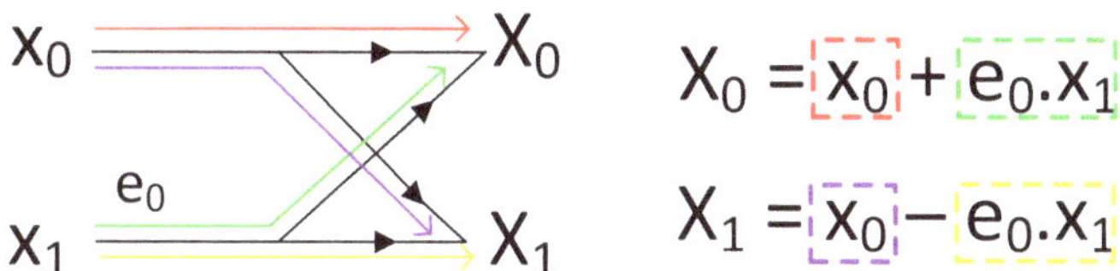

Figure 2. Simple butterfly operation.

Figure 3 shows the data flow of radix-2, decimation in time, 8-point Cooley-Tukey's FFT in three stages where each stage consists of four butterfly operations. After each stage, the partial outputs of previous stages are rearranged as an input of the next stage. From the AP-based point of view where each row can be regarded as a different processor with their own registers, two input and one exponential factor must be stored in the same row to perform a butterfly operation. However, after completion of a butterfly stage, the output of the current stage must be rearranged for the next stage since the computation pattern changes and the AP can perform the butterfly operation if and only if the operands (i.e., two inputs and coefficient) are in the same row. The exponential coefficients (e_{xy}) can be placed to the CAM arrays before the operations. For an n-point FFT operation, the overall system requires $log_2(n)$ APs and each AP requires $\frac{n}{2}$ rows. For example, the system requires 10 APs and 512-rows in each AP for 1024-point FFT operation. Since this is an in-memory FFT processor, the memory requirement is higher than the traditional FFT processors (e.g., [33–35]). On the other hand, the proposed processor does not need any traditional logic circuit, therefore provides an overall area efficiency. Section 4 discusses the comparison in detail.

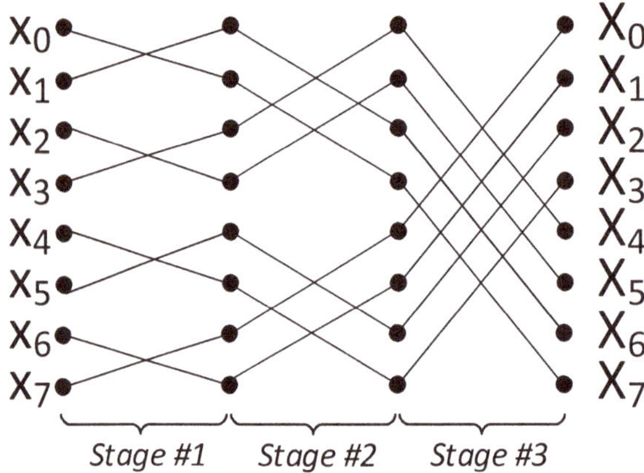

Figure 3. 8-point traditional FFT.

3.2. Data Movement

To process the data inside the AP accelerator, the outer system (i.e., processor) needs to communicate well enough with the accelerator (i.e., FFT processor). To feed the input data and retrieve the output data, the processor should have access to the data of the CAMs as row addressed. The main reason for this is that the traditional processors process the data as row-wise on the contrary of APs where data is processed as column-wise (see Section 2). Additionally, the sensors sample the data in time as sequential and provide it in this manner. The Figure 1 shows this hierarchy where the address decoder handles the communication between the AP and the processor. This decoder activates the specific row of the AP as described in the address input. The previous studies on associative computing [27,36] also provide a decoder mechanism for this purpose. In such an architecture, every row of the AP becomes addressable by the outer processor. On the other hand, the main purpose of in-memory accelerators is parallelizing the jobs done on large chunks of data where the sequential access to the individual memory locations is not much necessary during the operation. It is only needed during the initialization of the CAM array where the processor feeds the data as serial. However, even for this purpose, the random-access feature is still not much needed since this copy operation are done in order from the first line until the end. Therefore, the decoder circuit provides over functionality to the overall system which has no additional benefit.

Instead of using an address decoder, the shift register mechanism is introduced for the sake of area, performance, and energy efficiencies. Figure 4 shows the proposed in-memory FFT architecture explicitly where the costly decoder mechanism is replaced with the shift register-based approach. In this approach, a shift register is placed as vertical to the rows of the AP. The shift register has the same number of registers (flip-flops) as the number of rows in the AP. The outputs of each shift register are connected to the activation input of the corresponding rows. In this case, if the register outputs a logic-1, the row becomes activated while the logic-0 deactivates the corresponding row. The data movement operation from the processor to the AP is performed as follows; First, the processor selects the location of the AP's columns to which data is written by setting the corresponding mask registers.

The processor also asserts the *init pin* of the shift register to initiate the bulk data movement, so that the first register in the shift register becomes logic-1 in the next clock cycle. Therefore, in the first cycle, the first row is activated and ready to be written. At the same time, the outer processor synchronously provides the input data that is written to the selected columns of the first row. In the second cycle, shift register content is shifted by a single bit and the second row is activated and write operation is done for this row. At every time, the activated row by the shift register is written. The processor feeds the data as synchronized with the shift register, so they must be clocked by the same source. In this manner, the write operation for each row continues until reaching to the end row of the AP. To initialize an AP with n rows, $n + 1$ cycles are required. In this case, even though the inter-communication between the APs is column-wise through the switching matrix, the communication between the processor and the AP is handled as row-wise but more efficiently. After processing the data in the AP accelerator, the data can be retrieved by the processor in the same manner where the processor reads the data of the activated row from the bit lines as serial. The same shift register can be used for both writing and reading. When compared with the complexity of a decoder circuitry which needs n-1 1-to-2 demultiplexers for n-row CAM, the shift register approach requires n flip-flops only.

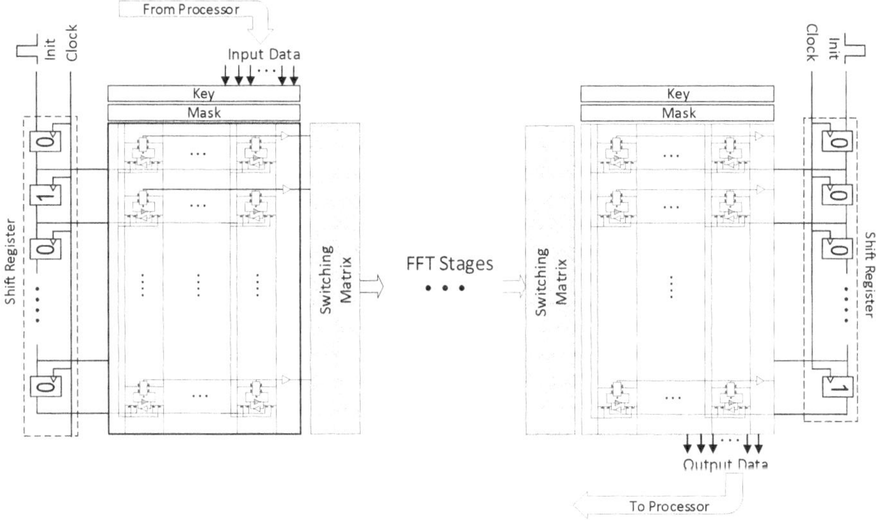

Figure 4. Pipelined in-memory FFT processor architecture.

3.3. Area-Optimized Architecture

For the in-memory FFT processors, two different architectures are proposed which are throughput and area-optimized, respectively. The throughput-optimized architecture performs each stage of the FFT in an AP-CAM as shown in Figure 4. The communication patterns between the APs can be fixed since the FFT size is fixed to 1024-point and it is known as a priori. On the other hand, the communication pattern varies with respect to the current stage as seen in Figure 3. Even though this architecture provides high-throughput in-memory FFT, it needs to replacement of AP-CAMs 10 times (i.e., $\log_2(n)$). An area-efficient alternative can be possible through the reconfigurable switching matrix where the results of a single AP stage are feedbacked back to the AP itself (see Figure 5). After completion of a butterfly stage, the reconfigurable switching matrix can be configured according to the next stage. However, this approach requires additional area and control costs. If the number of rows of a CAM array (n) is more than the number of columns (m) in an AP which is generally so since parallelism is obtained as row-wise, the area complexity of a reconfigurable switching matrix ($n \times n$) becomes more than CAM itself ($n \times m$). For instance, to perform a 1024-point FFT on 12-bit data,

132×512-bit CAM array is required. On the other hand, it requires a 512×512-bit reconfigurable switching matrix. Even the CAM cell size is assumed as $2\times$ of the traditional memory, the switching matrix requires about $1.94\times$ more area. Furthermore, the control over the switching matrix becomes intractable also since every cell must be controlled individually.

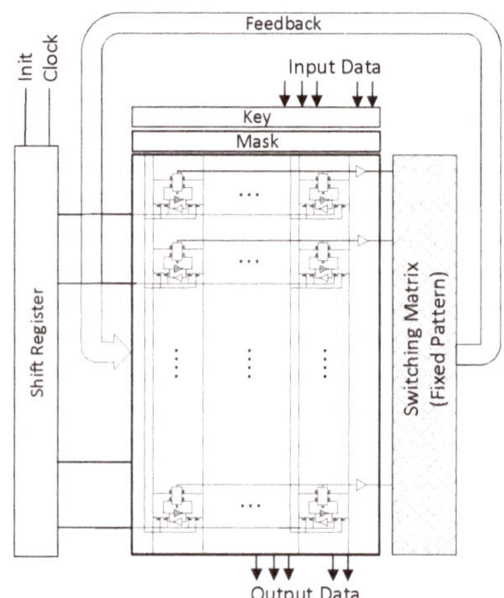

Figure 5. The ultra-area-efficient FFT processor based on singleton's FFT and feedback.

There are many algorithmic implementation of FFT (e.g., prime-factor FFT [37], Krukar's FFT [38], and Bluestein's FFT [39]) where some of them are optimized for specific input types (e.g., prime sizes, powers of two). Singleton's FFT [40] is an approach for performing FFT in the same manner and operational complexity as Cooley-Tukey FFT in traditional computers. On the other hand, it provides an incomparable advantage for APs. Even though the traditional FFT requires the change in the communication pattern where each FFT stage requires different input pairs, Singleton's FFT fixes the pattern of the data flow between the butterfly stages. For the visualization, Figure 6 shows an 8-point FFT using Singleton's method where the input x_i of every step goes into butterfly with input $x_{i+n/2}$ where n is the FFT size. Even though variable computation pattern is not an issue for general-purpose processors or ASICs which always have a random-access memory structure, it provides a vital advantage for parallel in-memory processing systems detailed as follows.

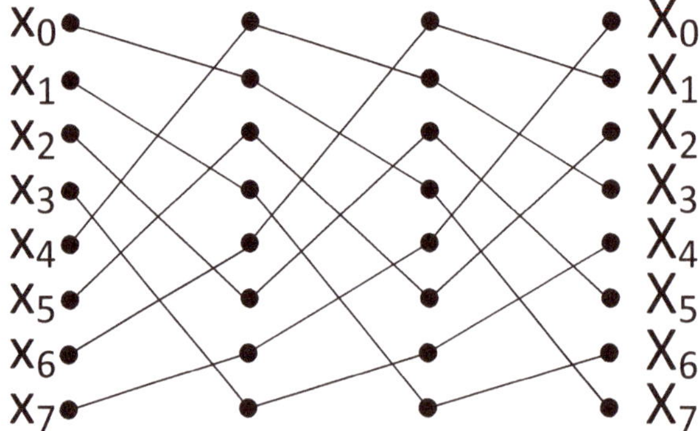

Figure 6. 8-point Singleton's FFT.

The area-optimized architecture exploits the Singleton's FFT to fix the inter-communication pattern between the stages. In this way, the whole FFT computation can be performed by using a single FFT stage. If the FFT implementation was the traditional one (i.e., Cooley-Tukey), the switching matrix would have to be reconfigurable. Figure 5 exploits the proposed area-optimized FFT architecture. To move the data from/to processor, a single shift register is used as described above. The switching matrix has a fixed pattern feedbacked to the AP itself so that every FFT stages are performed on the same AP. One drawback of this architecture is that after every computation, the new twiddle factors of the corresponding stage must be loaded to the APs from the outer processor by using the proposed shift register-based data movement approach. On the other hand, the cost of this overhead seems negligible compared with the whole butterfly operation on 1K data.

3.4. Dual-Issue Butterfly Operation

For the further optimization on the performance, the data flow diagram of a single butterfly operation on the AP (i.e., A, B = butterfly(e, a, b)) are inspected. Figure 7 shows the corresponding directed acyclic graph (DAG) of a butterfly operation on the AP where each box corresponds to an operation described inside and the lines show the data dependencies (flow of the data). Since AP performs a complex multiplication operation as four real multiplications, the diagram shows the operations on the real and imaginary parts with subscripts r and i respectively. At the first insight, it is obvious that the operations show a perfectly symmetric flow. For example, at the beginning while multiplying e_r with b_r, the same multiplication operation of $b_i \times e_i$ are performed. The set of instructions for performing these operations are the same, therefore can be performed as parallel. At that point, an AP row can be divided into two parts to perform the operations as parallel by adding extra matching circuit. Figure 8 shows the modified architecture for dual-way issue AP. The proposed modification does not require any additional cost to the controller part since the performed operations are identical, so the generated signals for the key and mask registers are exactly the same. At some point, if any operations needs to be performed between the operands on these two parts (e.g., t_i computation), the switch between them can be closed and it behaves as a single row. While this modification requires an 10% area overhead to the overall system because of the additional matching circuit, it provides around $1.9\times$ speedup due to the parallel execution of the costly multiplication operations.

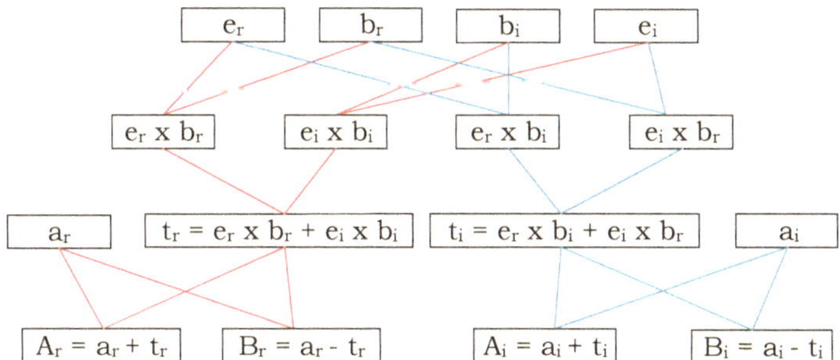

Figure 7. Directed acyclic graph of a butterfly operation.

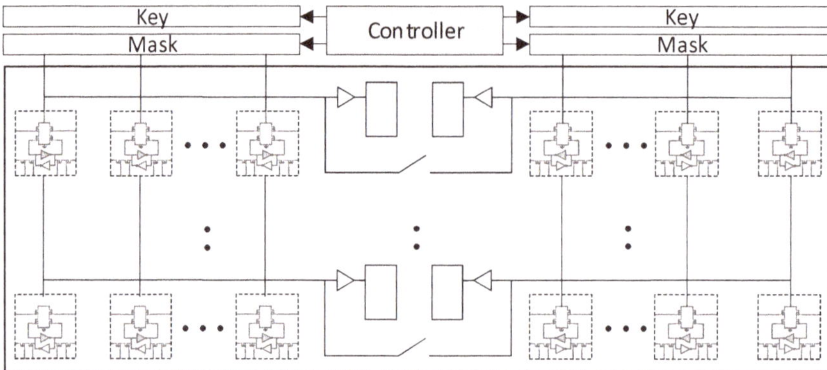

Figure 8. Dual-issue FFT on the AP.

4. Evaluation

For the evaluation of the proposed in-memory FFT processors (both area-optimized and throughput-optimized), the simulator in [29] are used to perform both system-level and circuit-level (pre-layout) simulations in Matlab and HSPICE, respectively. For the transistor model, the Predictive Technology Models (PTM) [41] is used to simulate high-density memories with 65 nm feature sizes [42]. Even though the used technology is 65 nm, the CAM cells are custom designed to decrease the current leakage and therefore energy consumption since traditional ternary CAM functionality is not needed for APs. The area of the cell design is calculated by referencing the fabricated SRAM and CAM designs in 65 nm [43,44]. The parasitic effects such as the line resistances are taken into account during the circuit simulation to obtain the accurate results [45]. Performance metrics and results are obtained by cross-checking the output of both Matlab and HSPICE simulations. For the sense amplifier, a low-power, sub-ns amplifier design in [46] is employed in the circuit. While comparing the results with the previous studies in the literature, the processors that are in the same category are taken into account. For example, for the data type, only fixed-point FFT processors are compared since it is not fair to compare a fixed-point processor with floating point one.

Table 3 shows the comparison of two in-memory FFT processors with other state-of-the-art FFT processors. The table includes both area-efficient (feedbacked) and throughput-efficient (pipelined) versions of the AP processors indicated as AP (F) and AP (P) respectively. In the AP, all butterfly operations on a CAM are performed simultaneously, so the running time of one stage does not depend on the number of samples if it fits into the memory. On the other hand, the word-length of the FFT operands affects the effective throughput since the operations are done as bitwise. The table shows that the proposed feedbacked in-memory FFT processor has the smallest area. Actually, to store the m-bit FFT operands (i.e., complex numbers) for n-point FFT, 6m × n bits memory is needed. On the other hand, the feedbacked FFT processor performs both storage and computation by using about 11m × n bits memory. When the area of a CAM cell is assumed as 2× of a normal memory cell, this leads to an inference that both computation and storage can be done in around 3.6× of the overall storage area. According to the normalized power results, the proposed processor shows a fair performance. On the other hand, the figure of merit (FOM), an overall evaluation metric of (FFT/s/Energy/Area) shows the best result within the others since the proposed FFT processor provides ultra-area efficiency.

One can put a single multiplier and adder and claim the invention of the smallest FFT processor. Therefore, the smallest area cannot be the sole claim. For this reason, while reporting the results, the GSample/s per area (GS/s/mm^2) are provided. Figure 9 proves the overall claim of the study which is proposing an ultra-area-efficient FFT processor. According to the figure, the proposed processors shows the best area efficiency in terms of GSample/s/area when compared with the other processors. In other words, the in-memory FFT processors exhibit the best FFT performance per unit area. The recent study in [34] claims the better normalized throughput per unit area than the state-of-the-art available designs. Beyond this study, the proposed design accomplishes a 33.2% improvement over their reported results.

Table 3. Comparison of FFT Processors without normalization.

Specification	AP (F)	AP (P)	[47]	[33]	[48]	[35]	[34]
FFT Size (N)	1024	1024	1024	256	2048	1024	4096
Technology	65 nm	65 nm	65 nm	90 nm	65 nm	65 nm	65 nm
V_{dd}	0.45 V	0.45 V	0.27 V	1 V	0.45 V	0.6 V	1.2 V
Word-length	12-bit	12-bit	16-bit	10-bit	12-bit	32-bit *	14-bit
Area	0.099 mm^2	0.99 mm^2	8.29 mm^2	5.1 mm^2	1.37 mm^2	3.6 mm^2	1.46 mm^2
Power	12 mW	123 mW	4.15 mW	165 mW	1.01 mW	60.3 mW	68.6 mW
Throughput/Area (GS/s/mm^2)	0.89	0.89	0.03	0.47	0.015	0.22	0.67
FOM (FFT/Energy/Area)	70.4	7.09	6.82	15.3	7.04	3.60	2.37

* The bitwidth of the architecture is variable over the FFT stages and the maximum one is 32-bit.

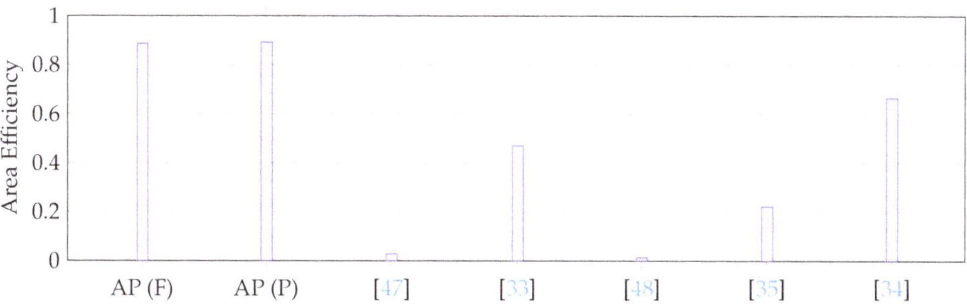

Figure 9. Area efficiencies of FFT processors (GS/s/mm^2).

In some cases, custom FFT processors can be used as directly coupled with an outer data source (sensor, channel, etc.) without any intermediate processor. If there is no outer processor, the coupled system must generate the address while sending the data. A basic counter can be used for this purpose. The proposed methodology of shift register also eliminates this need where the requirement can be fulfilled with a basic shift register. The shift register-based approach can also support multiple writings at the same time (i.e., multi-row activation); however, this is not necessary for the current content. According to the comparison between shift register and address decoder approaches for 1K-FFT processor, the synthesized design on Cadence shows that the shift register consists of fewer flip-flops and logic gates, and hence takes up 25% less area. Furthermore, the shift register is also shown to be more energy efficient which consumes around 0.4× of the address decoder.

For a further inspection on the designed architecture, a design space exploration is performed on the architecture with different operand bit widths (12-32 bits) and FFT sizes (128-4K). Figure 10 shows the energy/FFT and throughput results of the area-efficient FFT processors (feedbacked) normalized to 12-bit 1K-point FFT proposed above. Since proposed architecture performs the operations as bitwise, both throughput and energy are highly correlated with it, therefore decreases as bitwidth increases. On the other hand, if the FFT data can fit inside the memory, the throughput of a single butterfly stage increases as $\mathcal{O}(n)$. Overall, FFT throughput depends on the total number of stages as well which is formulated as $\mathcal{O}(\log_2 n)$. In overall, the normalized throughput with respect to FFT size changes by $\mathcal{O}(n/\log_2 n)$. In traditional FFT architectures, the throughput of a single butterfly stage decreases as FFT size increases since it needs to use the available resources sequentially, therefore, overall throughput changes by $\mathcal{O}(1/(n \times \log_2 n))$. Figure 11 shows the energy/FFT results for both the proposed FFT and the architectures from [48,49] where the FFT size changes between 128–2048 points. The architectures in [48,49] can be configured to perform 128, 256, 512, 1024, or 2048-point FFTs. The result demonstrates that proposed AP-based FFT shows better scaling in terms of energy/FFT with respect to increasing FFT sizes. Since the need for higher point FFT increases in the domains such as MRI which also requires parallel computation on the data coming from many receivers [50], the in-memory FFT architecture can propose an efficient solution together with the high-speed data placement through the proposed shift register-based approach.

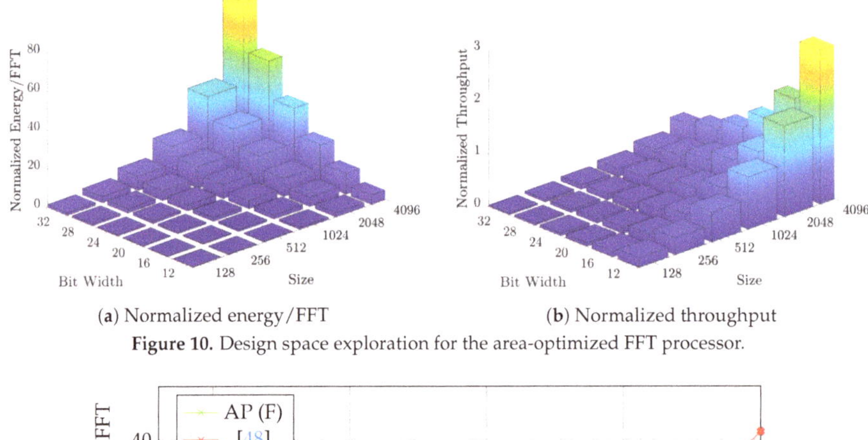

(**a**) Normalized energy/FFT   (**b**) Normalized throughput

Figure 10. Design space exploration for the area-optimized FFT processor.

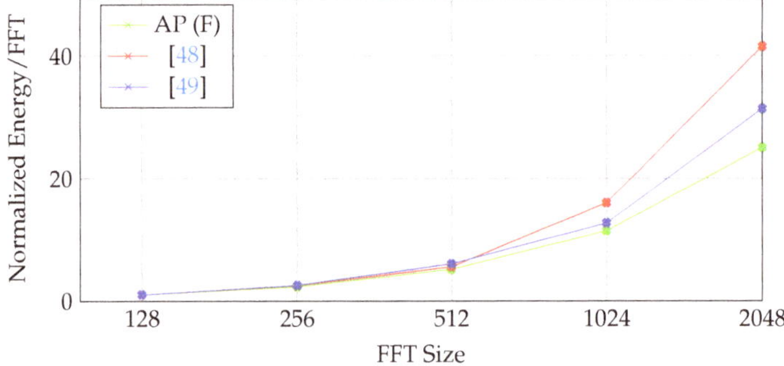

Figure 11. Comparison of normalized Energy/FFT scaling with respect to FFT size.

Even though the proposed FFT processor achieves a great deal of area efficiency due to the dense structure of the memory arrays, another paradigm that can be beneficial on this architecture is approximate computing. Approximate computing is a popular computing paradigm that relaxes the correctness constraints of a system for the sake of energy and performance improvement [51,52]. The paradigm can be applied to the error-tolerant applications. APs facilitate the approximate computing inherently since the operations are performed bit-by-bit basis [28]. As an example case, the proposed architecture can be evaluated for communication applications in which the bitwidth of the FFT processor can be adjusted dynamically during the run time concerning the estimated channel signal-to-noise ratio (SNR), aiming at achieving the desired performance at a reduced energy consumption [32]. Figure 12 shows an example case for 1K FFT where the change in average peak signal-to-noise ratio (PSNR) and error rate with respect to the bitwidth are shown where the reference is 32-bit FFT. When interpreted with Figure 10 where the normalized energy and throughput results are presented with respect to bitwidth, the approximate in-memory FFT can be performed dynamically by adjusting the bitwidth during runtime to obtain the optimum energy consumption together with the required throughput.

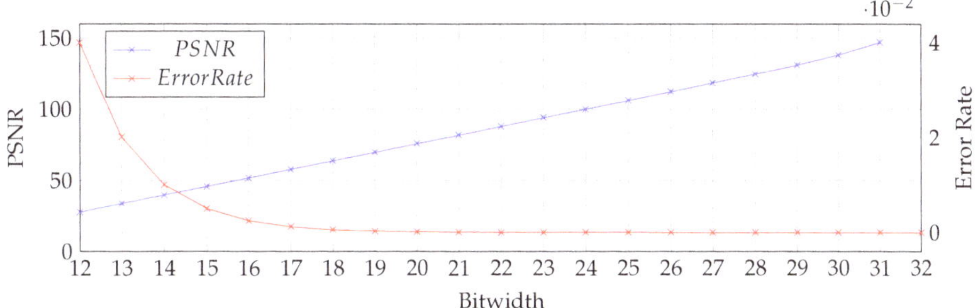

Figure 12. Bitwidth vs. average PSNR and error rate of 1024-point FFT.

5. Conclusions

In this study, an ultra-area-efficient FFT processor through the in-memory associative processor is introduced. The proposed processor performs FFT directly inside the memory. For better communication with the external systems, the traditional accelerator architecture is improved by proposing a better data moving mechanism specific to the AP-based accelerators. Furthermore, the study introduces a dual-way associative processing methodology to perform the symmetric tasks of the butterfly operation at nearly $2\times$ speed without any cost to the controller. The proposed design has the smallest area occupancy reported until now. The efficiency of the proposed architecture is proven by comparing it with the state-of-the-art FFT processors in terms of performance, power, and area. Beyond the smallest reported area, the proposed processor achieves the best area efficiency (normalized throughput per area) within its own class of FFT processors. It means that the proposed architecture delivers the best performance in a given area.

Author Contributions: Conceptualization, H.E.Y.; investigation, H.E.Y. and W.G.; methodology, H.E.Y. and K.N.S.; project administration, K.N.S.; software, H.E.Y. and W.G.; supervision, K.N.S., F.J.K., and A.M.E.; validation, H.E.Y. and W.G.; writing—original draft, H.E.Y., K.N.S., and W.G.

Funding: This research received no external funding

Conflicts of Interest: The authors declare no conflict of interest.

References

1. Mutlu, O.; Ghose, S.; Gómez-Luna, J.; Ausavarungnirun, R. Processing data where it makes sense: Enabling in-memory computation. *Microprocess. Microsyst.* **2019**, *67*, 28–41. [CrossRef]
2. Big data needs a hardware revolution. *Nature* **2018**, *554*, 145. [CrossRef] [PubMed]
3. Dally, W.J. Challenges for Future Computing Systems. In Proceedings of the 2015 Amsterdam Conference, Amsterdam, The Netherlands, 19–21 January 2015.
4. Ghose, S.; Hsieh, K.; Boroumand, A.; Ausavarungnirun, R.; Mutlu, O. Enabling the Adoption of Processing-in-Memory: Challenges, Mechanisms, Future Research Directions. *arXiv* **2018**, arXiv:1802.00320.
5. Kozyrakis, C.E.; Perissakis, S.; Patterson, D.; Anderson, T.; Asanovic, K.; Cardwell, N.; Fromm, R.; Golbus, J.; Gribstad, B.; Keeton, K.; et al. Scalable processors in the billion-transistor era: IRAM. *Computer* **1997**, *30*, 75–78. [CrossRef]
6. Gokhale, M.; Lloyd, S.; Macaraeg, C. Hybrid Memory Cube Performance Characterization on Data-centric Workloads. In Proceedings of the 5th Workshop on Irregular Applications: Architectures and Algorithms, Austin, TX, USA, 15 November 2015; ACM: New York, NY, USA, 2015; pp. 7:1–7:8. [CrossRef]
7. Ghose, S.; Hsieh, K.; Boroumand, A.; Ausavarungnirun, R.; Mutlu, O. The Processing-in-Memory Paradigm: Mechanisms to Enable Adoption. In *Beyond-CMOS Technologies for Next Generation Computer Design*; Topaloglu, R.O., Wong, H.S.P., Eds.; Springer International Publishing: Cham, Switzerland, 2019; pp. 133–194.
8. Kanev, S.; Darago, J.P.; Hazelwood, K.; Ranganathan, P.; Moseley, T.; Wei, G.Y.; Brooks, D. Profiling a Warehouse-scale Computer. In Proceedings of the 42nd Annual International Symposium on Computer Architecture, Portland, OR, USA, 13–17 June 2015; ACM: New York, NY, USA, 2015; pp. 158–169. [CrossRef]

9. Seshadri, V.; Kim, Y.; Fallin, C.; Lee, D.; Ausavarungnirun, R.; Pekhimenko, G.; Luo, Y.; Mutlu, O.; Gibbons, P.B.; Kozuch, M.A.; et al. RowClone: Fast and energy-efficient in-DRAM bulk data copy and initialization. In Proceedings of the 2013 46th Annual IEEE/ACM International Symposium on Microarchitecture (MICRO), Davis, CA, USA, 7–11 December 2013; pp. 185–197.

10. Mittal, S. A Survey of ReRAM-Based Architectures for Processing-In-Memory and Neural Networks. *Mach. Learn. Knowl. Extr.* **2018**, *1*, 75–114. [CrossRef]

11. Ielmini, D.; Wong, H.S.P. In-memory computing with resistive switching devices. *Nat. Electron.* **2018**, *1*, 333–343. [CrossRef]

12. Li, S.; Xu, C.; Zou, Q.; Zhao, J.; Lu, Y.; Xie, Y. Pinatubo: A Processing-in-memory Architecture for Bulk Bitwise Operations in Emerging Non-volatile Memories. In Proceedings of the 53rd Annual Design Automation Conference, Austin, TX, USA, 5–9 June 2016; ACM: New York, NY, USA; pp. 173:1–173:6. [CrossRef]

13. Sim, J.; Imani, M.; Choi, W.; Kim, Y.; Rosing, T. LUPIS: Latch-up based ultra efficient processing in-memory system. In Proceedings of the 2018 19th International Symposium on Quality Electronic Design (ISQED), Santa Clara, CA, USA, 13–14 March 2018; pp. 55–60. [CrossRef]

14. Chen, B.; Cai, F.; Zhou, J.; Ma, W.; Sheridan, P.; Lu, W.D. Efficient in-memory computing architecture based on crossbar arrays. In Proceedings of the 2015 IEEE International Electron Devices Meeting (IEDM), Washington, DC, USA, 7–9 December 2015; pp. 17.5.1–17.5.4. [CrossRef]

15. Imani, M.; Gupta, S.; Rosing, T. Ultra-Efficient Processing In-Memory for Data Intensive Applications. In Proceedings of the 54th Annual Design Automation Conference 2017, Austin, TX, USA, 18–22 June 2017; ACM: New York, NY, USA, 2017; pp. 6:1–6:6. [CrossRef]

16. Chi, P.; Li, S.; Xu, C.; Zhang, T.; Zhao, J.; Liu, Y.; Wang, Y.; Xie, Y. PRIME: A Novel Processing-in-Memory Architecture for Neural Network Computation in ReRAM-Based Main Memory. In Proceedings of the 2016 ACM/IEEE 43rd Annual International Symposium on Computer Architecture (ISCA), Seoul, korea, 18–22 June 2016; pp. 27–39. [CrossRef]

17. Stone, H.S. A Logic-in-Memory Computer. *IEEE Trans. Comput.* **1970**, *C-19*, 73–78. [CrossRef]

18. Santoro, G.; Turvani, G.; Graziano, M. New Logic-In-Memory Paradigms: An Architectural and Technological Perspective. *Micromachines* **2019**, *10*, 368. [CrossRef] [PubMed]

19. Cofano, M.; Vacca, M.; Santoro, G.; Causapruno, G.; Turvani, G.; Graziano, M. Exploiting the Logic-In-Memory paradigm for speeding-up data-intensive algorithms. *Integration* **2019**, *66*, 153–163. [CrossRef]

20. Chua, L. Memristor-The missing circuit element. *IEEE Trans. Circuit Theory* **1971**, *18*, 507–519. [CrossRef]

21. Apalkov, D.; Khvalkovskiy, A.; Watts, S.; Nikitin, V.; Tang, X.; Lottis, D.; Moon, K.; Luo, X.; Chen, E.; Ong, A.; et al. Spin-transfer Torque Magnetic Random Access Memory (STT-MRAM). *J. Emerg. Technol. Comput. Syst.* **2013**, *9*, 13:1–13:35. [CrossRef]

22. Hennig, J.; Nauerth, A.; Friedburg, H. RARE imaging: A fast imaging method for clinical MR. *Magn. Reson. Med.* **1986**, *3*, 823–833. [CrossRef] [PubMed]

23. Li, L.; Wyrwicz, A.M. Parallel 2D FFT implementation on FPGA suitable for real-time MR image processing. *Rev. Sci. Instrum.* **2018**, *89*, 093706. [CrossRef] [PubMed]

24. Shi, L.; Andronesi, O.; Hassanieh, H.; Ghazi, B.; Katabi, D.; Adalsteinsson, E. Mrs sparse-fft: Reducing acquisition time and artifacts for in vivo 2d correlation spectroscopy. In Proceedings of the International Society for Magnetic Resonance in Medicine Annual Meeting and Exhibition (ISMRM'13), Salt Lake City, UT, USA, 20–26 April 2013.

25. Potter, J.L. *Associative Computing: A Programming Paradigm for Massively Parallel Computers*; Perseus Publishing: New York, NY, USA, 1991.

26. Foster, C.C. *Content Addressable Parallel Processors*; John Wiley & Sons, Inc.: New York, NY, USA, 1976.

27. Yavits, L.; Morad, A.; Ginosar, R. Computer Architecture with Associative Processor Replacing Last-Level Cache and SIMD Accelerator. *IEEE Trans. Comput.* **2015**, *64*, 368–381. [CrossRef]

28. Yantir, H.E.; Eltawil, A.M.; Kurdahi, F.J. Approximate Memristive In-memory Computing. *ACM Trans. Embed. Comput. Syst.* **2017**, *16*, 129:1–129:18. [CrossRef]

29. Yantır, H.E.; Eltawil, A.M.; Kurdahi, F.J. A Hybrid Approximate Computing Approach for Associative In-Memory Processors. *IEEE J. Emerg. Sel. Top. Circuits Syst.* **2018**, *8*, 758–769. [CrossRef]

30. Fourier, J. Chapter 26—Joseph Fourier, Théorie analytique de la chaleur (1822). In *Landmark Writings in Western Mathematics 1640–1940*; Grattan-Guinness, I., Cooke, R., Corry, L., Crépel, P., Guicciardini, N., Eds.; Elsevier Science: Amsterdam, The Netherlands, 2005; pp. 354–364, ISBN 978-0-444-50871-3.

31. Cooley, J.; Tukey, J. An Algorithm for the Machine Calculation of Complex Fourier Series. *Math. Comput.* **1965**, *19*, 297–301. [CrossRef]

32. Abdelaal, R.A.; Yantır, H.E.; Eltawil, A.M.; Kurdahi, F.J. Power Performance Tradeoffs Using Adaptive Bit Width Adjustments on Resistive Associative Processors. *IEEE Trans. Circuits Syst. Regul. Pap.* **2019**, *66*, 302–312. [CrossRef]

33. Chen, Y.; Lin, Y.W.; Tsao, Y.C.; Lee, C.Y. A 2.4-Gsample/s DVFS FFT Processor for MIMO OFDM Communication Systems. *IEEE J. -Solid-State Circuits* **2008**, *43*, 1260–1273. [CrossRef]

34. Liu, S.; Liu, D. A High-Flexible Low-Latency Memory-Based FFT Processor for 4G, WLAN, and Future 5G. *IEEE Trans. Very Large Scale Integr. Syst.* **2018**, 1–13. [CrossRef]

35. Ba, N.L.; Kim, T.T. An Area Efficient 1024-Point Low Power Radix-22FFT Processor With Feed-Forward Multiple Delay Commutators. *IEEE Trans. Circuits Syst. I Regul. Pap.* **2018**, *65*, 3291–3299. [CrossRef]

36. Guo, Q.; Guo, X.; Patel, R.; Ipek, E.; Friedman, E.G. AC-DIMM: Associative Computing with STT-MRAM. *SIGARCH Comput. Archit. News* **2013**, *41*, 189–200. [CrossRef]

37. Good, I.J. The Interaction Algorithm and Practical Fourier Analysis. *J. R. Stat. Soc. Ser. B* **1958**, *20*, 361–372. [CrossRef]

38. Rader, C.M. Discrete Fourier transforms when the number of data samples is prime. *Proc. IEEE* **1968**, *56*, 1107–1108. [CrossRef]

39. Bluestein, L. A linear filtering approach to the computation of discrete Fourier transform. *IEEE Trans. Audio Electroacoust.* **1970**, *18*, 451–455. [CrossRef]

40. Singleton, R. A method for computing the fast Fourier transform with auxiliary memory and limited high-speed storage. *IEEE Trans. Audio Electroacoust.* **1967**, *15*, 91–98. [CrossRef]

41. Arizona State University. *Predictive Technology Model (PTM)*; Arizona State University: Tempe, AZ, USA, 2011.

42. Sinha, S.; Yeric, G.; Chandra, V.; Cline, B.; Cao, Y. Exploring sub-20nm FinFET design with Predictive Technology Models. In Proceedings of the DAC Design Automation Conference 2012, San Francisco, CA, USA, 3–7 June 2012; pp. 283–288. [CrossRef]

43. Zhang, K.; Bhattacharya, U.; Chen, Z.; Hamzaoglu, F.; Murray, D.; Vallepalli, N.; Wang, Y.; Zheng, B.; Bohr, M. SRAM design on 65nm CMOS technology with integrated leakage reduction scheme. In Proceedings of the 2004 Symposium on VLSI Circuits. Digest of Technical Papers, Honolulu, HI, USA, 17–19 June 2004; pp. 294–295. [CrossRef]

44. Hayashi, I.; Amano, T.; Watanabe, N.; Yano, Y.; Kuroda, Y.; Shirata, M.; Dosaka, K.; Nii, K.; Noda, H.; Kawai, H. A 250-MHz 18-Mb Full Ternary CAM With Low-Voltage Matchline Sensing Scheme In 65-nm CMOS. *IEEE J. -Solid-State Circuits* **2013**, *48*, 2671–2680. [CrossRef]

45. Wilson, L. *International technology roadmap for semiconductors (ITRS)*; Semiconductor Industry Association: Washington, DC, USA, 2013; Volume 1–17.

46. Schinkel, D.; Mensink, E.; Klumperink, E.; van Tuijl, E.; Nauta, B. A Double-Tail Latch-Type Voltage Sense Amplifier with 18ps Setup+Hold Time. In Proceedings of the 2007 IEEE International Solid-State Circuits Conference, San Francisco, CA, USA, 11–15 February 2007; Digest of Technical Papers; pp. 314–605. [CrossRef]

47. Seok, M.; Jeon, D.; Chakrabarti, C.; Blaauw, D.; Sylvester, D. A 0.27 V 30 MHz 17.7 nJ/transform 1024-pt complex FFT core with super-pipelining. In Proceedings of the 2011 IEEE International Solid-State Circuits Conference Digest of Technical Papers (ISSCC), San Francisco, CA, USA, 20–24 February 2011; pp. 342–344. [CrossRef]

48. Yang, C.; Yu, T.; Markovic, D. Power and Area Minimization of Reconfigurable FFT Processors: A 3GPP-LTE Example. *IEEE J. -Solid-State Circuits* **2012**, *47*, 757–768. [CrossRef]

49. Guichang, Z.; Fan, X.; Willson, A.N. A power-scalable reconfigurable FFT/IFFT IC based on a multi-processor ring. *IEEE J. -Solid-State Circuits* **2006**, *41*, 483–495. [CrossRef]

50. McDougall, M.P.; Wright, S.M. 64-channel array coil for single echo acquisition magnetic resonance imaging. *Magn. Reson. Med.* 2005, *54*, 386–392. [CrossRef] [PubMed]

51. Mittal, S. A Survey of Techniques for Approximate Computing. *ACM Comput. Surv.* **2016**, *48*, 62:1–62:33. [CrossRef]
52. Agrawal, A.; Choi, J.; Gopalakrishnan, K.; Gupta, S.; Nair, R.; Oh, J.; Prener, D.A.; Shukla, S.; Srinivasan, V.; Sura, Z. Approximate computing: Challenges and opportunities. In Proceedings of the 2016 IEEE International Conference on Rebooting Computing (ICRC), San Diego, CA, USA, 17–19 October 2016; pp. 1–8. [CrossRef]

 *micromachines*

Article

Speeding Up the Write Operation for Multi-Level Cell Phase Change Memory with Programmable Ramp-Down Current Pulses

Chenchen Xie [1,2], Xi Li [2,*], Houpeng Chen [2], Yang Li [1,2], Yuanguang Liu [1,2], Qian Wang [2], Kun Ren [2] and Zhitang Song [2]

[1]   Schools of Microelectronics, University of Chinese Academy of Sciences, Beijing 100049, China
[2]   State Key Laboratory of Functional Materials for Informatics; Shanghai Institute of Microsystem and Information Technology, Chinese Academy of Sciences, Shanghai 200050, China
*   Correspondence: ituluck@mail.sim.ac.cn; Tel.: +86-021-6251-1070

Received: 5 June 2019; Accepted: 5 July 2019; Published: 8 July 2019

Abstract: Multi-level cell (MLC) phase change memory (PCM) can not only effectively multiply the memory capacity while maintaining the cell area, but also has infinite potential in the application of the artificial neural network. The write and verify scheme is usually adopted to reduce the impact of device-to-device variability at the expense of a greater operation time and more power consumption. This paper proposes a novel write operation for multi-level cell phase change memory: Programmable ramp-down current pulses are utilized to program the RESET initialized memory cells to the expected resistance levels. In addition, a fully differential read circuit with an optional reference current source is employed to complete the readout operation. Eventually, a 2-bit/cell phase change memory chip is presented with a more efficient write operation of a single current pulse and a read access time of 65 ns. Some experiments are implemented to demonstrate the resistance distribution and the drift.

Keywords: multi-level cell; phase change memory; programmable ramp-down current pulses

---

1. Introduction

Data is the most competitive resource in the twenty first century and its heat has never been cut down. Especially with the advent of the big-data era and artificial intelligence, a massive amount of data needs to be processed and saved, which undoubtedly brings unprecedented challenges to the memory market. Phase change memory (PCM), one of the most promising novel non-volatile memories, attracts much attention due to its prominent performances. Compared with the mainstream flash memory, PCM has an excellent reliability below 20 nm technology [1] and its scaling is more favorable when the NMOS (N-Metal-Oxide-Semiconductor) devices are replaced by the FinFETs [2]. What is more, the large resistance contrast between amorphous and crystalline states (typically three or four orders of magnitude) in the memory cell means that PCM has more potential in multi-level cell (MLC) storage, which is a crucial feature for reducing the cost-per-bit and increasing the memory capacity. The MLC PCM can also be used in artificial neural networks as synapses, which provides a promising solution for energy-efficient artificial neural networks (ANNs) [3,4]. Therefore, the research on multi-level phase change memory cell storage is of great significance to the future development of the non-volatile memory market.

However, the realization of MLC PCM still faces several challenges. First of all, new program and read schemes should be specifically proposed since the intermediate states that represent the extra bits are avoided as much as possible in conventional phase change memory. Then, the corresponding circuits need to be well-designed, taking both performance and efficiency into consideration. Finally, as a novel storage technology that improves the capacity at the expense of reliability, a physical issue called "resistance drift" may produce severe reliability problems as it reduces the separation between adjacent levels.

Previous research has made some progress in multi-level cell phase change memory technology. T. Nirschl et al. came up with a novel multi-level program algorithm based on write and verify cycles to produce highly optimized resistance distributions in PCM [5]. G. F. Close et al. analyzed the impact of noise in multi-bit PCM from different levels [6]. N. Papandreou et al. introduced advanced iterative programing schemes for multilevel storage in PCM to achieve a high robustness to cell variability and low latency [7]. A new cell-state metric was proposed by N. Papandreou et al. to obtain larger level contrast in PCM and reduce the sensitivity to drift [8]. A 256-Mcell PCM chip operating at 2+ bit/cell, which means that the actual capacity can reach 512 Mb, was presented by Gael F. Close et al. [2]. Milos Stanisavljevic et al. discussed the storage and retention of data in MLC PCM at elevated temperatures [9].

This paper starts with the principle of multi-level cell storage in phase change memory and explores the relationship between the resistance distribution of a memory cell and the program current pulses. Then, a PCM memory chip that demonstrates an MLC operation at 2-bit/cell is presented. The entire work involves the program scheme of multi-level storage, chip structure, circuit realization, and the results of the simulation and experiments. Eventually, a 4-Mcell PCM is expanded to an 8 Mb capacity by multi-level storage technology.

The remainder of this paper is organized as follows: Section 2 briefly introduces the basic characteristics of PCM and discusses the fundamental principles of MLC PCM; chip architecture, specific write-read schemes combined with the circuit implement are demonstrated respectively in Section 3; Section 4 presents the results of experiments and the simulation; and conclusions are drawn in Section 5.

2. Phase Change Storage Technology

2.1. Basic Characteristics of Phase Change Memory

The basic principle of phase change storage is the chalcogenide phase change materials' (typical $Ge_2Sb_2Te_5$, GST) reversible transformation between two different phases (amorphous and crystalline phase) by internal structure changes [10,11]. The great difference in electrical properties between two phases makes it possible to store binary data: the amorphous phase with a high resistance usually represents '0' and the crystalline one represents '1', with a lower resistance.

Figure 1 shows the storage array of phase change memory and the transmission electron microscope (TEM) image of a PCM cell. Each cell consists of a layer of phase change material sandwiched between a top and bottom electrode and an access device, which is typically a MOSFET (Metal-Oxide-Semiconductor Field-Effect Transistor). Phase transformation is usually performed by applying programming pulses (voltage or current) to the bit line of the selected phase change memory cell. The Joule heat generated by the current flowing through the phase change memory cell causes the phase change material to melt and quench, thus producing mushroom-shaped amorphous phase in the crystalline phase, as shown in Figure 1.

Figure 2a shows the program and read pulses of PCM. The RESET program operation from crystalline to amorphous phase is usually performed by a rectangular current pulse with a large amplitude and narrow width. In order to make the phase change material quench to amorphous phase, the RESET pulse must have an abrupt trailing edge. As for SET operation, a wider current pulse with a lower amplitude is usually used to heat the cell to its crystallization temperature until it becomes crystalline phase. The typical current –voltage (I–V) characteristics of a PCM cell are shown in Figure 2b. With the increase of the voltage applied to the memory cell, the current flowing through the amorphous phase cell increases slowly. Until the voltage reaches a certain value V_{th}, however, the resistance of the phase change memory cell drops sharply, which is known as the threshold switching phenomenon of chalcogenide compounds. Therefore, during the reading process, the voltage applied to the addressed cell must be kept well below V_{th} to ensure the accuracy of the read-out data.

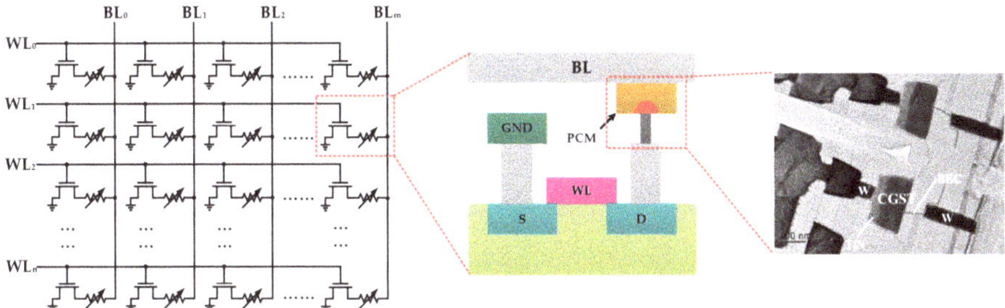

Figure 1. The storage array of phase change memory and the TEM image of a phase change memory (PCM) cell.

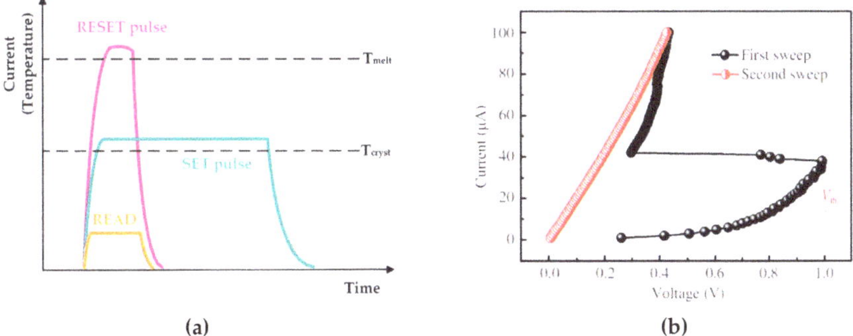

(a)   (b)

Figure 2. (**a**) Program and read operations of phase change memory (PCM); (**b**) the threshold switching phenomenon of chalcogenide compounds.

2.2. Multilevel Cell Storage

In PCM, the essential difference between two opposite phases is that the amorphous degree of the phase change material layer is different; in other words, the amorphous region and its thickness are different. The electrical resistance of the cell is only utilized to measure these differences. In conventional applications, intermediate states are usually avoided in the PCM cell to guarantee the accuracy of data storage. However, by changing some parameters, like the amplitude, of programming pulses, the PCM cell can be stabilized in the intermediate state, which is the basic state for multilevel storage in PCM [8]. What is more, the large resistance contrast, which is around three to four orders, between amorphous and crystalline phase leaves a sufficient margin for the realization of intermediate states. Figure 3 shows the sectional view of the phase change material layer with different amorphous regions.

When studying the programming conditions for realizing intermediate states, the initial state of the PCM cell should be considered. Figure 4a,b show the characteristic programming curve of the PCM cell resistance as a function of pulse amplitude. For the case where the initial state is high resistance and the programming operations are performed with SET pulses of different amplitudes, with the increase of the pulse amplitude, the resistance of the PCM cell first decreases and then increases. Taking 0.35 mA as the demarcation point, the curves before and after it both show some linearity. As for the other case, the overall curve does not show linearity, but it increases monotonously with the increase of the SET pulse amplitude. However, if the curve is piecewise analyzed, the part whose amplitude is between 0.42 mA and 0.7 mA also has a certain linearity. As shown in Figure 4c,d, whether in terms of the resistance distribution range or its consistency, using the SET operation to program memory cells is a better scheme for PCM multilevel storage.

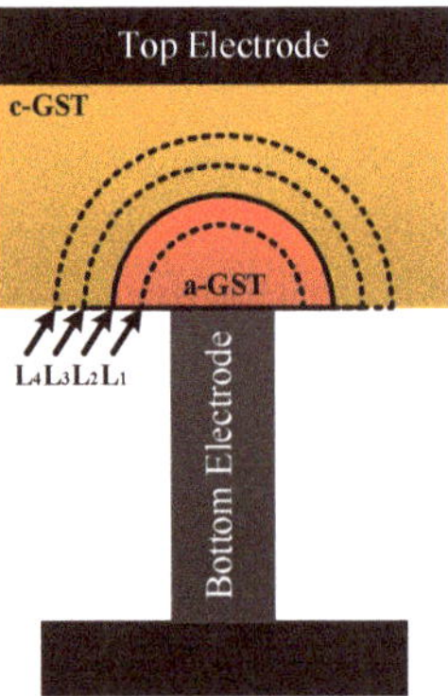

Figure 3. The sectional view of the phase change material layer with different amorphous regions.

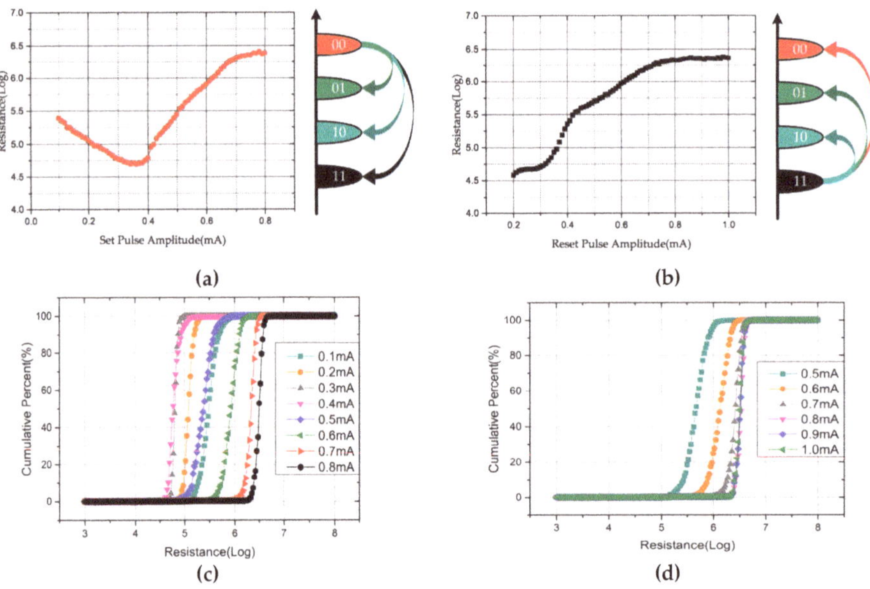

Figure 4. The characteristic programming curve of the phase change memory (PCM) cell resistance as a function of pulse amplitude for (**a**) RESET initialization and (**b**) SET initialization; resistance distribution for (**c**) RESET initialization and (**d**) SET initialization.

3. Multilevel Cell Phase Change Memory Chip

3.1. Chip Architecture

The overall framework of the 4 M 2-bit/cell phase change memory chip, which is shown in Figure 5, includes the following modules: PCM Storage Array, Row Decoder, Column Decoder, Column Selector, BandGap, Writer Driver, Voltage Controlled Oscillator (VCO), Pulse Control, Sense Amplifier, Logic Control, Address Buffer and Latch, Data Input/Output Buffer et al. The entire PCM Storage Array is divided into four 1 M cell blocks. The Row and Column Decoders locate the addressed memory cells according to the address signal saved in the Address Latch. BandGap and VCO generate the corresponding reference and clock signal on the basis of configuration parameters. Then, the Logic Control Module converts the external control signals, such as CS_, WE_, and OE_, into the internal read-write command to control the Write Driver and Sense Amplifier. Finally, the written and readout data interact with peripheral devices through the Data I/O Interface. Figure 6 shows the layout of the chip. Compared with traditional phase change memory, the biggest difference of MLC PCM lies in the read-write scheme and the specific circuit implementation, which will be covered in the flowing two subsections.

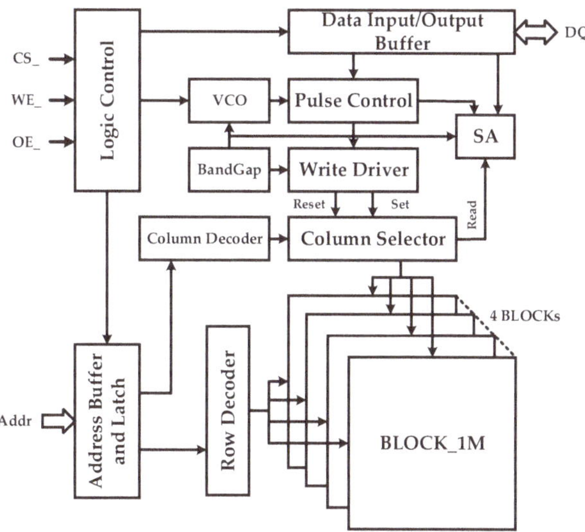

Figure 5. The architecture framework of the 4 M 2-bit/cell phase change memory chip.

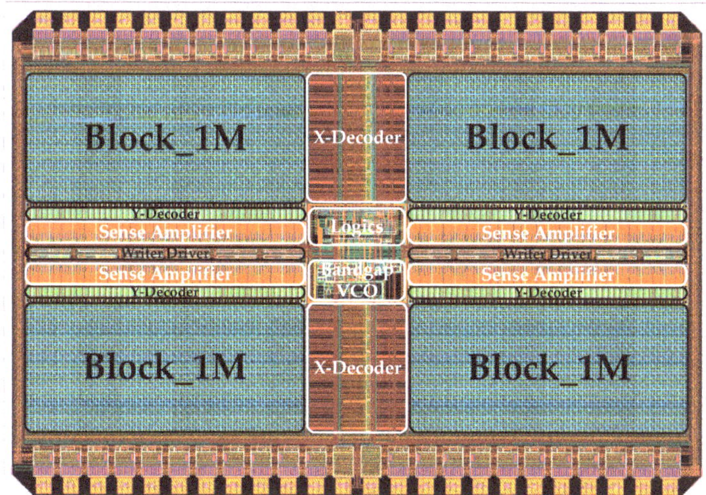

Figure 6. The layout of the 4 M 2-bit/cell phase change memory chip.

3.2. Program Scheme and Circuit

From the analysis in the previous section, it can be seen that the broader resistance distribution can be obtained if the high-resistance PCM cells are operated with rectangular current pulses of different amplitudes. However, due to the process mismatch and energy loss in the bit line, the memory cells in the array may not achieve the same resistance level under the same pulse operation. To minimize the impact caused by cell variety, Samsung and STMicroelectronics propose "ASQ technology" [11] and "SET-Sweep Programming" [12], respectively, both of which are designed to extend the crystalline time of the PCM cells. Based on the same principle, a programmable ramp-down current pulse scheme is adopted to achieve a better cell resistance distribution.

As shown in Figure 7, the descending edge of the slope current is achieved by constructing a finite number of ramp current pulses. Furthermore, in order to further obtain the optimal operating parameters of PCM cells related to the process, the initial height, initial width, and number and width of ramp current pulses are all adjustable.

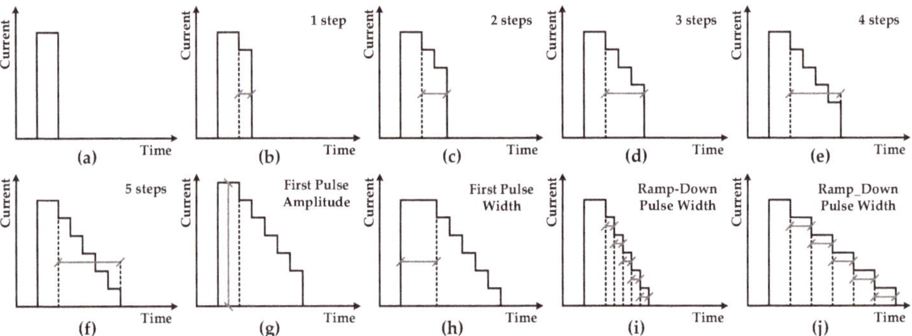

Figure 7. Programmable ramp-down current pulse for SET operation: (**a**) is a rectangular SET pulse without a ramp-down edge; (**b–f**) are ramp-down pulses with 1/2/3/4/5 steps, respectively; (**g**) is a five-step ramp-down pulse with a larger initial amplitude; (**h**) is a five-step ramp-down pulse with a larger initial width; (**i,j**) are five-step ramp-down pulses with different widths.

To achieve the above scheme, the ramp-down current pulse generator circuit designed in this paper is shown in Figure 8a. The generator consists of eight current mirrors. During the SET programming process, the control switches S<0>~S<7> are turned on or off sequentially according to a certain order, and the SET current pulse with a specific shape can then be generated. The slope of the descent edge can be changed by controlling the opening time of each current source. In addition, in order to facilitate adjustment, a number of switches are designed in each current source, as shown in the lower half of Figure 8a. Four different amplitudes can be obtained by adjusting the combination of signal S0H <1:0>, and the height of each pulse in the ramp-down current can then be adjusted. Considering the high voltage on the bit line during the write operation, the transmission gate is implemented by a single PMOS, which can reduce the wiring of the layout and save the area at the same time. Figure 8b shows the control circuit block diagram of the pulse generator. The external signals are transformed into three kinds of control signals: RDPulse, RSPulse, and ST<5:0>, corresponding to READ, RESET, and SET operations, respectively.

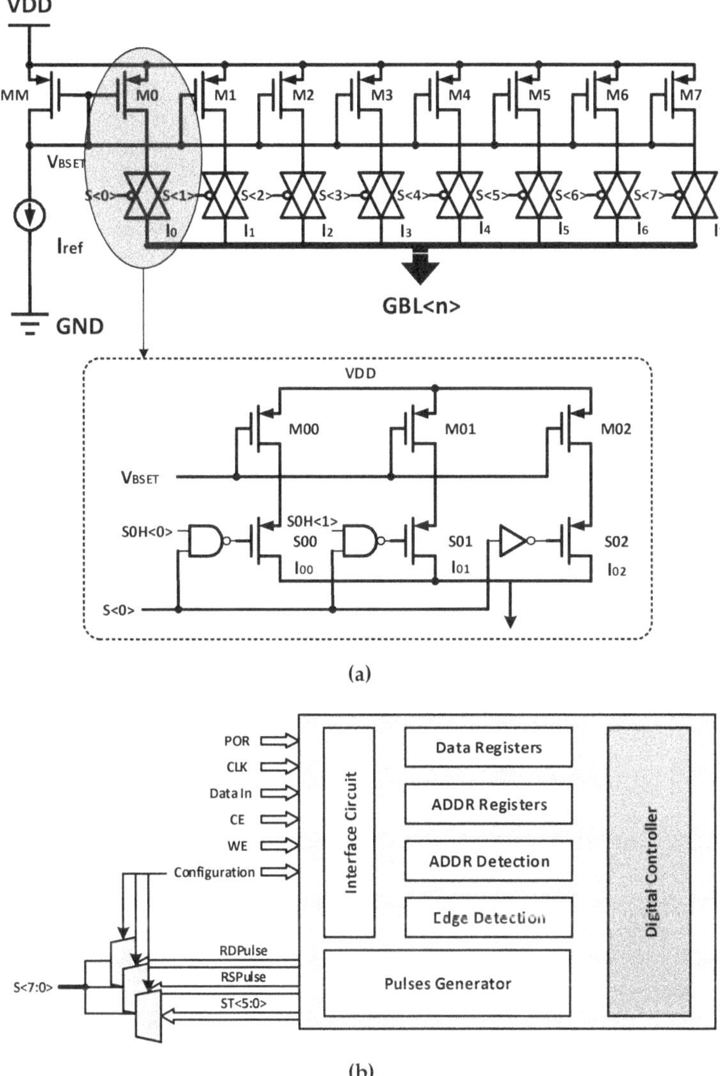

Figure 8. (**a**) Ramp-down current pulse generator circuit; (**b**) control circuit block diagram of the pulse generator.

3.3. Readout Scheme and Circuit

The readout scheme of phase change memory is essentially adopted to utilize a specific circuit to measure the resistance of the memory cell. When the cell resistance is greater or less than the specific resistance value R_H or R_L, the readout circuit outputs different digital levels respectively. The resistance interval $R_L \sim R_H$ is called the readout window of PCM. Generally, we choose $R_{REF} = (R_L + R_H)/2$ as the reference resistance of the readout circuit. For MLC PCM with multi-bit stored in each cell, more readout windows need to be set up. In this paper, a readout scheme for 2-bit/cell phase change memory with an optional reference source is proposed, and the whole readout process is divided into two read operations: high-bit and low-bit readouts.

According to Ohm's law, the resistance value of the PCM cell can be distinguished by two kinds of readout circuits: a current-bias voltage readout circuit and voltage-bias current readout circuit. By applying a constant current to the memory cell, the current-bias voltage readout circuit generates a reading voltage according to the cell resistance value. The voltage comparator then compares the reading voltage with the reference voltage to complete the cell resistance discrimination and output the logic level "0" or "1". Correspondingly, the current-bias voltage readout circuit applies a certain voltage to the memory cell, and then compares the generated current with the reference current and outputs the logic level. However, due to the threshold effect of PCM and parasitic capacitance of the storage array, the realization of the current-bias voltage readout circuit is not realistic in practical applications.

Figure 9 shows the fully differential high-speed readout circuit, which is based on the voltage-bias current readout scheme, included in this paper. The whole readout circuit can be divided into five parts: Clamp Circuit, Fully Differential Current Comparator, Optional I_{ref}, Self-bias Voltage Comparator, and Readout Inverter. The Clamp Circuit controls the bit line voltage to $V_{clamp}-V_{th0}$ with a single transistor NM0. By setting V_{clamp} and V_{th0} reasonably, the bit line voltage can be limited under the threshold voltage of the PCM cell. This approach has a great bandwidth and can provide a fast clamping operation. Furthermore, in order to avoid the effect of path parasitic charge on the PCM cells during the whole read operation, a discharge transistor NM5 is added to the readout circuit. The fully differential current comparator is composed of two sets of current mirrors which are cross-coupled. It can quickly respond to the difference between I_{read} and I_{ref} and amplify them into differential voltage signals V_1 and V_2. Since there are multiple readout windows when reading each cell, the reference current source is designed to be optional. Firstly, three standard reference currents that can be changed by adjusting configuration parameters are generated by the bias circuit module inside the chip. Then, the high-bit reference current source is selected for the first read operation, and the reference current is mirrored into the current comparator through the current mirror composed of PM5. Finally, the selection of the current reference source during the second read operation is determined by the logic circuit controlled by the first readout result. The generated differential voltage signals V_1 and V_2 are then delivered to the Self-bias Voltage Comparator. It consists of two inverters and a pair of complementary MOSFET. The inverters composed of PM7 and NM7 are used to invert differential voltage signal V_1. Additionally, the inverted V_1 shifts the threshold of the second inverter, which is composed of PM8 and NM8, to the opposite direction by controlling the working state of PM6 and NM6. Then, the second inverter can respond more quickly to the change of differential voltage signal V_2 and output the final result. The Readout Inverter is used to reverse the output of the Self-bias Voltage Comparator and recover the electrical level of the output signal.

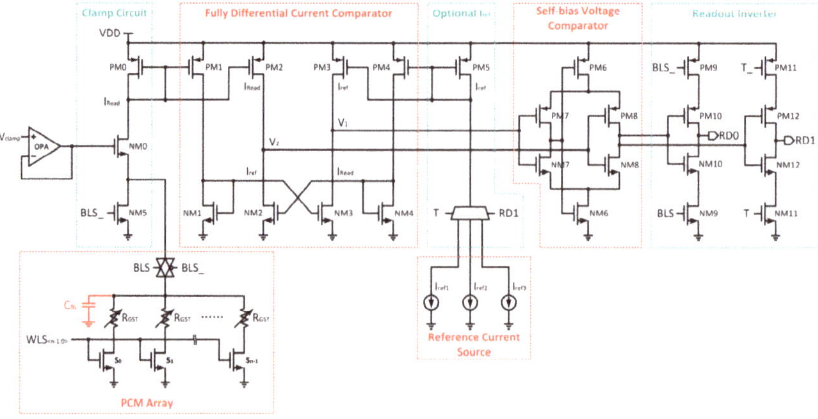

Figure 9. Fully differential high-speed readout circuit with an optional reference current source.

4. Experimental Results

In this section, the experiment results of the 4 M 2-bit/cell PCM chip with the assistance of automatic test equipment (ATE) are presented. A brief discussion of the different program pulses for four resistance levels and the comparison with the result of write and verify scheme are then given. Following this, the Resistance Drift, which is the most dominant issue that hinders MLC functionality in PCM, is demonstrated on the basis of the test results. Finally, some simulation diagrams of the program and readout circuits are displayed.

4.1. The Resistance Distribution of 2-Bit/Cell Phase Change Memory

Figure 10 shows the resistance distribution of four states in PCM cells and the corresponding program pulses. After RESET initialization, the PCM cells are programmed with different shaped current pulses, including rectangular and ramp-down current pulses, by adjusting the configuration parameters. Almost all the resistance distribution within the range of PCM cell resistance variation can be obtained through this approach. As shown in Figure 10b, the optimal RESET pulse is a rectangular current pulse with an amplitude of 0.9 mA and a width of 52 ns. A current pulse with a larger amplitude cannot increase the resistance of the "00" state, but will result in more power consumption. Additionally, the width of 52 ns is sufficient enough to operate all well-performing cells to "00". To program the RESET cell to the "01" state, a rectangular current pulse with a smaller amplitude and larger width is performed. For the two states with a lower resistance value, complete crystallization of the PCM cells can be achieved with ramp-down current pulses of different amplitudes. In fact, ramp down pulses with four steps are enough to program the memory cells to their states and the extra two steps are added to achieve a better consistency.

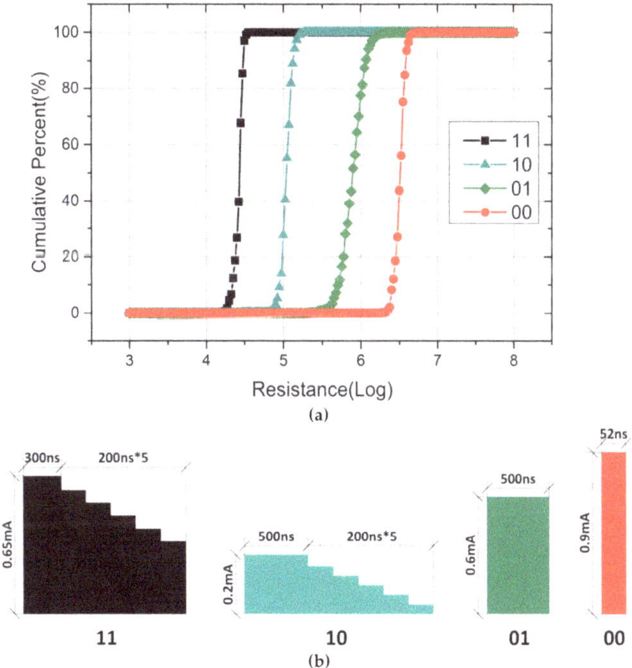

Figure 10. (**a**) Resistance distribution of four states in phase change memory (PCM) cells with a ramp-down current pulse scheme; (**b**) the corresponding program pulses.

As a contrast, another program scheme based on write and verify is processed with the assistance of ATE. Unlike the previous scheme, this approach starts with an SET operation and then melting rectangular pulses of varying amplitudes in the partial-RESET regime are utilized to increase the resistance. After each program operation, the cell resistance will be readout to verify. If the cell resistance has reached the expected level, the program operation is completed. Otherwise, a rectangular current pulse with a larger amplitude will be used to program until the cell resistance reaches the expected range. In order to compare the two schemes, the resistance range of four states is set as shown in Figure 10a. Figure 11 displays the resistance distribution and the iteration times of four states. To make sure that the cell resistance reaches the expected range accurately, the amplitude increment of the current pulse in each iteration cannot be too large. Consequently, the number of iterations is positively related to the target resistance. For the "00" state with a high resistance, there are over 70 iterations. Note that each iteration includes a read and write operation. Therefore, even though the write and verify scheme improves the consistency of the resistance distribution, the cost of operation time and power consumption is unacceptable. In addition, if the whole scheme is integrated in the chip, the design of the circuit will become more complicated. In conclusion, the scheme of a single-pulse program is preferred in terms of the operation time, power consumption, and cost.

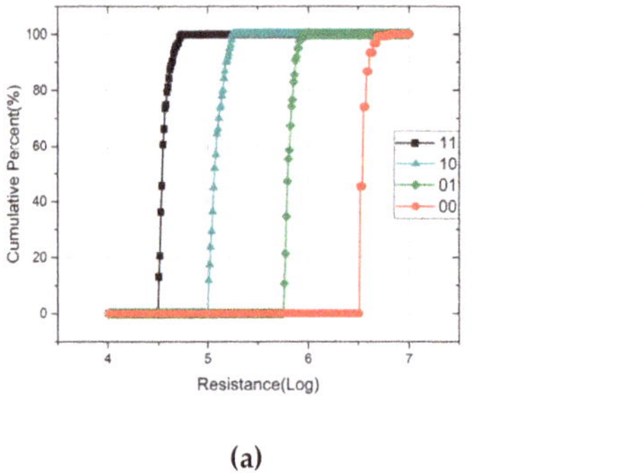

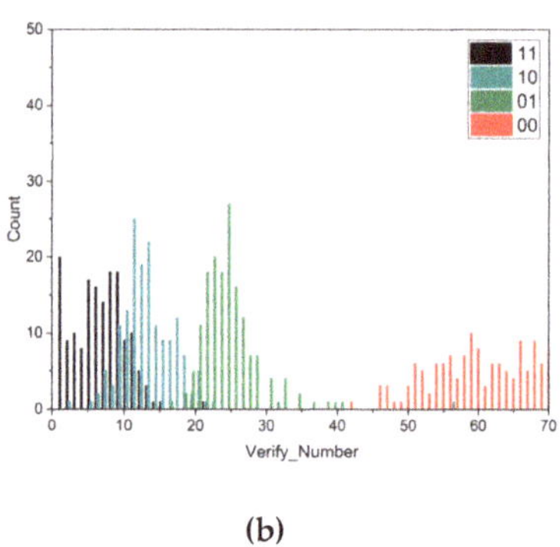

(a) **(b)**

Figure 11. (**a**) Resistance distribution of four states in phase change memory (PCM) cells with the write and verify scheme; (**b**) the number of iterations.

4.2. Resistance Drift

Amorphous materials are known to display structure relaxation (SR), which is the atomistic-scale rearrangement of an amorphous structure. The amorphous GST in PCM cells also suffers from this phenomenon, resulting in an increase of the electrical resistance with time [13]. As a novel storage technology that improves the capacity at the expense of performance, MLC storage in PCM faces reliability problems as resistance drift reduces the separation between adjacent levels. To study the effect of resistance drift on data retention in memory cells, the resistance variation of PCM cells is recorded within 1000 s after programming. As shown in Figure 12, resistance drift mainly occurs within 100 s after programming. After that, the resistance still increases a little with time, but the separation is enough to distinguish four states.

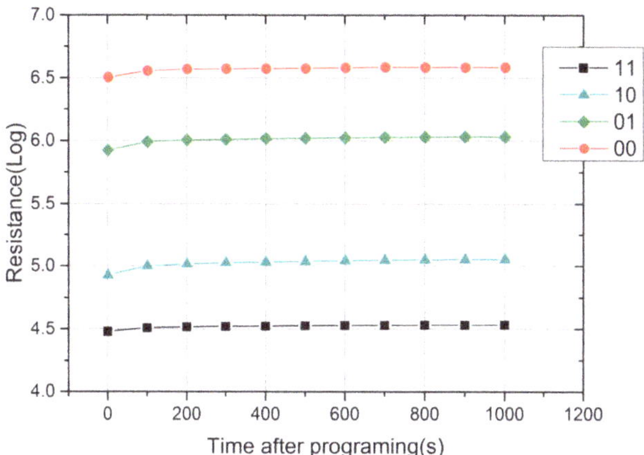

Figure 12. Measurements showing the multilevel drift behavior for a 1000 s time frame.

4.3. Simulation Results

Figure 13a shows the simulation graphs of the ramp-down current pulse generator circuit, and the shape of each pulse corresponds to the design scheme in Figure 7. Figure 13b,c show the readout simulation results of four states in the PCM cell during the two read operations. Taking the worst case into consideration, the final readout time is 65 ns.

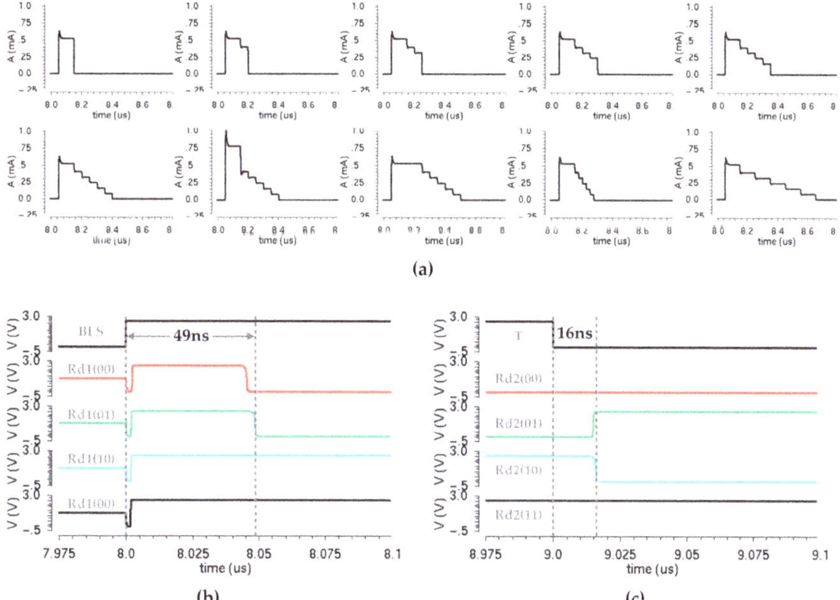

Figure 13. (**a**) Simulation graphs of the ramp-down current pulse generator circuit; (**b**,**c**) readout simulation results of four states in the phase change memory (PCM) cell during the two read operations.

5. Conclusions

A 2-bit/cell phase change memory chip is presented in this paper with a speed-up write operation. The program scheme adopted in this paper is started with the initialization of memory cells. Then, different shaped pulses, which are produced by the programmable ramp-down current pulse generator, are applied to the addressed cells and program them to the target level. The read operation of the 2-bit/cell is accomplished by a specially designed fully differential read circuit with an optional reference current source. The final results of the simulation and experiment verify the feasibility of the scheme and the functionality of multi-level storage in PCM.

As a comparison, Table 1 summarizes some information and the performance of the chips proposed in this paper and [2]. Our work improves the write and read speed for 2-bit MLC PCM by 6.25 times and 4.9 times, respectively, and decreases the write time from 9.7 μs to <1.6 μs and read time from 320 ns to 65 ns. The omission of the write & verify process reduces not only the number of generated pulses for each bit, but also the power consumption during the programming. Furthermore, the ADCs (analog-to-digital converters) and DACs (digital-to-analog converters) that are necessary for the chip in [2] are dismissed in the new scheme, which greatly cuts down the complexity and cost of the chip design. Therefore, compared with the write and verify scheme, the scheme proposed in this paper is more attractive because of its advantages in speed, power consumption, and cost.

Table 1. Summary of the chips proposed in this paper and [2].

Chips	Chip Proposed in This Paper	Chip Proposed in [2]
	CMOS Technology	
Node	SMIC 40 nm	90 nm
Supply Voltage	2.5 V	Digital: 1.2 V Phase change memory (PCM) and analog: 2.5–3.0 V
	PCM Cell Array	
Material	C-GST	Doped GST
Access Device	NMOS	NMOS
Cells	4 M cells, 16 accessed in parallel	256 M cells, 16 accessed in parallel
	Write	
Access Time	RESET 52 ns+SET 1.5 μs @ 2 bits/cell	9.7 μs @ 2 bits/cell
Program Scheme	Programmable ramp down current pulse	Open-loop single shot, or closed-loop write and verify with one ADC and two DACs integrated in the chip
	Readout	
Access Time	65 ns @ 2 bits/cell	320 ns @ 2 bits/cell
Read Scheme	Fully differential read circuit with optional reference current source	1 bit range+6-bit ADC

Author Contributions: X.L., C.X., and H.C. proposed the scheme and designed the chip; K.R. improved the materials and process; Q.W. designed the layout; C.X. and X.L. conceived and designed the experiments; C.X. and Y.L. performed the experimental testing; Y.L. provided the data for comparative experiments; C.X. analyzed the data and wrote the paper; Z.S. was in charge of the entire project.

Funding: This work is supported by the National Key Research and Development Program of China (2017YFA0206101, 2017YFB0701703, 2017YFA0206104, 2017YFB0405601, 2018YFB0407500), National Natural Science Foundation of China (61874178, 61874129), Science and Technology Council of Shanghai (17DZ2291300), and Shanghai Sailing Program (19YF1456100).

Conflicts of Interest: The authors declare no conflicts of interest.

References

1. Kim, I.S.; Cho, S.L.; Im, D.H.; Cho, E.H.; Kim, D.H.; Oh, G.H.; Chung, C.H. High performance PRAM cell scalable to sub-20nm technology with below 4F2 cell size, extendable to DRAM applications. In Proceedings of the 2010 Symposium on VLSI Technology, Honolulu, HI, USA, 15–17 June 2010; pp. 203–204. [CrossRef]

2. Close, G.F.; Frey, U.; Morrish, J.; Jordan, R.; Lewis, S.C.; Maffitt, T.; Eleftheriou, E. A 256-Mcell Phase-Change Memory Chip Operating at 2+Bit/Cell. In *IEEE Transactions on Circuits and Systems I: Regular Papers*; IEEE: Piscataway, NJ, USA, 2013; Volume 60, pp. 1521–1533.

3. Nandakumar, S.R.; Boybat, I.; le Gallo, M.; Sebastian, A.; Rajendran, B.; Eleftheriou, E. Supervised learning in spiking neural networks with MLC PCM synapses. In Proceedings of the 2017 75th Annual Device Research Conference (DRC), South Bend, IN, USA, 25–28 June 2017; pp. 1–2. [CrossRef]

4. Lee, J.; Lim, D.; Jeong, H.; Ma, H.; Shi, L. Exploring Cycle-to-Cycle and Device-to-Device Variation Tolerance in MLC Storage-Based Neural Network Training. In *IEEE Transactions on Electron Devices*; IEEE: Piscataway, NJ, USA, 2019; Volume 66, pp. 2172–2178.

5. Nirschl, T.; Philipp, J.B.; Happ, T.D.; Burr, G.W.; Rajendran, B.; Lee, M.H.; Joseph, E. Write Strategies for 2 and 4-bit Multi-Level Phase-Change Memory. In Proceedings of the 2007 IEEE International Electron Devices Meeting, Washington, DC, USA, 10–12 December 2007; pp. 461–464. [CrossRef]

6. Close, G.F.; Frey, U.; Breitwisch, M.; Lung, H.L.; Lam, C.; Hagleitner, C.; Eleftheriou, E. Device, circuit and system-level analysis of noise in multi-bit phase-change memory. In Proceedings of the 2010 International Electron Devices Meeting, San Francisco, CA, USA, 6–8 December 2010; pp. 29.5.1–29.5.4. [CrossRef]

7. Papandreou, N.; Pozidis, H.; Pantazi, A.; Sebastian, A.; Breitwisch, M.; Lam, C.; Eleftheriou, E. Programming algorithms for multilevel phase-change memory. In Proceedings of the 2011 IEEE International Symposium of Circuits and Systems (ISCAS), Rio de Janeiro, Brazil, 15–18 May 2011; pp. 329–332. [CrossRef]

8. Papandreou, N.; Sebastian, A.; Pantazi, A.; Breitwisch, M.; Lam, C.; Pozidis, H.; Eleftheriou, E. Drift-resilient cell-state metric for multilevel phase-change memory. In Proceedings of the 2011 International Electron Devices Meeting, Washington, DC, USA, 5–7 December 2011; pp. 3.5.1–3.5.4. [CrossRef]

9. Stanisavljevic, M.; Athmanathan, A.; Papandreou, N.; Pozidis, H.; Eleftheriou, E. Phase-change memory: Feasibility of reliable multilevel-cell storage and retention at elevated temperatures. In Proceedings of the 2015 IEEE International Reliability Physics Symposium, Monterey, CA, USA, 19–23 April 2015; pp. 5B.6.1–5B.6.6. [CrossRef]

10. Raoux, S.; Burr, G.W.; Breitwisch, M.J.; Rettner, C.T.; Chen, Y.C.; Shelby, R.M.; Lam, C.H. Phase-change random access memory: A scalable technology. *IBM J. Res. Dev.* **2019**, *52*, 465–479. [CrossRef]

11. Lee, K.J.; Cho, B.H.; Cho, W.Y.; Kang, S.; Choi, B.G.; Oh, H.R.; Park, M.H. A 90nm 1.8V 512Mb Diode-Switch PRAM with 266MB/s Read Throughput. In Proceedings of the 2007 IEEE International Solid-State Circuits Conference. Digest of Technical Papers, San Francisco, CA, USA, 11–15 February 2007; pp. 472–616. [CrossRef]

12. Bedeschi, F.; Boffmo, C.; Bonizzoni, E.; Resta, C.; Torelli, G.; Zella, D. Set-sweep programming pulse for phase-change memories. In Proceedings of the 2006 IEEE International Symposium on Circuits and Systems, Island of Kos, Greece, 21–24 May 2006. [CrossRef]

13. Ielmini, D.; Lavizzari, S.; Sharma, D.; Lacaita, A.L. Physical interpretation, modeling and impact on phase change memory (PCM) reliability of resistance drift due to chalcogenide structural relaxation. In Proceedings of the 2007 IEEE International Electron Devices Meeting, Washington, DC, USA, 10–12 December 2007; pp. 939–942. [CrossRef]

Article

Effect of Annealing Temperature for Ni/AlO$_x$/Pt RRAM Devices Fabricated with Solution-Based Dielectric

Zongjie Shen [1,2], Yanfei Qi [1,3], Ivona Z. Mitrovic [2], Cezhou Zhao [1,2], Steve Hall [2], Li Yang [4,5], Tian Luo [1,2], Yanbo Huang [1,2] and Chun Zhao [1,2,*]

[1] Department of Electrical and Electronic Engineering, Xi'an Jiaotong-Liverpool University, Suzhou 215123, China
[2] Department of Electrical Engineering and Electronics, University of Liverpool, Liverpool L69 3BX, UK
[3] School of Electronic and Information Engineering, Xi'an Jiaotong University, Xi'an 710061, China
[4] Department of Chemistry, Xi'an Jiaotong-Liverpool University, Suzhou 215123, China
[5] Department of Chemistry, University of Liverpool, Liverpool L69 3BX, UK
* Correspondence: chun.zhao@xjtlu.edu.cn; Tel.: +86-(0)512-8816-1402

Received: 10 May 2019; Accepted: 28 June 2019; Published: 2 July 2019

Abstract: Resistive random access memory (RRAM) devices with Ni/AlO$_x$/Pt-structure were manufactured by deposition of a solution-based aluminum oxide (AlO$_x$) dielectric layer which was subsequently annealed at temperatures from 200 °C to 300 °C, in increments of 25 °C. The devices displayed typical bipolar resistive switching characteristics. Investigations were carried out on the effect of different annealing temperatures for associated RRAM devices to show that performance was correlated with changes of hydroxyl group concentration in the AlO$_x$ thin films. The annealing temperature of 250 °C was found to be optimal for the dielectric layer, exhibiting superior performance of the RRAM devices with the lowest operation voltage (<1.5 V), the highest ON/OFF ratio (>10^4), the narrowest resistance distribution, the longest retention time (>10^4 s) and the most endurance cycles (>150).

Keywords: bipolar resistive switching characteristics; annealing temperatures; solution-based dielectric; resistive random access memory (RRAM)

1. Introduction

As one of the promising candidates for next-generation nonvolatile memories, resistive random access memory (RRAM) has received considerable attention due to significant advantages concerning simplicity of structure, low power consumption, fast read & write speed, high scalability and 3-D integration feasibility compared to the industry standard silicon-based flash memories [1–7]. Current candidate materials for the resistive switching (RS) layer of RRAM devices include perovskite, ferromagnetic and metal oxide-based materials [1,3–5,8–11]. In particular, metal oxide-based materials such as AlO$_x$, NiO$_x$, TiO$_x$ and HfO$_x$ are currently extensively discussed because of the simplicity of the material [10,12–14]. Among these materials, AlO$_x$ has been widely applied in gate insulator layers [15–18] and has attracted extensive attention in the RRAM field owing to its wide band gap (~8.9 eV), high thermal stability with Si and Pt, high dielectric constant (~8) and large breakdown electric field [10,14,19–22] as Kim et al. has reported [19,20,23–26]. In addition, the superior elasticity [27] and high toughness [28] make it possible for AlO$_x$ to be applied under various conditions including vibration and pressure environments [29–31]. Cano et al. reported that AlO$_x$-based dielectric layer showed superior stability under environments with hydrofluoric acid pressure [29] and Choi et al. reported large-scale flexible electronics application with AlO$_x$ thin film [31], which have demonstrated that the AlO$_x$ thin film has great potential as a metal oxide layer in RRAM devices.

A number of fabrication methods for incorporation of a metal oxide RS layer in AlO_x-based RRAM devices have been investigated. Methods based on solution processes for metal oxide thin films have been extensively considered, namely spin [32–34] and dip coating [35–37], drop casting [34,36–38] and different printing methods. Compared with traditional fabrication methods such as atomic-layer-deposition (ALD) [17,39,40] and magnetron sputtering [28,40,41], the solution-based method has advantages of low fabrication cost with the elimination of vacuum deposition processes [42], ease of preparation for precursor materials [39,43,44] and high efficiency of device throughput [27], which reveals the promising prospect of solution-based methods in RS layer fabrication. Several factors including plasma cleaning time, deposition gaseous environment and annealing temperature are considered to influence the performance of solution-based metal oxide thin films. A limited number of investigations have been reported regarding the relationship between annealing temperature and performance of RRAM device with solution-based RS layer [10,38].

In this work, the AlO_x thin film was deposited with a spin-coating method and then annealed at temperatures of 200 °C to 300 °C, in increments of 25 °C. The RRAM devices with solution-based AlO_x thin film were characterized electrically in terms of operation voltage, ON/OFF ratio between the high resistance state (HRS) and low resistance state (LRS), resistance distribution, retention time and endurance cycles. X-ray photoelectron spectroscopy (XPS) results indicate that these performance metrics are associated with different gradients of hydroxyl group (-OH) concentrations in the AlO_x thin films with different annealing temperatures. Devices with AlO_x thin films annealed at 250 °C demonstrated superior performance with the lowest operation voltage (<1.5 V), the highest ON/OFF ratio ($>10^4$), the narrowest resistance distribution, the longest retention time ($>10^4$ s) and the most endurance cycles (>150).

2. Device Fabrication

The fabricated Ni(top)/AlO_x/Pt(bottom) memory device structure with dimensions 2 mm × 2 mm is shown in Figure 1a. Firstly, the substrate comprising layers Pt (200 nm)/Ti/SiO_2/Si was ultrasonically cleaned in acetone, ethanol and deionized (DI) water, sequentially. Then an aluminum nitrate nonahydrate ($Al(NO_3)_3·9H_2O$) solution consisting of ~9.353 g $Al(NO_3)_3·9H_2O$ and 10 mL deionized water was prepared as the 2.5 M AlO_x precursor. The precursor solution was stirred vigorously for 20 min under ambient air conditions. The Pt substrate surface layer was given a hydrophilic treatment in a plasma cleaner in an atmospheric environment. The AlO_x precursor solution, filtered through a 0.45 μm polyether sulfone (PES) syringe, was spin-coated onto the substrate at a spin rate of 4500 rpm for 40 s and subsequently annealed at the different desired temperatures of 200 °C, 225 °C, 250 °C, 275 °C and 300 °C for 60 min under ambient conditions. A ~40 nm-thick top electrode (TE) layer of Ni and a ~40 nm-thick capping layer of Al were both deposited by e-beam evaporation. Figure 1b shows a scanning electron microscope (SEM) cross-sectional image of the device, confirming the target thicknesses of ~40 nm, ~30 nm and ~100 nm for Ni, AlO_x and Pt layers respectively.

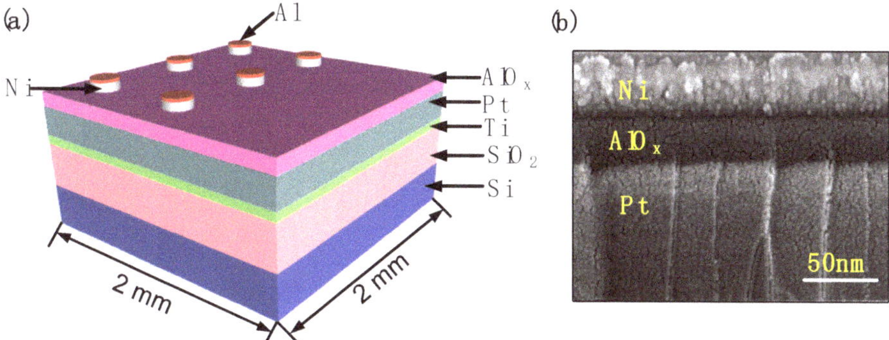

Figure 1. (**a**) Schematic of an Al/Ni/solution-based AlO$_x$/Pt RRAM device; (**b**) a scanning electron microscope (SEM) cross-sectional image of the Al/Ni/solution-based AlO$_x$/Pt RRAM device.

An Agilent B1500A high-precision semiconductor analyzer (Agilent Santa Rosa, CA, USA) was employed to measure the I-V characteristics with a two-probe configuration. All electrical measurements were performed in the dark and at room temperature within a Faraday cage. In addition, to investigate the effect of annealing temperatures on device performance, X-ray photoelectron spectroscopy (XPS) spectra of constituent Al and O core level (CL) elements were measured.

3. Results and Discussion

3.1. Memoristic Characteristics Based on Al/Ni/Solution-Based AlO$_x$/Pt RRAM

The RRAM devices were operated under 1 mA compliance current (CC) and observed to exhibit typical bipolar RS behavior, as illustrated by the I-V characteristics in Figure 2. The devices with the dielectric layer annealed at 200 °C exhibit typical RRAM breakdown characteristics at very low voltage <0.3 V while breakdown characteristics of 300 °C annealed devices are not usually observed even for voltages higher than 18 V, which is of course, unsuitable for RRAM device application [45,46]. Therefore, RRAM devices with dielectric layers annealed at 225 °C, 250 °C, 275 °C were considered for further evaluation. Compared with unipolar I-V characteristics of other RRAM devices [47], all RRAM devices with Al/Ni/solution-based AlO$_x$/Pt structure demonstrate typical bipolar I-V characteristics without forming operation. The current compliance (CC) is set at 1 mA to prevent catastrophic breakdown of the RRAM devices. During cycling, the HRS was transferred to LRS abruptly in the SET process and the resistance of the LRS began to increase abruptly toward HRS in the RESET process. The SET and RESET process controls the RRAM device transition to ON and OFF states. It is observed that the majority of values of SET voltages (V_{SET}) for the three samples are around 1.5 V while some are up to 4 V. In the RESET process, nearly all RESET voltages (V_{RESET}) are around −1 V approximately. As illustrated in Figure 3, in the SET operation, the average values of V_{SET} are around 3.2 V, 1.0 V and 2.4 V at 225 °C, 250 °C and 275 °C, respectively. RRAM devices with dielectric layer annealed at 250 °C exhibit the lowest SET voltages (Figure 3a) with the highest ON/OFF ratio (>10^4) between LRS (ON state) and HRS (OFF state). Similar results can be observed in the RESET operation (Figure 3b) although the variation of V_{RESET} average values is not as obvious as that of V_{SET}. Figure 2d shows the cumulative probability for resistance distribution of the RRAM devices annealed at various temperatures. All values of memory resistance at HRS (R_{HRS}) and LRS (R_{LRS}) of consecutive forming-free DC switching cycles were read at 0.1 V. As illustrated in Figure 2d, curves of resistance distribution almost overlap at LRS, indicating that no significant dependence on annealing temperature is apparent at LRS. However, an obvious variation can be observed at R_{HRS}. The uniformity and narrowness of the resistance distribution are key metrics for stability and quality of RRAM devices. A narrow resistance distribution is considered to be a good demonstration of the stability and performance of devices [7,48–50]. In this work, the narrowest resistance distribution of Al/Ni/solution-based AlO$_x$/Pt RRAM devices is found for the 250 °C annealing temperature, which therefore presents the best uniformity of the devices.

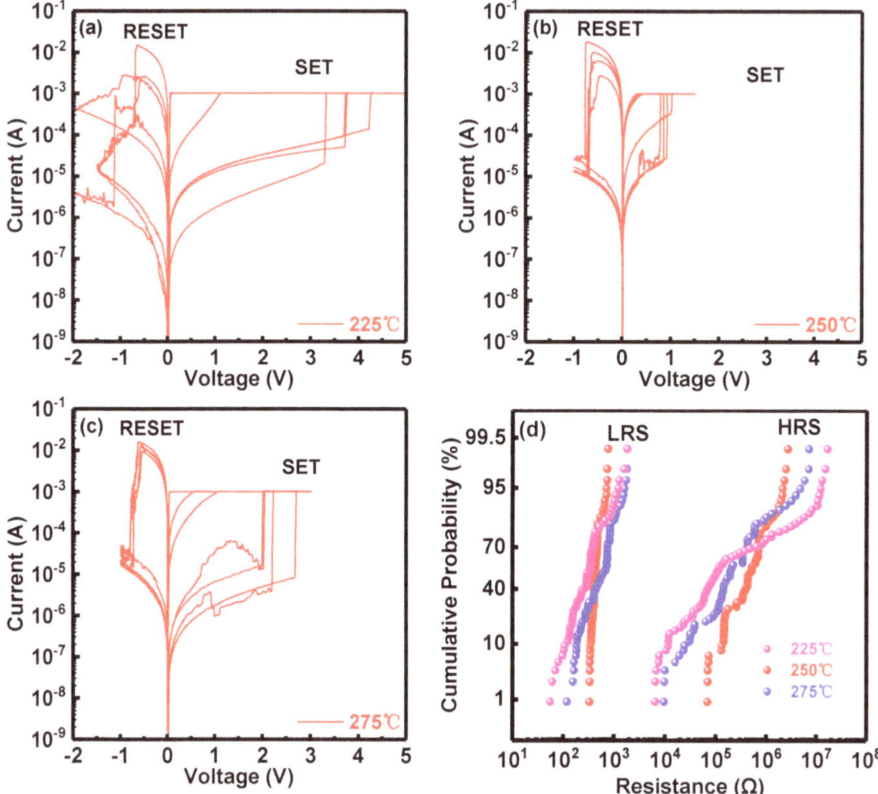

Figure 2. I-V curves of Al/Ni/solution-based AlO$_x$/Pt RRAM devices with (resistive switching) RS layer annealed at (**a**) 225 °C; (**b**) 250 °C and (**c**) 275 °C. (**d**) Resistance distribution of Al/Ni/solution-based AlO$_x$/Pt RRAM device with RS layer deposited at various temperatures.

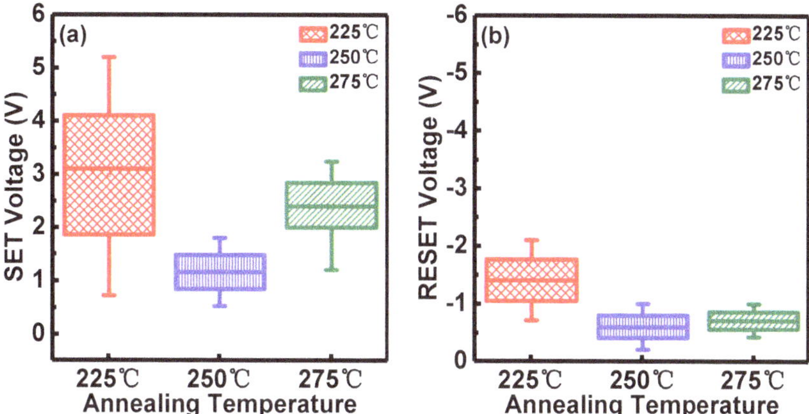

Figure 3. Voltage distribution of (**a**) SET operation and (**b**) RESET operation for Al/Ni/solution-based AlO$_x$/Pt RRAM devices with RS layer annealed at different temperatures.

3.2. Endurance and Retention Properties of Al/Ni/Solution-Based AlO$_x$/Pt RRAM

Figure 4 demonstrates the retention and endurance properties at HRS and LRS for the RRAM devices with RS layers annealed at various temperatures. With the results of resistance distribution above, the resistance values of retention and endurance belong to the range of HRS and LRS values in Figure 2d. Resistance values both at HRS and LRS are read at 0.2 V. Figure 4a–c show DC cycles vs resistance at 1 mA CC of devices annealed at 225 °C, 250 °C and 275 °C, which show similar characteristics to those observed in the resistance distribution of Figure 2d. The best resistance distribution can be observed in 250 °C annealed RRAM devices and the worst uniformity of resistance can be observed in 225 °C annealed RRAM devices. Similarly, the endurance property with the best uniformity is demonstrated in the RRAM device annealed at 250 °C while the worst performance is observed in the RRAM device annealed at 225 °C. The same retention property can be observed in Figure 4d, which shows that the device can sustain data for more than 10^4 s.

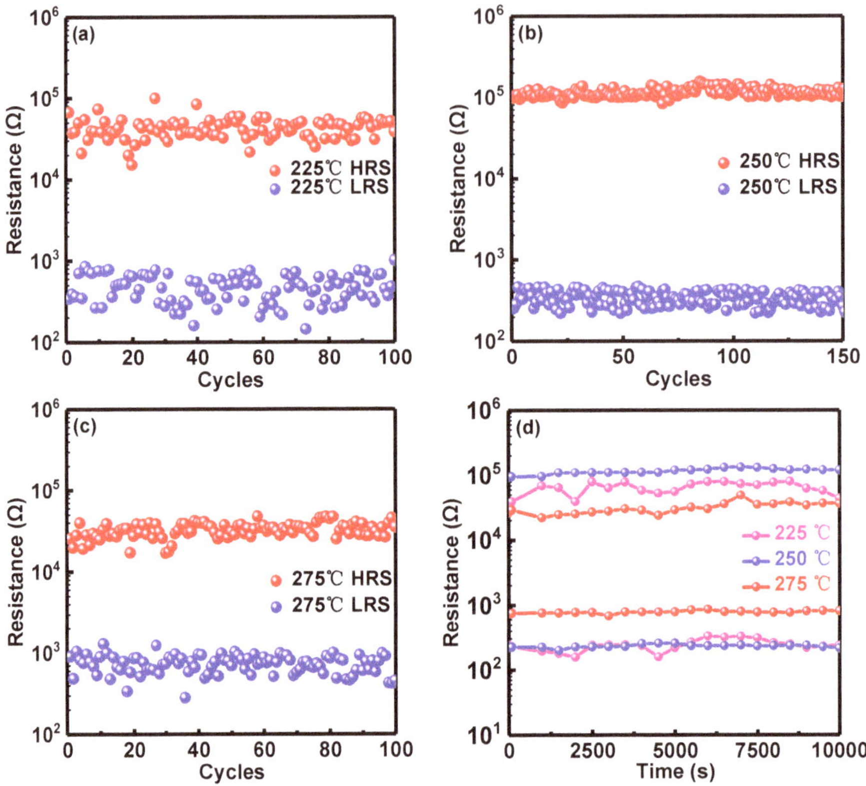

Figure 4. Endurance property of Al/Ni/solution-based AlO$_x$/Pt RRAM devices with RS layer annealed at (**a**) 225 °C; (**b**) 250 °C and (**c**) 275 °C. (**d**) Retention property of RRAM devices annealed at various temperatures.

The best performance was found for an annealing temperature of 250 °C with the lowest operation voltage (<1.5 V), the highest ON/OFF ratio (>10^4), the narrowest resistance distribution, the longest retention time (>10^4 s) and the most endurance cycles (150).

3.3. Switching Mechanism of Al/Ni/Solution-Based AlO$_x$/Pt RRAM

With typical bipolar RS performance demonstrated by Al/Ni/solution-based AlO$_x$/Pt RRAM devices, the RS modeling with fitting curves (250 °C annealed devices) illustrated in Figure 5a is used to investigate the conduction mechanism. Figure 5a shows evidence for space-charge limited current (SCLC) as the dominant conduction mechanism in 250 °C annealed devices. The fitting results show positive and negative bias regions of I-V characteristics in double logarithmic plots. A large area overlap of SET and RESET can be observed due to the approximately equal values of CC and RESET current. The currents are seen to follow Ohmic conduction (I ∝ V) in the low voltage regime [51,52]. At higher bias voltages, the OFF-state slope shows a transition to about 2.0, consistent with Child' s square law [53,54]. By further increasing the applied voltage, the slope increased to approximately 8.7, again consistent with the SCLC mechanism [53–56].

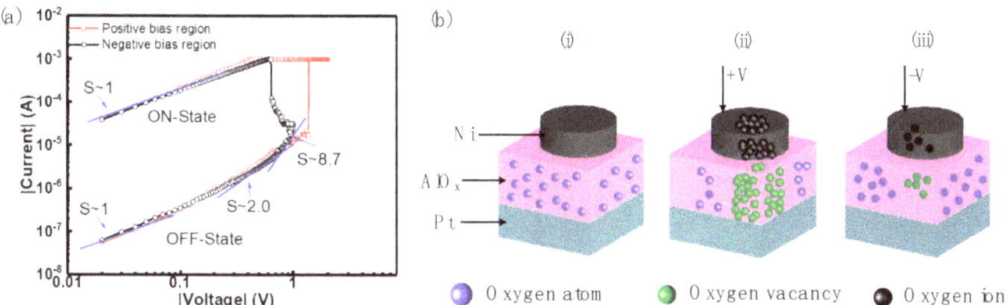

Figure 5. (**a**) Curve fitting of I-V characteristics for Al/Ni/solution-based AlO$_x$/Pt RRAM devices indicating SCLC conduction. (**b**) Diagrams to describe the switching mechanism of Al/Ni/solution-based AlO$_x$/Pt RRAM devices at (i) the initial state, (ii) the ON state and (iii) the OFF state, respectively.

Bipolar RS performance of all RRAM devices with different annealing temperatures are considered to be associated with the formation and rupture process of conductive filaments (CF) associated with oxygen vacancies, in the SET/RESET process [2–4,15,57]. Figure 5b shows a schematic representation of this process consisting of ON and OFF states, which is considered as the switching mechanism of these devices. The formation and rupture process of CF is associated with the distribution of oxygen ions and oxygen vacancies in the TE and RS layer [22,48,57–59]. Figure 5b(i) shows the initial state of RRAM devices without applied voltage, indicating oxygen atoms present in the AlO$_x$ thin film. With application of a positive voltage to the Ni electrode in the SET operation, electrons are captured by oxygen atoms in the AlO$_x$ thin film [15,27,60–62], to yield oxygen ions which drift to TE. The generation process of oxygen ions can be represented as:

$$O + 2e^- \rightarrow O^{2-}$$

The oxygen vacancies remain in the AlO$_x$ thin film and constitute the dominant components of CF. This formation process of CF consisting of oxygen vacancies in the AlO$_x$ thin film is considered to be responsible for the resistance state transition (HRS to LRS) of RRAM devices at the ON state, as depicted in Figure 5b [48,58,60]. Conversely, in the RESET operation, with a negative voltage applied to TE, oxygen ions stored in the electrode drift back to the AlO$_x$ thin film under the influence of the negative electrical field and therefore reduce the density of oxygen vacancies in the AlO$_x$ thin film [18,63]. This action dominates the rupture process of CF [15,22,48] and the RRAM devices perform at the OFF state (LRS to HRS).

The formation and rupture mechanism of CF is confirmed to be associated with the characteristics of the RS layer in filamentary RRAM devices with the dependency on film thickness, measurement temperature and deposition temperature [64–67]. In this work, the device performance is found to be dependent on annealing temperature of the dielectric layer and the best performance is observed in the device with a dielectric layer annealed at 250 °C.

Physical characterization was undertaken using XPS. Figure 6a–c show XPS spectra of O 1s core levels for the AlO$_x$ thin films annealed at 225 °C, 250 °C and 275 °C. The O 1s CL spectrum can be de-convoluted into two sub-peaks with binding energies located at 531.1 eV (O$_1$) and 532.2 eV (O$_2$) [40,64–67]. The O$_1$ and O$_2$ peaks are associated with the metal-oxygen bonds (O$_1$) and hydroxyl group (O$_2$), respectively [5,66,67]. As illustrated in Figure 6a–c, the hydroxyl-related peak (O$_2$) increased with annealing temperatures from 225 °C to 250 °C and decreased from 250 °C to 275 °C. Similar behavior has been observed by Xu et al. [68]. The highest and the lowest concentration of the hydroxyl group is found for samples annealed at 250 °C and 225 °C, respectively. Figure 6d shows the integrated intensity of the two sub-peaks referring to the concentration of hydroxyl group (M-OH) and metal-oxygen bonds (M-O) for the three samples. The observed variation in concentration of hydroxyl group has been found to show strong correlation to RRAM device performance. The best performing

RRAM device annealed at 250 °C has the highest concentration of hydroxyl group, while the worst performance is observed for device annealed at 225 °C which exhibits the lowest concentration of hydroxyl group.

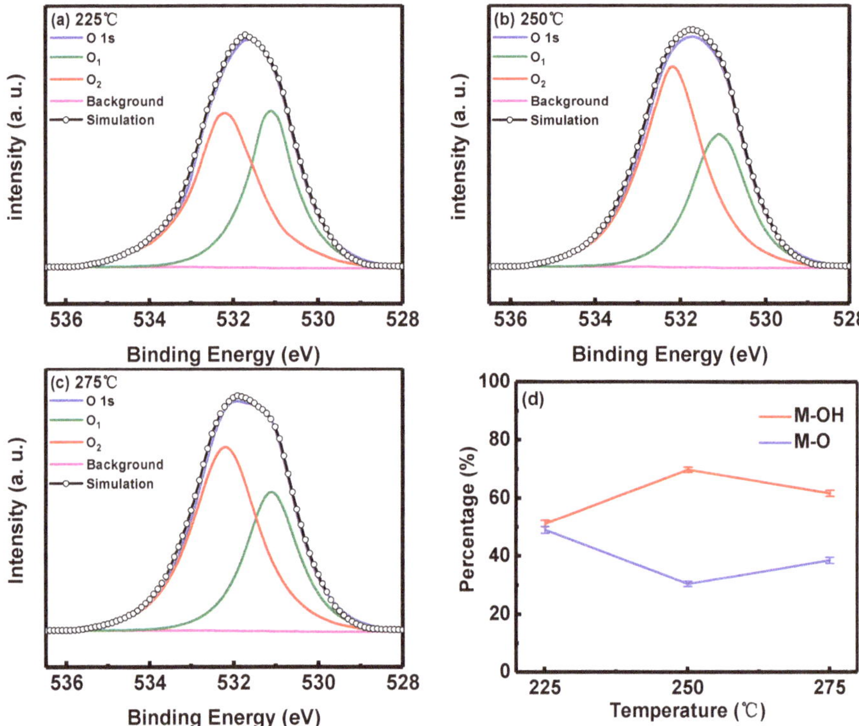

Figure 6. XPS spectra of O 1s CLs for Al/Ni/solution-based AlO$_x$/Pt RRAM devices annealed at (**a**) 225 °C; (**b**) 250 °C and (**c**) 275 °C. (**d**) Integrated intensities of O 1s CL sub-peak referring to M-OH bond and M-O bond for solution-based AlO$_x$ layers annealed at different temperatures.

With the different concentrations of M-O and M-OH in the dielectric layer, two main species of compositions, namely AlO$_x$ and Al(OH)$_x$, play dominant roles in switching behavior. We now propose a hypothesis for the relationship between composition and surface roughness of the dielectric layer. The more complex the compositions of the dielectric layer, the higher surface roughness will be present [69–71]. The surface roughness assessed by Atomic Force Microscope (AFM) of dielectric layers annealed at 225 °C, 250 °C and 275 °C are 0.682 nm, 0.230 nm and 0.524 nm, respectively. In 225 °C annealed devices, the similar concentration (~50%) of M-O and M-OH can be detected in the film indicating that the concentration of AlO$_x$ and Al(OH)$_x$ are almost equal. Hence the dielectric layer performance might be affected concurrently by two main compositions. A smooth surface of the dielectric layer is essential to achieve low leakage current and the realization of high-performance dielectric thin films. A higher concentration of M-OH is observed in the 250 °C annealed AlO$_x$ thin film, which indicates that Al(OH)$_x$ has a more dominant influence on the layer properties. Compared with Al(OH)$_x$, the influence of AlO$_x$ is less significant, which results in a lower surface roughness. In addition, the existence of the hydroxyl group in the dielectric layer is associated with water absorption, which affects the permittivity of AlO$_x$ with a slight fluctuation (~9.3–~11.5) and hence the capacitance associated with the dielectric thin film. This part will be submitted to further investigation.

4. Conclusions

RRAM devices with Al/Ni/AlO$_x$/Pt structure were fabricated by a solution-based process with the RS layer annealed at 200 °C, 225 °C, 250 °C, 275 °C and 300 °C. The effect on RRAM device performance for annealing temperatures of 225 °C, 250 °C, 275 °C was investigated in terms of the operation voltages of RS characteristics, resistance distribution, endurance cycles and retention uniformity. The worst device performance was observed for an annealing temperature of 225 °C and the better performance was demonstrated in the device annealed at 275 °C. The best performance was found for an annealing temperature of 250 °C with the lowest operation voltage (<1.5 V), the highest ON/OFF ratio (>10^4), the narrowest resistance distribution, the longest retention time (>10^4 s) and the most endurance cycles (150), which indicates the lowest energy consumption and the excellent stability of the RRAM devices. An XPS study has been conducted to determine elements present in the AlO$_x$ thin films prepared at different annealing temperature with the aim of explaining the variation of associated RRAM devices performance. The device performance was considered to be related to the concentration gradient of hydroxyl groups in the solution-based AlO$_x$ thin films for different annealing temperatures.

Author Contributions: Conceptualization, Z.S., Y.Q. and C.Z. (Chun Zhao); Methodology, Z.S., Y.Q. and C.Z. (Chun Zhao); Software, Z.S., Y.Q., T.L. and Y.H.; Validation, Z.S., C.Z. (Chun Zhao), C.Z. (Cezhou Zhao) and S.H.; Formal Analysis, Z.S., Y.Q., I.Z.M., C.Z. (Cezhou Zhao), S.H. and C.Z. (Chun Zhao); Investigation, Z.S. and Y.Q.; Resources, C.Z. (Chun Zhao), C.Z. (Cezhou Zhao) and L.Y.; Data Curation, Z.S., Y.Q., T.L. and Y.H.; Writing-Original Draft Preparation, Z.S., Y.Q. and C.Z. (Chun Zhao); Writing-Review & Editing, Z.S., Y.Q., I.Z.M., C.Z. (Cezhou Zhao), S.H. and C.Z. (Chun Zhao); Visualization, Z.S. and Y.Q.; Supervision, Y.Q., I.Z.M., C.Z. (Cezhou Zhao), S.H. and C.Z. (Chun Zhao); Project Administration, C.Z. (Cezhou Zhao) and C.Z. (Chun Zhao); Funding Acquisition, I.Z.M., C.Z. (Cezhou Zhao) and C.Z. (Chun Zhao).

Funding: This research was funded in part by the National Natural Science Foundation of China (21503169, 2175011441,61704111), Key Program Special Fund in XJTLU (KSF-P-02, KSF-T-03, KSF-A-04, KSF-A-05, KSF-A-07). The author Ivona Z. Mitrovic acknowledges the British Council UKIERI project no. IND/CONT/G/17-18/18.

Conflicts of Interest: The authors declare no conflict of interest.

References

1. Ambrosi, E.; Bricalli, A.; Laudato, M.; Ielmini, D. Impact of oxide and electrode materials on the switching characteristics of oxide ReRAM devices. *Faraday Discuss.* **2019**, *213*, 87–98. [CrossRef] [PubMed]

2. Bang, S.; Kim, M.; Kim, T.; Lee, D.; Kim, S.; Cho, S. Gradual switching and self-rectifying characteristics of Cu/α-IGZO/p+-Si RRAM for synaptic device application. *Solid State Electron.* **2018**, *150*, 60–65. [CrossRef]

3. Bertolazzi, S.; Bondavalli, P.; Roche, S.; San, T.; Choi, S. Nonvolatile Memories Based on Graphene and Related 2D Materials. *Adv. Mater.* **2019**, *31*, e1806663. [CrossRef] [PubMed]

4. Bukke, R.N.; Avis, C.; Jang, J. Solution-Processed Amorphous In–Zn–Sn Oxide Thin-Film Transistor Performance Improvement by Solution-Processed Y$_2$O$_3$ Passivation. *IEEE Electron Devices Lett.* **2016**, *37*, 433–436. [CrossRef]

5. Bukke, R.N.; Mude, N.N.; Lee, J.; Avis, C.; Jang, J. Effect of Hf alloy in ZrO$_x$ gate insulator for solution processed a-IZTO thin film transistors. *IEEE Electron Devices Lett.* **2018**. [CrossRef]

6. Chen, P.-H.; Su, Y.-T.; Chang, F.-C. Stabilizing Resistive Switching Characteristics by Inserting Indium-Tin-Oxide Layer as Oxygen Ion Reservoir in HfO$_2$-Based Resistive Random Access Memory. *IEEE Trans. Electron Dev.* **2019**, *66*, 1276–1280. [CrossRef]

7. Arun, N.; Kumar, K.; Mangababu, A.; Rao, S.; Pathat, A. Influence of the bottom metal electrode and gamma irradiation effects on the performance of HfO$_2$-based RRAM devices. *Radiat. Eff. Defects Solids* **2019**, *174*, 66–75. [CrossRef]

8. Cazorla, M.; Aldana, S.; Maestro, M.; Gonzalez, M. Thermal study of multilayer resistive random access memories based on HfO$_2$ and Al$_2$O$_3$ oxides. *J. Vac. Sci. Technol. B* **2019**, *37*. [CrossRef]

9. Chen, J.; Yin, C.; Li, Y.; Qin, C. LiSiO$_x$-based Analog Memristive Synapse for Neuromorphic Computing. *IEEE Electron Devices Lett.* **2019**. [CrossRef]

10. Duan, W.; Tang, Y.; Liang, X.; Rao, C.; Chu, J.; Wang, G.; Pei, Y. Solution processed flexible resistive switching memory based on Al-In-O self-mixing layer. *J. Appl. Phys.* **2018**, *124*. [CrossRef]

11. Bartlett, P.; Berg, A.I.; Bernasconi, M.; Brown, S.; Burr, G.; Foroutan-Nejad, C.; Gale, E.; Huang, R.; Ielmini, D.; Kissling, G.; et al. Phase-change memories (PCM)-Experiments and modelling: General discussion. *Faraday Discuss.* **2019**, *213*, 393–420. [CrossRef] [PubMed]

12. Gao, S.; Yi, X.; Shang, J.; Liu, G.; Li, R.-W. Organic and hybrid resistive switching materials and devices. *Chem. Soc. Rev.* **2019**, *48*, 1531–1565. [CrossRef] [PubMed]

13. Han, P.; Lai, T.-C.; Wang, M.; Zhao, X.-R.; Cao, Y.-Q.; Wu, D.; Li, A.-D. Outstanding memory characteristics with atomic layer deposited $Ta_2O_5/Al_2O_3/TiO_2/Al_2O_3/Ta_2O_5$ nanocomposite structures as the charge trapping layer. *Appl. Surf. Sci.* **2019**, *467–468*, 423–427. [CrossRef]

14. He, Z.-Y.; Wang, T.-Y.; Chen, L.; Zhu, H.; Sun, Q.-Q.; Ding, S.-J.; Zhang, D. Atomic Layer-Deposited $HfAlO_x$-Based RRAM with Low Operating Voltage for Computing In-Memory Applications. *Nanoscale Res. Lett.* **2019**, *14*, 51. [CrossRef] [PubMed]

15. Tian, M.; Zhong, H. Effects of Electrode on the Performance of Al_2O_3 Based Metal-Insulator-Metal Antifuse. *ECS J. Solid State Sci. Technol.* **2019**, *8*, N32–N35. [CrossRef]

16. Hur, J.H.; Kim, D. A study on mechanism of resistance distribution characteristics of oxide-based resistive memory. *Sci. Rep.* **2019**, *9*, 302. [CrossRef] [PubMed]

17. Kadhim, M.S.; Yang, F.; Sun, B.; Hou, W.; Peng, H.; Hou, Y.; Jia, Y.; Yuan, L.; Yu, Y.; Zhao, Y. Existence of Resistive Switching Memory and Negative Differential Resistance State in Self-Colored MoS_2/ZnO Heterojunction Devices. *ACS Appl. Electron. Mater.* **2019**, *1*, 318–324. [CrossRef]

18. Kang, K.; Ahn, H.; Song, Y.; Lee, W.; Kim, J.; Kim, Y.; Yoo, D.; Lee, T. High-Performance Solution-Processed Organo-Metal Halide Perovskite Unipolar Resistive Memory Devices in a Cross-Bar Array Structure. *Adv. Mater.* **2019**, *31*, e1804841. [CrossRef]

19. Kim, G.; Kornijcuk, V.; Kim, D.; Kim, I.; Hwang, C.; Jesong, D. Artificial Neural Network for Response Inference of a Nonvolatile Resistance-Switch Array. *Micromachines (Basel)* **2019**, *10*, 219. [CrossRef]

20. Kim, S.; Chen, J.; Chen, Y.C.; Kim, M.H.; Kim, H.; Kwon, M.W.; Hwang, S.; Ismail, M.; Li, Y.; Miao, X.-S.; et al. Neuronal dynamics in HfO_x/AlO_y-based homeothermic synaptic memristors with low-power and homogeneous resistive switching. *Nanoscale* **2018**, *11*, 237–245. [CrossRef]

21. Kim, T.-H.; Kim, S.; Kim, H.; Kim, M. Highly uniform and reliable resistive switching characteristics of a $Ni/WO_x/p + -Si$ memory device. *Solid State Electron.* **2018**, *140*, 51–54. [CrossRef]

22. Ram, J.; Kumar, R. Effect of Annealing on the Surface Morphology, Optical and and Structural Properties of Nanodimensional Tungsten Oxide Prepared by Coprecipitation Technique. *J. Electron. Mater.* **2018**, *48*, 1174–1183. [CrossRef]

23. Kang, X.; Guo, J.; Gao, Y.; Ren, S.; Chen, W. NiO-based resistive memory devices with highly improved uniformity boosted by ionic liquid pre-treatment. *Appl. Surf. Sci.* **2019**, *480*, 57–62. [CrossRef]

24. Le, P.Y.; Tran, H.; Zhao, Z.; Mckenzie, D. Tin oxide artificial synapses for low power temporal information processing. *Nanotechnology* **2019**. [CrossRef] [PubMed]

25. Lee, B.R.; Park, J.; Lee, T.; Kim, T. Highly Flexible and Transparent Memristive Devices Using Cross-Stacked Oxide/Metal/Oxide Electrode Layers. *ACS. Appl. Mater. Interfaces* **2019**, *11*, 5215–5222. [CrossRef] [PubMed]

26. Lübben, M.; Valov, I. Active Electrode Redox Reactions and Device Behavior in ECM Type Resistive Switching Memories. *Adv. Electron. Mater.* **2019**. [CrossRef]

27. Zhang, R.; Huang, H.; Xia, Q.; Ye, C.; Wei, X. Role of Oxygen Vacancies at the TiO_2/HfO_2 Interface in Flexible Oxide-Based Resistive Switching Memory. *Adv. Electron. Mater.* **2019**, *5*. [CrossRef]

28. Moussa, S.; Mauzeroll, J. Review—Microelectrodes: An Overview of Probe Development and Bioelectrochemistry Applications from 2013 to 2018. *J. Electrochem. Soc.* **2019**, *166*, G25–G38. [CrossRef]

29. Cano, A.M.; George, S.; Marquardt, A.; DuMont, D. Effect of HF Pressure on Thermal Al_2O_3 Atomic Layer Etch Rates and Al_2O_3 Fluorination. *J. Phys. Chem. C* **2019**, *123*, 10346–10355. [CrossRef]

30. Lin, C.-Y.; Wang, J.-C.; Chen, T.-C. Analysis of suspension and heat transfer characteristics of Al_2O_3 nanofluids prepared through ultrasonic vibration. *Appl. Energy* **2011**, *88*, 4527–4533. [CrossRef]

31. Zawrah, M.F.; Khattab, R.M.; Girgis, L.G.; Daidamony, H. Stability and electrical conductivity of water-base Al_2O_3 nanofluids for different applications. *HBRC J.* **2019**, *12*, 227–234. [CrossRef]

32. Wang, H.; Zhang, H.; Liu, J.; Xue, D.; Liang, H. Hydroxyl Group Adsorption on GaN (0001) Surface: First Principles and XPS Studies. *J. Electron. Mater.* **2019**, *48*, 2430–2437. [CrossRef]

33. Li, L. Ternary Memristic Effect of Trilayer-Structured Graphene-Based Memory Devices. *Nanomaterials (Basel)* **2019**, *9*, 518. [CrossRef] [PubMed]

34. Niu, G.; Calka, P.; Huang, P.; Sharath, S.; Petzold, S.; Gloskovskii, A.; Zhao, Y.; Kang, J. Operando diagnostic detection of interfacial oxygen 'breathing' of resistive random access memory by bulk-sensitive hard X-ray photoelectron spectroscopy. *Mater. Res. Lett.* **2019**, *7*, 117–123. [CrossRef]

35. Li, L.; Li, G. High-Performance Resistance-Switchable Multilayers of Graphene Oxide Blended with 1,3,4-Oxadiazole Acceptor Nanocomposite. *Micromachines (Basel)* **2019**, *10*, 140. [CrossRef] [PubMed]

36. Petzold, S.; Sharath, S.; Lemke, J.; Hildebrandt, E.; Trautmann, C. Heavy Ion Radiation Effects on Hafnium Oxide based Resistive Random Access Memory. *IEEE Trans. Nucl. Sci.* **2019**. [CrossRef]

37. Russo, P.; Xiao, M.; Zhou, N. Electrochemical Oxidation Induced Multi-Level Memory in Carbon-Based Resistive Switching Devices. *Sci. Rep.* **2019**, *9*, 1564. [CrossRef]

38. Liu, A.; Noh, Y.-Y.; Zhu, H.; Sun, H.; Xu, Y. Solution Processed Metal Oxide High-kappa Dielectrics for Emerging Transistors and Circuits. *Adv. Mater.* **2018**, e1706364. [CrossRef]

39. Liu, T.; Wu, W.; Liao, K.; Sun, Q.; Gong, X. Fabrication of carboxymethyl cellulose and graphene oxide bio-nanocomposites for flexible nonvolatile resistive switching memory devices. *Carbohydr. Polym.* **2019**, *214*, 213–220. [CrossRef]

40. Wu, H.; Yao, P.; Zhao, M.; Liu, Y.; Xi, Y. Reliability Perspective on Neuromorphic Computing Based on Analog RRAM. In Proceedings of the 2019 IEEE International Reliability Physics Symposium (IRPS), Monterey, CA, USA, 23 May 2019.

41. Mao, J.-Y.; Zhou, L.; Ren, Y.; Yang, J.; Chang, C. A bio-inspired electronic synapse using solution processable organic small molecule. *J. Mater. Chem. C* **2019**, *7*, 1491–1501. [CrossRef]

42. Nenning, A.; Fleig, J. Electrochemical XPS investigation of metal exsolution on SOFC electrodes: Controlling the electrode oxygen partial pressure in ultra-high-vacuum. *Surf. Sci.* **2019**, *680*, 43–51. [CrossRef]

43. Yi, X.; Yu, Z.; Niu, X.; Shang, J.; Mao, G.; Yin, T.; Yang, H.; Xue, W.; Dhanapal, P.; Qu, S.; et al. Intrinsically Stretchable Resistive Switching Memory Enabled by Combining a Liquid Metal-Based Soft Electrode and a Metal-Organic Framework Insulator. *Adv. Electron. Mater.* **2019**, *5*. [CrossRef]

44. Yen, T.J.; Wang, X.; Li, J.; Cho, K. All Non metal Resistive Random Access Memory. *Sci. Rep.* **2019**, *9*, 6144. [CrossRef] [PubMed]

45. Long, S.; Lian, X.; Cagli, C.; Perniola, L. A Model for the Set Statistics of RRAM Inspired in the Percolation Model of Oxide Breakdown. *IEEE Electron Dev. Lett.* **2013**, *34*, 999–1001. [CrossRef]

46. Sun, C.; Lu, S.; Jin, F.; Mo, W.; Song, J.; Dong, K. The Resistive Switching Characteristics of TiN/HfO$_2$/Ag RRAM Devices with Bidirectional Current Compliance. *J. Electron. Mater.* **2019**, *48*, 2992–2999. [CrossRef]

47. Chen, Y.-C.; Chen, Y.-F.; Wu, X.; Zhou, F.; Guo, M.; Lin, C.-Y.; Hsieh, C.-C.; Fowler, B.; Chang, T.-C.; Lee, J.; et al. Dynamic conductance characteristics in HfO$_x$-based resistive random access memory. *RSC Adv.* **2017**, *7*, 12984–12989. [CrossRef]

48. Qi, Y.; Zhao, C.; Zhao, C.; Xu, W.; Shen, Z.; He, J.; Zhao, T.; Fang, Y.; Liu, Q.; Yi, R.; et al. Enhanced resistive switching performance of aluminum oxide dielectric with a low temperature solution-processed method. *Solid State Electron.* **2019**, *158*, 28–36. [CrossRef]

49. Lee, D.; Kim, S.; Cho, K. Integration of 4F2 selector-less crossbar array 2Mb ReRAM based on transition metal oxides for high density memory applications. In Proceedings of the 2012 Symposium on VLSI Technology (VLSIT), Honolulu, HI, USA, 12–14 June 2012.

50. He, Y.; Ma, G.; Zhou, X.; Cai, H.; Liu, C.; Zhang, J.; Wang, H. Impact of chemical doping on resistive switching behavior in zirconium-doped CH$_3$NH$_3$PbI$_3$ based RRAM. *Org. Electron.* **2019**, *68*, 230–235. [CrossRef]

51. Chang, Y.-F.; Fowler, B.; Chen, Y.-C.; Chen, Y.-T.; Wang, Y.; Xue, F.; Zhou, F.; Lee, J. Intrinsic SiO$_x$-based unipolar resistive switching memory. II. Thermal effects on charge transport and characterization of multilevel programing. *J. Appl. Phys.* **2014**, *116*. [CrossRef]

52. Lin, C.-Y.; Wu, C.-Y.; Wu, C.-Y.; Hu, C.; Tsenga, T.-Y. Bistable Resistive Switching in Al$_2$O$_3$ Memory Thin Films. *J. Electrochem. Soc.* **2007**, *154*, 4. [CrossRef]

53. Liu, Z.; Chen, P.; Liu, Y.; Yang, M.; Wong, J.; Cen, Z.; Zhang, S. Temperature-Dependent Charge Transport in Al/Al Nanocrystal Embedded Al$_2$O$_3$ Nanocomposite/p-Si Diodes. *ECS Solid State Lett.* **2012**, *1*, Q4–Q7. [CrossRef]

54. Kim, Y.; Ohmi, S.; Tsutsui, K.; Iwai, H. Space-Charge-Limited Currents in La$_2$O$_3$ Thin Films Deposited by E-Beam Evaporation after Low Temperature Dry-Nitrogen Annealing. *Jpn. J. Appl. Phys.* **2005**, *44*, 4032–4042. [CrossRef]

55. Chuang, K.-C.; Chu, C.; Zhang, H.; Luo, J.; Li, W.; Li, Y. Impact of the Stacking Order of HfO_x and AlO_x Dielectric Films on RRAM Switching Mechanisms to Behave Digital Resistive Switching and Synaptic Characteristics. *IEEE J. Electron Devices Soc.* **2019**. [CrossRef]

56. Rodrigues, A.N.; Santos, Y.P.; Rodrigues, C.L.; Macedo, M.A. Al_2O_3 thin film multilayer structure for application in RRAM devices. *Solid State Electron.* **2018**, *149*, 1–5. [CrossRef]

57. Rodriguez-Fernandez, A.; Aldana, S.; Campabadalm, F.; Sune, J. Resistive Switching with Self-Rectifying Tunability and Influence of the Oxide Layer Thickness in $Ni/HfO_2/n+$-Si RRAM Devices. *IEEE Trans. Electron Dev.* **2017**, *64*, 3159–3166. [CrossRef]

58. Wu, L.; Dong, C.; Wang, X.; Li, J.; Li, M. Annealing effect on the bipolar resistive switching memory of NiZn ferrite films. *J. Alloys Compd.* **2019**, *779*, 794–799. [CrossRef]

59. Gao, L.; Li, Y.; Li, Q.; Song, Z.; Ma, F. Enhanced resistive switching characteristics in Al_2O_3 memory devices by embedded Ag nanoparticles. *Nanotechnology* **2017**, *28*, 215201. [CrossRef] [PubMed]

60. Wu, X.; Cha, D.; Bosman, M.; Raghavan, N.; Migas, D.; Borisenko, V.; Zhang, X.; Li, K.; Pey, K. Intrinsic nanofilamentation in resistive switching. *J. Appl. Phys.* **2013**, *113*. [CrossRef]

61. Lee, J.; Schell, W.; Zhu, X.; Lu, W. Charge Transition of Oxygen Vacancies during Resistive Switching in Oxide-Based RRAM. *ACS. Appl. Mater. Interfaces* **2019**, *11*, 11579–11586. [CrossRef]

62. Sarkar, B.; Lee, B.; Misra, V. Understanding the gradual reset in $Pt/Al_2O_3/Ni$ RRAM for synaptic applications. *Semicond. Sci. Technol.* **2015**, *30*. [CrossRef]

63. Qi, Y.; Zhao, C.; Liu, C.; Fang, Y.; He, J.; Luo, T.; Yang, L.; Zhao, C. Comparisons of switching characteristics between $Ti/Al_2O_3/Pt$ and $TiN/Al_2O_3/Pt$ RRAM devices with various compliance currents. *Semicond. Sci. Technol.* **2018**, *33*. [CrossRef]

64. Cook, S.; Dylla, M.; Rosenberg, A.; Mansley, Z.; Fong, D. The Vacancy-Induced Electronic Structure of the $SrTiO_{3-\delta}$ Surface. *Adv. Electron. Mater.* **2019**, *5*. [CrossRef]

65. Lyu, F.; Bai, Y.; Wang, Q.; Wang, L.; Zhang, X.; Yin, Y. Coordination-assisted synthesis of iron-incorporated cobalt oxide nanoplates for enhanced oxygen evolution. *Mater. Today Chem.* **2019**, *11*, 112–118. [CrossRef]

66. Mefford, J.T.; Kurilovich, A.; Saunders, J.; Hardin, W.; Abakumov, A.; Forslund, R. Decoupling the roles of carbon and metal oxides on the electrocatalytic reduction of oxygen on $La_{1-x}Sr_xCoO_3$-delta perovskite composite electrodes. *Phys. Chem. Chem. Phys.* **2019**, *21*, 3327–3338. [CrossRef] [PubMed]

67. Sun, B.; Qian, Y.; Liang, Z.; Guo, Y.; Xue, Y.; Tian, J.; Cui, H. Oxygen vacancy-rich BiO_{2-x} ultra-thin nanosheet for efficient full-spectrum responsive photocatalytic oxygen evolution from water splitting. *Sol. Energy Mater. Sol. Cells* **2019**, *195*, 309–317. [CrossRef]

68. Xu, W.; Wang, H.; Xie, F.; Chen, J.; Cao, H.; Xu, J. Facile and environmentally friendly solution-processed aluminum oxide dielectric for low-temperature, high-performance oxide thin-film transistors. *ACS Appl. Mater. Interfaces* **2015**, *7*, 5803–5810. [CrossRef]

69. Juárez-Moreno, J.A.; Ávila-Ortega, A.; Oliva, A.I.; Avilés, F. Effect of wettability and surface roughness on the adhesion properties of collagen on PDMS films treated by capacitively coupled oxygen plasma. *Appl. Surf. Sci.* **2015**, *349*, 763–773.

70. Allahbakhsh, A.; Sharif, F.; Mazinani, S. The Influence of Oxygen-Containing Functional Groups on the Surface Behavior and Roughness Characteristics of Graphene Oxide. *Nano* **2013**, *8*. [CrossRef]

71. Boronat, M.; Corma, A.; Illas, F.; Radilla, J. Mechanism of selective alcohol oxidation to aldehydes on gold catalysts: Influence of surface roughness on reactivity. *J. Catal.* **2011**, *278*, 50–58. [CrossRef]

Article

Fine-Grained Power Gating Using an MRAM-CMOS Non-Volatile Flip-Flop

Jaeyoung Park * and Young Uk Yim

School of Computer Science and Electrical Engineering, Handong Global University, Pohang-si 37554, Korea; zero12@gmail.com
* Correspondence: jaeyoung.park@handong.edu; Tel.: +1-858-658-2657

Received: 28 April 2019; Accepted: 17 June 2019; Published: 20 June 2019

Abstract: An area-efficient non-volatile flip flop (NVFF) is proposed. Two minimum-sized Metal-Oxide-Semiconductor Field-Effect Transistor (MOSFET) and two magnetic tunnel junction (MTJ) devices are added on top of a conventional D flip-flop for temporary storage during the power-down. An area overhead of the temporary storage is minimized by reusing a part of the D flip-flop and an energy overhead is reduced by a current-reuse technique. In addition, two optimization strategies of the use of the proposed NVFF are proposed: (1) A module-based placement in a design phase for minimizing the area overhead; and (2) a dynamic write pulse modulation at runtime for reducing the energy overhead. We evaluated the proposed NVFF circuit using a compact MTJ model targeting an implementation in a 10 nm technology node. Results indicate that area overhead is 6.9% normalized to the conventional flip flop. Compared to the best previously known NVFFs, the proposed circuit succeeded in reducing the area by 4.1× and the energy by 1.5×. The proposed placement strategy of the NVFF shows an improvement of nearly a factor of 2–18 in terms of area and energy, and the pulse duration modulation provides a further energy reduction depending on fault tolerance of programs.

Keywords: STT-MRAM; flip-flop; power gating; low-power

1. Introduction

Power gating has been researched as an effective energy-reduction technique [1–3]. This reduces static power consumption by shutting power off. However, data needs to be transferred to another storage component before the power-off and restored after the power-on [4,5]. Such transfers of data introduce energy and area overheads. Therefore, it is important to develop a low overhead temporary storage component and an efficient strategy of the use of the storage components. Off-chip memories have been used for the temporary storage [2,3]; however, a complex interface between the off-chip memories and a chip stands in the way of wide adoption of such off-chip memories in power gating scheme.

Embedded non-volatile flip-flops (NVFFs) are promising enablers to fine-grained power gating because these do not require a complex interface to transfer state from/to the external storage. One critical issue is an overhead to store/restore data onto non-volatile temporary storage of the NVFF before/after the power-down. It is universally the case when adding a new feature (e.g., non-volatility) to the existing flip-flop. However, what is the best way to build a low overhead NVFF? A magnetic tunnel junction (MTJ) of a spin torque transfer magnetic memory (STT-MRAM) is a candidate for the non-volatile storage of the NVFF because the MTJ does not occupy the silicon area; the MTJ is placed between metal layers. However, the area and energy overheads can be significant if write and read circuits for the MTJ are not carefully optimized.

In this paper, we propose an area-efficient MRAM-Complementary Metal-Oxide-Semiconductor (CMOS) hybrid non-volatile flip-flop. We reutilize the existing CMOS flip-flop for transferring data to/from MTJs to reduce the area overhead. Only two minimum-sized transistors are added. In addition, the energy overhead for storing data onto MTJs is reduced by 50% by reusing a write current to write two MTJs. We evaluated the proposed NVFF circuit using a compact MTJ model and a 10 nm Predictive Technology Model (PTM) MOSFET model [6–8]. The proposed NVFF has an improvement by a factor of 4–23 in terms of the area over state-of-the-art circuits. In addition, energy for the storing operation is reduced by 1.5× compared to the best previously proposed NVFF circuits.

We also propose optimization strategies for the use of the proposed NVFF. Because the proposed NVFF introduces a non-negligible area overhead compared to the conventional D flip-flop (FF), it is important to use the proposed NVFF carefully to minimize the area overhead. Replacing all the conventional FFs in a design with NVFFs imposes a large area penalty. Therefore, we first analyze a circuit and then place the NVFFs only for a selected module which can minimize the area penalty. We place the proposed NVFF in a module that has a low ratio of the flop area to total. Because a module contains flops, combinational logic circuits, and passive devices, the area penalty by the NVFF is minimized where FFs occupy the relatively small area to the total. In the other words, the area increase by the NVFF becomes relatively small if the other components in the same modules occupy the greater portion of the module.

In addition, the write pulse duration for the NVFF needs to be carefully optimized due to the stochastic nature of the MTJ write. The MTJ write process itself is fundamentally stochastic and the actual time to the completion varies dramatically with the distribution having a very long tail [9–11]. This means that write energy also varies quite significantly and the write energy can be wasteful if the applied write pulse duration is not carefully selected. Instead of using the conventional deterministic strategy with a fixed pulse duration that guarantee a target error probability, we exploit this stochastic property to save more energy by adjusting the pulse duration adaptively. A key insight is that high fault-tolerance programs can endure more error from the NVFF so that we can reduce the write pulse duration for the programs to save write energy even if the NVFF itself introduces higher error probability.

We demonstrate the optimization strategies on an OpenSPARC design which is an open-source version of UltraSPARC processor [12]. Four programs—matrix multiply, sort, bzip2, and prime—are also selected for this experiment [13]. Analysis indicates the placement shows an improvement of nearly a factor of 2–18 in terms of area and energy and the pulse duration modulation maximizes energy savings of the proposed NVFF for programs have high fault tolerance. The detailed analysis are presented in the following sections.

2. MRAM-CMOS Non-Volatile Flip-Flops

2.1. State-of-the-Art MRAM-CMOS NVFFs

MRAM-CMOS NVFFs typically need extra circuits for writing and reading MTJs. Multiple realizations of the extra circuits, which use additional write drivers and sense amplifiers, have been proposed [14–17]. In [15], two NAND gates, seven inverters, and three NMOS switch transistors are used for the external write driver and the sense amplifier with a significant reduction of D-Q delay. In [16], four NOR gates, four inverters, and 16 NMOS transistors are used to reduce C-Q delay and sensing currents.

In [14], only three extra transistors are added for writing and reading MTJs because the existing cross-coupled inverter pair of the conventional D-FFs is used to assist with storing and restoring operations of MTJs. Figure 1 shows the storing and restoring operations of the NVFF [14]. For the storing operation, MTJ_A is written to *antiparallel* (AP) state by lowering Reset-ENable (REN) signal when QS is logical 'H' for Q = 1. MTJ_B is written to *parallel* (P) state by raising the REN and Set-ENable (SEN) signals in the second write phase. The restoring operation is achieved by the regenerative feedback of the inverter pair because a different voltage is developed between QS and QSb nodes when MTJ_A and MTJ_B have different resistances. This is why the NVFF requires only three additional transistors. However, sizable transistors are needed to drive sufficient current with low Vgs because the Vgs is dropped by IR drop through an MTJ (Vgs = Vdd − I × R_{MTJ}). Moreover, the storing energy is doubled because two MTJs need to be written in different phases.

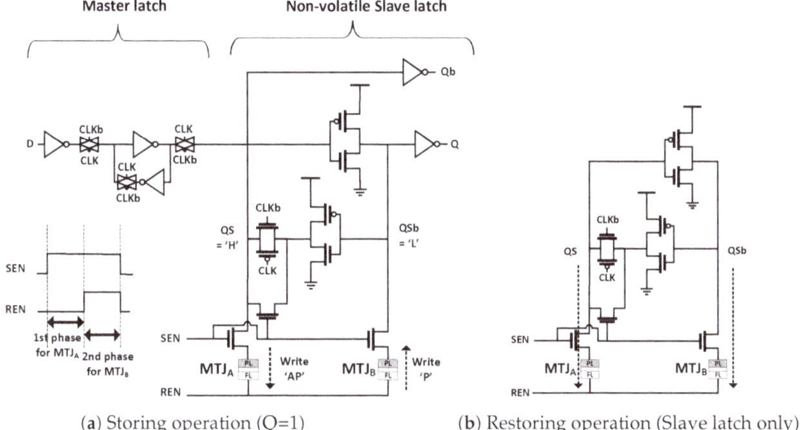

(**a**) Storing operation (Q=1)   (**b**) Restoring operation (Slave latch only)

Figure 1. Schematic of an non-volatile flip flop (NVFF) of Yamamoto et al. [14].

2.2. Current Reutilization NVFF

We propose a current reutilization technique to reduce energy and area overheads. The main idea that a single write current for an MTJ can be used to write another MTJ. Instead of applying two separate current pulses to write two MTJs at different phases, we can write two MTJs using a single write current at the same time. The current reutilization should not introduce large area overhead. We achieved this by inserting one minimum-sized NMOS transistor because two MTJs can be placed on the same current path via the NMOS transistor (M1) as shown in Figure 2. A write current is passed through MTJ_A, M1, and MTJ_B when a switch transistor (M1) is turned on for Q = 1. The MTJ_A is written to the AP state because the current direction is from the pinned layer (PL) to the free layer (FL) of the MTJ, and MTJ_B becomes the P state because the current direction is reversed (FL→PL). For storing Q = 0, a write current goes through MTJ_B, M1, and MTJ_A; therefore, the situation is reversed (MTJ_B = AP, MTJ_A = P). The proposed current reutilization technique allows for writing both MTJs using one write current at the same time. Thus, we could reduce the write current by 50%, resulting in a half storing energy. In contrast to an NVFF of Yamamoto et al. [14], an inverter pair drives a write current and a minimum-sized NMOS transistor is only used as a switch. In addition, a full Vdd is applied to a gate of the inverter pair during the storing operation.

The restoring operation is achieved by another minimum-sized transistor (M2). This reutilizes the inverter pair of the existing D-FF. Different voltages are developed between QS and QSb of the slave latch by two MTJs that have different resistances when M1 and M2 are turned on after the power-up.

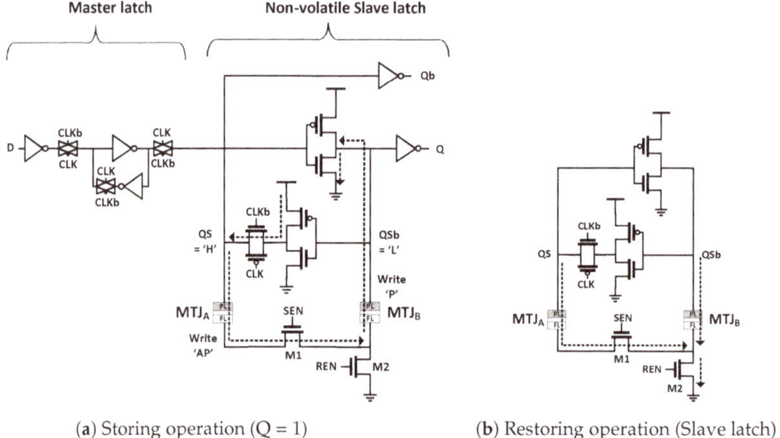

(**a**) Storing operation (Q = 1)   (**b**) Restoring operation (Slave latch)

Figure 2. Schematic of the proposed current reutilization NVFF.

2.3. Evaluation of the Proposed NVFF

We designed the proposed NVFF using a 10 nm predictive technology model (PTM) MOSFET model and a compact MTJ model [7,8]. Key parameters of the perpendicular MTJ is described in Table 1. Figure 3 shows Simulation Program with Integrated Circuit Emphasis (SPICE) simulation results of the proposed NVFF using the models. The proposed NVFF operates as a conventional D-FF in normal operations. On top of the D-FF, non-volatile operations are added. The storing operation is performed before the power-down. The output Q is stored in MTJs when SEN is raised. MTJ_A is written to the AP state and MTJ_B is the P state for Q = 1. During the power-down mode, the output Q is lowered. The Q is restored when power is up again at 28.6 ns (restoring operation).

Table 1. Key parameters of perpendicular magnetic tunnel junction (MTJ) [6–8].

Parameter	Value	Unit
Intrinsic critical current	24	μA
Thermal stability factor	58	
Tunnel Magnetoresistance ratio (TMR)	∼100	%
Diameter of MTJ	20	nm
Out-of-plane magnetic field	0.4	T

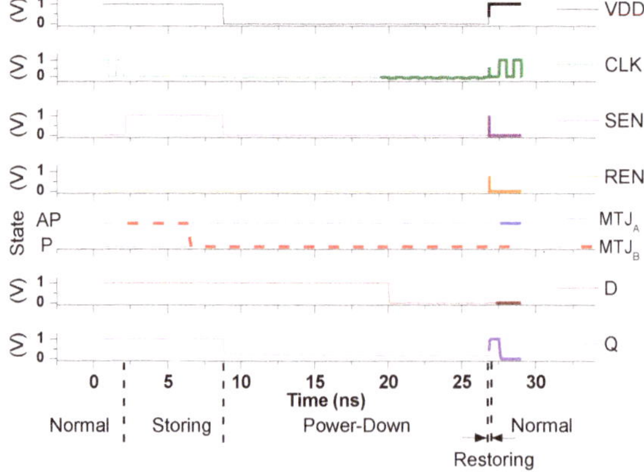

Figure 3. Waveforms of each node of the proposed NVFF. (Output '1' is stored and restored. *x*-axis denotes time and the *y*-axes indicates voltage (V) or states of MTJs.)

We compare the proposed NVFF with the state-of-the-art NVFF circuits as shown in Table 2. The proposed NVFF shows an improvement of nearly a factor of 2–17 in terms of restoring energy compared to the state-of-the-art NVFF circuits. Note that the restoring energy is reduced by 50% if MTJ and CMOS devices are the same as an NVFF [14]. The proposed NVFF implemented in an advance technology node and the greater part of the storing energy reduction comes from the technology scaling.

The relative area increase is only 6.9% (2/29) because only two minimum-sized transistors are added to the conventional D flip-flop (FF) that has 29 transistors. We did not directly compare the area because technologies of the reference circuits are different, and the actual area strongly depends on the layout optimization. Thus, we used a relative area overhead to the D-FF of each technology for this comparison. Note that the size of the PMOS transistor in the inverter is assumed to be $2\times$ NMOS transistor. The relative area overheads of state-of-the-art NVFF architectures are from 28.0% to 160.0%. Therefore, the proposed NVFF has an improvement of nearly a factor of 4–23 in terms of the area overhead compared to state-of-the-art NVFF circuit.

Table 2. Performance summary and comparison with the state-of-the-art NVFFs.

		MRAM-Based							FeRAM-	ReRAM-
		This Work	[17]	[16]	[15]	[18]	[14]	[19] [a]	[20]	[21]
Technology node (nm)		10	90	45	45	90	65	65	130	65
Area overhead [b] (%)		6.9	131.0	160.0	120.0	103.4	109.0	28.0	64.0	32.0
Energy (pJ)	Storing	0.2	175.5	1.9	1.6	0.3	5.0	0.5	2.4	-
	Restoring	0.002	-	0.171	0.007	-	0.349	0.197	-	-
Delay (ns)	Storing	6.6	-	-	-	10.0	29.5	6.4	1640.0	-
	Restoring	0.01	0.169 [c]	2.0	0.184	1.0	2.0	2.0	1230.0	16.0
C-Q delay (ps)		43.8	318.1 [c]	68.8	186.2	67.2	73.8	-	-	<1 ns
Power-Delay Product (fJ)		0.3	2.8 [c]	1.1	2.3	0.7	1.4	-	-	-

[a] Spin Hall Effect MTJ, [b] normalized to the conventional D flip-flop, [c] Data from [22]. FeRAM-Ferroelectric RAM, ReRAM-Resistive RAM.

A simulated delay of the restoring operation is 10 ps and a storing time is set to 6.6 ns to have a sufficiently low error probability. We computed the error probability of the proposed NVFF. We used the following probability model derived in [11] using a Neel–Brown relaxation formula to compute error probability. The model describes the switching probability $P_{SW}(t, I)$, which is the probability of switching occurring for a pulse duration t at current I:

$$P_{SW}(t, I) = 1 - e^{-\frac{t}{\tau_0 e^{\Delta(1 - I/I_{c0})}}},$$ (1)

where Δ is the thermal stability factor and τ_0 is is the inverse of the thermal attempt frequency that has a typical value of 1 ns [10,11]. I_{C0} is a critical current and I is an applied current to write. A computed write error probability is 1.5×10^{-13} where an average write current is 24.6 μA and a storing time is 6.6 ns.

3. Optimization Strategies for the Proposed NVFF

In this section, we describe optimization strategies of the use of the proposed NVFF. Because the proposed NVFF introduces a non-negligible area overhead compared to the conventional D-FF, it is important to use the proposed NVFF carefully to minimize the area overhead. In addition, the MTJ write process itself is fundamentally stochastic and the actual time to completion varies dramatically with the distribution having a very long tail [9–11]. This means that write energy also varies quite significantly and the write energy can be wasted if the applied pulse duration is not carefully adjusted.

We propose a two-phase optimization strategy: (1) a static NVFF placement in a design phase and (2) dynamic pulse width modification at runtime. The proposed two-phase flow is illustrated in Figure 4. In a design phase, we place the NVFF only in a module that is able to maximize the benefit of the NVFF. At runtime, we dynamically adjust the write pulse duration to save more energy for a program that has high fault tolerance. The details are described as follows.

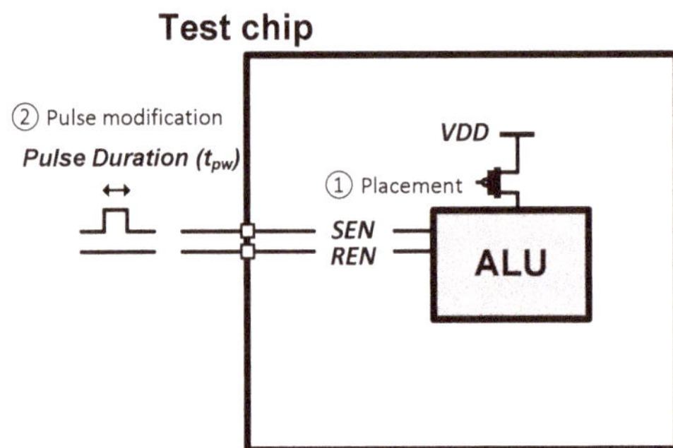

Figure 4. Overview of the proposed two-phase optimization flow.

3.1. Pre-Fabrication Optimization: A Module-Based Placement

A key question for the optimization is where the NVFF is placed to reduce static power while minimizing the area overhead. Replacing all the conventional FFs with NVFFs imposes a large area penalty. Therefore, we first analyze a circuit and then place the NVFFs only for a selected module. This is a fine-grained (or cluster-based) power gating approach. We characterize a circuit using two metrics, static power and a ratio of the FF area to the total, and then place the NVFFs in a module that has high static power and low area ratio. Because the area penalty is minimized where FFs occupy the relatively small area to the total. In addition, more static power can be saved if the module itself consumes high static power. Because the power gating can reduce static (leakage) power by shutting off power supply and it has no impact on dynamic power, placing the NVFFs in a high-static-power module can save more static power. Because a module generally contains not only FFs but also has combinational logic circuits and passive devices, the power gating can also reduce the static power of the combinational logic circuits and passive devices in the same module too.

We demonstrate the proposed optimization strategy in designing OpenSPARC T1 core, which is an open-source version of UltraSPARC processor. We first synthesized all modules and performed the placement and routing using *Synopsys* 32 nm EDK standard cell library [23]. We used *Synopsys Design Compiler, IC Compiler, and Primetime* for synthesizing, placement and routing, and static timing and power analysis, respectively [24–26]. We then selected seven high computational modules, and analyzed area and static power. As shown in Table 3, ALU (exu_alu) and decoder (ifu_dec) modules have fewer FFs than the other five modules. This results in lower area ratio to total. The increased area is less than 1% if the conventional FFs in the modules are replaced with the proposed NVFFs. Among two modules, the static power of the ALU is higher than that of the decoder. Therefore, we selected the ALU for a module to place the proposed NVFF. The placement shows an improvement of nearly a factor of 2–18 in terms of area and energy compared to the other modules.

All performances of seven modules are summarized in Table 3. The area and power are computed using *Synopsys Primetime*. Storing and restoring energy from the 10 nm PTM model are scaled up based on a constant field scaling method [27] because 32 nm standard cell library is used for the placement and routing of the OpenSPARC core. A break-even time ($T_{breakeven}$) is determined when energy saving by the power-gating is equal to the energy overhead by storing and restoring operations.

Table 3. Performance summary of seven modules.

	FF Area (μm^2)	Total Area (μm^2)	FF/Total (%)	NVFF Area (μM^2)	Increased Area (%)	P_{static} (mW)	$E_{storing}$ + $E_{restoring}$ (pJ)	$T_{breakeven}$ (ns)
exu_alu	429.5	15,022.5	2.9	459.1	0.2	1.8	72.6	40.8
exu_div	3714.6	12,218.2	30.4	3970.9	2.1	0.2	628.0	3924.0
exu_ecl	2319.3	6869.5	33.8	2479.4	2.3	0.1	392.1	4292.9
exu_rml	1729.2	4340.0	39.8	1848.5	2.7	0.4	733.1	1929.6
ifu_dec	277.5	4049.1	6.8	296.7	0.5	0.4	46.9	119.7
ifu_fcl	1785.6	5991.8	29.8	1908.8	2.1	0.5	301.9	616.6
ffu_dp	5466.6	13,722.1	39.8	5843.8	2.7	1.3	924.2	716.1

3.2. Post-Fabrication Optimization: A Pulse Width Modulation

We now describe a post-fabrication optimization strategy. The main idea is that a write pulse duration can be adaptively adjusted to reduce the write energy overhead for programs which have high fault tolerance. Because of a trade-off between energy and error probability of the propose NVFF, the write energy can be reduced by sacrificing error probability. This is true where each program has its unique fault tolerance even if the hardware design remains unchanged. In other words, some programs can tolerate more error so that we can use a shorter pulse duration for the programs to save more energy.

To implement this idea, we first examine the fault tolerance of programs to validate whether the fault tolerance varies over programs. Four programs—matrix multiply, sort, bzip2, and prime—are selected for this experiment. Gate-level simulations of the programs are performed on an OpenSPARC core using *Synopsys VCS* to inject faults and monitor the final outputs [28]. The fault injection process is based on a gate-level simulation that is halted at a randomly-determined cycle. The gate-level simulator extracts outputs of the combinational blocks and flip-flops for the cycle of interest. We inject faults (e.g., flipped value) on the flops based on the probabilities of their occurrence. After the injection, the analysis continues to the end of programs to determine whether the fault has been masked or a system failure has occurred. Outcomes from the fault injection are compared to a golden fault-free run. System failures by the fault injection are categorized as one of the following: detected unrecoverable error (DUE), Output match, silent data corruption (SDC), Hung, or Masked. We did this fault injection process for four programs. As shown in Figure 5, the most frequent category is Masked (above 90% of all cases). The second highest category is DUE, followed by SDC and Output match. The Hung case is not observed in the simulation. Among the four programs, bzip2 shows the lowest system failure rate, 1.2%. The DUE is only 0.6%, whereas the other programs are above 2.2%. This clearly shows that bzip2 has better tolerance in this experiment; therefore, a shorter pulse duration can be used for the program to save more energy.

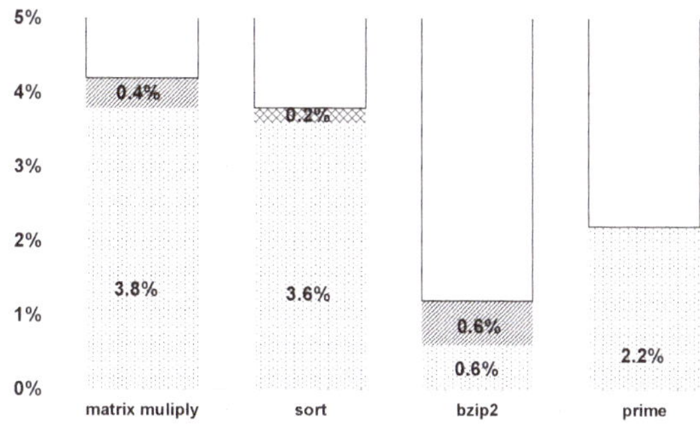

Figure 5. System failure results by fault injection.

We also examine how much energy we can save by adjusting the pulse duration. Figure 6 shows the error rate and the expected energy of a flop at different pulse duration. The error probability is inversely proportional to the write pulse duration as Equation (1), and the expected energy is linearly increased by the pulse duration while the error rate is exponentially decreased. At 6.6 ns, a write error probability is 1.5×10^{-13} and energy for a storing and restoring operations is 0.2 pJ. For short pulses such as 3.3 ns, the computed error probability is increased to 3.9×10^{-7} while the energy consumption is reduced by half. Because of such trade-off, the applied pulse duration for each program needs to be carefully selected based on the target error probability of the system even if the pulse duration modulation strategy maximizes energy savings of the proposed NVFF.

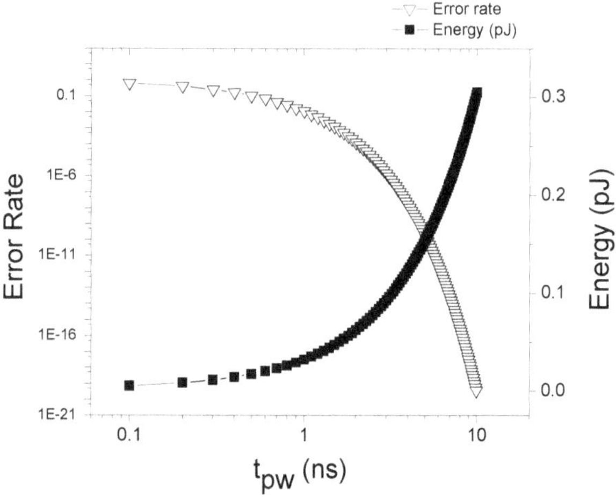

Figure 6. Error rate and the expected energy of the proposed NVFF at different write pulse duration.

In order to control the pulse duration, a control circuit is necessary. However, the area overhead per flop would be negligible because one circuit can control all FFs in a chip. In addition, the pulse duration is selected at software-level because the program information is needed.

4. Conclusions

In this paper, a novel area efficient NVFF is proposed. The relative area overhead is 6.9%, and the proposed NVFF shows an improvement of nearly a factor of 4–23 in terms of area overhead compared to state-of-the-art NVFF designs. The write current for the restoring operation is reduced by 50% using the proposed current-reuse technique. To our knowledge, the proposed NVFF enables a fine-grained power gating without significant area overhead. Compared to the best previously known NVFFs, the proposed NVFF succeeds in reducing the area by 4.1× and the energy by 1.5×.

Two optimization strategies for reducing area and energy overheads are also proposed: NVFF placement and pulse duration modulation strategies. We demonstrated the placement strategy on an OpenSPARC T1 core design. Analysis indicates that the placement on the ALU shows an improvement of nearly a factor of 2–18 in terms of area and energy compared to the other modules. We also demonstrated the fault tolerance variation over programs and the adaptive pulse duration strategy for the energy savings.

Author Contributions: J.P. contributed for methodology, validation, data curation, and writing—original draft preparation. Y.U.Y. contributed for writing—review and editing and supervision.

Funding: This work was supported by the National Program for Excellence in Software at Handong Global University (2017-0-00130) funded by the Ministry of Science and ICT.

Conflicts of Interest: The authors declare no conflict of interest.

Abbreviations

The following abbreviations are used in this manuscript:

STT-MRAM	spin torque transfer magnetic RAM
MTJ	magnetic tunnel junction
FF	flip flop
NVFF	non-volatile flip flop
PTM	predictive technology model
MUX	multiplexer
EDK	educational design kit
VCS	verilog compiler and simulator
ALU	arithmetic logic unit
DUE	detected unrecoverable error
SDC	silent data corruption

References

1. Shin, Y.; Seomun, J.; Choi, K.M.; Sakurai, T. Power gating: Circuits, design methodologies, and best practice for standard-cell VLSI designs. *ACM Trans. Design Autom. Electron. Syst.* **2010**, *15*, 28. [CrossRef]
2. Jeong, K.; Kahng, A.B.; Kang, S.; Rosing, T.S.; Strong, R. MAPG: Memory access power gating. In Proceedings of the Conference on Design, Automation and Test in Europe, Dresden, Germany, 12–16 March 2012; pp. 1054–1059.
3. Chiou, D.S.; Chen, S.H.; Chang, S.C.; Yeh, C. Timing driven power gating. In Proceedings of the 43rd Annual Design Automation Conference, San Francisco, CA, USA, 24–28 July 2006; pp. 121–124.
4. Sorin, D.J.; Martin, M.M.; Hill, M.D.; Wood, D.A. SafetyNet: Improving the availability of shared memory multiprocessors with global checkpoint/recovery. In Proceedings of the 29th Annual International Symposium on Computer Architecture, Anchorage, AK, USA, 25–29 May 2002; pp. 123–134.
5. Prvulovic, M.; Zhang, Z.; Torrellas, J. ReVive: Cost-effective architectural support for rollback recovery in shared-memory multiprocessors. In Proceedings of the 29th Annual International Symposium on Computer Architecture, Anchorage, AK, USA, 25–29 May 2002; pp. 111–122.
6. Sato, H.; Yamamoto, T.; Yamanouchi, M.; Ikeda, S.; Fukami, S.; Kinoshita, K.; Matsukura, F.; Kasai, N.; Ohno, H. Comprehensive study of CoFeB-MgO magnetic tunnel junction characteristics with single-and double-interface scaling down to 1X nm. In Proceedings of the 2013 IEEE International Electron Devices Meeting, Washington, DC, USA, 9–11 December 2013; Volume 3, pp. 1–3.
7. Zhang, Y.; Zhao, W.; Lakys, Y.; Klein, J.; Kim, J.V.; Ravelosona, D.; Chappert, C. Compact Modeling of Perpendicular-Anisotropy CoFeB/MgO Magnetic Tunnel Junctions. *IEEE Trans. Electron Devices* **2012**, *59*, 819–826. [CrossRef]
8. Zhao, W.; Cao, Y. Predictive technology model for nano-CMOS design exploration. *ACM J. Emerg. Technol. Comput. Syst.* **2007**, *3*, 1. [CrossRef]
9. Worledge, D.; Hu, G.; Trouilloud, P.; Abraham, D.; Brown, S.; Gaidis, M.; Nowak, J.; O'Sullivan, E.; Robertazzi, R.; Sun, J.; et al. Switching distributions and write reliability of perpendicular spin torque MRAM. In Proceedings of the 2010 International Electron Devices Meeting, San Francisco, CA, USA, 6–8 December 2010.
10. Raychowdhury, A.; Somasekhar, D.; Karnik, T.; De, V. Design space and scalability exploration of 1t-1stt mtj memory arrays in the presence of variability and disturbances. In Proceedings of the 2009 IEEE International Electron Devices Meeting (IEDM), Baltimore, MD, USA, 7–9 December 2009.
11. Diao, Z.; Li, Z.; Wang, S.; Ding, Y.; Panchula, A.; Chen, E.; Wang, L.C.; Huai, Y. Spin-transfer torque switching in magnetic tunnel junctions and spin-transfer torque random access memory. *J. Phys. Condens. Matter* **2007**, *19*, 165209. [CrossRef]
12. Microsystem, S. OpenSPARC™ T1 Microarchitecture Specification. 2009. Available online: https://www.oracle.com/technetwork/systems/opensparc/t1-01-opensparct1-micro-arch-1538959.html.
13. Dujmovic, J.J.; Dujmovic, I. Evolution and evaluation of SPEC benchmarks. *ACM Sigmetrics Perform. Eval. Rev.* **1998**, *26*, 2–9. [CrossRef]

14. Yamamoto, S.; Shuto, Y.; Sugahara, S. Nonvolatile delay flip-flop using spin-transistor architecture with spin transfer torque MTJs for power-gating systems. *Electron. Lett.* **2011**, *47*, 1027–1029. [CrossRef]

15. Jung, Y.; Kim, J.; Ryu, K.; Kim, J.P.; Kang, S.H.; Jung, S.O. An MTJ-based non-volatile flip-flop for high-performance SoC. *Int. J. Circuit Theory Appl.* **2014**, *42*, 394–406. [CrossRef]

16. Ryu, K.; Kim, J.; Jung, J.; Kim, J.P.; Kang, S.H.; Jung, S.O. A magnetic tunnel junction based zero standby leakage current retention flip-flop. *IEEE Trans. Very Large Scale Integr. (VLSI) Syst.* **2012**, *20*, 2044–2053. [CrossRef]

17. Zhao, W.; Belhaire, E.; Chappert, C. Spin-mtj based non-volatile flip-flop. In Proceedings of the 2007 7th IEEE Conference on Nanotechnology (IEEE NANO), Hong Kong, China, 2–5 August 2007; pp. 399–402.

18. Suzuki, D.; Hanyu, T. Magnetic-tunnel-junction based low-energy nonvolatile flip-flop using an area-efficient self-terminated write driver. *J. Appl. Phys.* **2015**, *117*, 17B504. [CrossRef]

19. Kwon, K.W.; Choday, S.H.; Kim, Y.; Fong, X.; Park, S.P.; Roy, K. SHE-NVFF: Spin Hall effect-based nonvolatile flip-flop for power gating architecture. *IEEE Electron Device Lett.* **2014**, *35*, 488–490. [CrossRef]

20. Kimura, H.; Fuchikami, T.; Maramoto, K.; Fujimori, Y.; Izumi, S.; Kawaguchi, H.; Yoshimoto, M. A 2.4 pJ ferroelectric-based non-volatile flip-flop with 10-year data retention capability. In Proceedings of the 2014 IEEE Asian Solid-State Circuits Conference (A-SSCC), KaoHsiung, Taiwan, 10–12 November 2014; pp. 21–24.

21. Lo, C.P.; Chen, W.H.; Wang, Z.; Lee, A.; Hsu, K.H.; Su, F.; King, Y.C.; Lin, C.J.; Liu, Y.; Yang, H.; et al. A ReRAM-based single-NVM nonvolatile flip-flop with reduced stress-time and write-power against wide distribution in write-time by using self-write-termination scheme for nonvolatile processors in IoT era. In Proceedings of the 2016 IEEE International Electron Devices Meeting (IEDM), San Francisco, CA, USA, 3–7 December 2016.

22. Na, T.; Ryu, K.; Kim, J.; Kang, S.H.; Jung, S.O. A comparative study of STT-MTJ based non-volatile flip-flops. In Proceedings of the 2013 IEEE International Symposium on Circuits and Systems (ISCAS), Beijing, China, 19–23 May 2013; pp. 109–112.

23. Goldman, R.; Bartleson, K.; Wood, T.; Kranen, K.; Melikyan, V.; Babayan, E. 32/28nm Educational Design Kit: Capabilities, deployment and future. In Proceedings of the 2013 IEEE Asia Pacific Conference on Postgraduate Research in Microelectronics and Electronics (PrimeAsia), Visakhapatnam, India, 19–21 December 2013; pp. 284–288.

24. Design Compiler; User Guide, Synopsys, 2000. Available online: https://www.synopsys.com/implementation-and-signoff/rtl-synthesis-test/design-compiler-graphical.html.

25. IC Compiler, User Guide, Synopsys, 2013. Available online: https://www.synopsys.com/implementation-and-signoff/physical-implementation/ic-compiler.html.

26. PrimeTime, User Guide version c-2009.06. Synopsys, June 2009. Available online: https://www.synopsys.com/content/dam/synopsys/implementation&signoff/datasheets/primetime-ds.pdf.

27. Borkar, S. Design challenges of technology scaling. *IEEE Micro* **1999**, *19*, 23–29. [CrossRef]

28. Verilog Compiler Simulator Synopsys, 2004. Avaliable online: https://www.synopsys.com/verification/simulation/vcs.html.

 *micromachines*

Article

Memristor Neural Network Training with Clock Synchronous Neuromorphic System

Sumin Jo [1], Wookyung Sun [1], Bokyung Kim [1], Sunhee Kim [2,*], Junhee Park [1,*] and Hyungsoon Shin [1,*]

[1] Department of Electronic and Electrical Engineering, Ewha Womans University, Seoul 03760, Korea; sumin5784@gmail.com (S.J.); wkyungsun@ewha.ac.kr (W.S.); bkkim0505@ewhain.net (B.K.)
[2] Department of System Semiconductor Engineering, Sangmyung University, Cheonan 31066, Korea
* Correspondence: happyshkim@smu.ac.kr (S.K.); junhee.park@ewha.ac.kr (J.P.); hsshin@ewha.ac.kr (H.S.)

Received: 1 May 2019; Accepted: 6 June 2019; Published: 8 June 2019

Abstract: Memristor devices are considered to have the potential to implement unsupervised learning, especially spike timing-dependent plasticity (STDP), in the field of neuromorphic hardware research. In this study, a neuromorphic hardware system for multilayer unsupervised learning was designed, and unsupervised learning was performed with a memristor neural network. We showed that the nonlinear characteristic memristor neural network can be trained by unsupervised learning only with the correlation between inputs and outputs. Moreover, a method to train nonlinear memristor devices in a supervised manner, named guide training, was devised. Memristor devices have a nonlinear characteristic, which makes implementing machine learning algorithms, such as backpropagation, difficult. The guide-training algorithm devised in this paper updates the synaptic weights by only using the correlations between inputs and outputs, and therefore, neither complex mathematical formulas nor computations are required during the training. Thus, it is considered appropriate to train a nonlinear memristor neural network. All training and inference simulations were performed using the designed neuromorphic hardware system. With the system and memristor neural network, the image classification was successfully done using both the Hebbian unsupervised training and guide supervised training methods.

Keywords: neuromorphic system; Hebbian training; guide training; memristor; image classification

1. Introduction

Neuromorphic hardware research has begun to develop new computing architectures [1–6]. From a broad point of view, neuromorphic research has two main streams [6]. One focuses on reproducing the exact biological phenomena that occur in the brain [3,6–10], while the other focuses on the development of a new computing device typically known as a neuromorphic chip. As the neuromorphic chip takes advantage of the biological neural network, it has several features such as massively parallel processing, local memory structure, high integrity, and low power consumption [4,5,11–22]. Neuromorphic hardware is especially efficient in terms of size and power consumption compared to typical Von Neumann architecture computing devices. The main difference between neuromorphic hardware and Von Neumann computers is the memory structure. In the human brain, the neural cell topology is determined by the connections between neurons (i.e., synaptic connectivity). This means that the biological neural network contains a memory device and a computing unit at the same time. On the contrary, the memory device and computing unit are separated in a typical Von Neumann computer. Most of the power is consumed from the data transfer between the memory device and computing unit. This power issue appears in extreme forms in recent data-intensive artificial intelligence (AI) applications.

There are approximately 100–500 trillion synapses in the adult human brain [23]. A memristor can be very densely integrated but remain energy efficient. Therefore, it has considerable potential to physically implement huge and complex network connectivity similar to the human brain [8,24–27]. In addition to this integration property, its I–V characteristic makes the memristor device an appropriate synapse device. It was first suggested and reported that the I–V characteristic is analogous to the behavior of biological synapses in [28]. Due to this device characteristic, memristors have been considered to have the potential to implement spike timing-dependent plasticity (STDP) in hardware. Neural networks can learn by themselves based on given information (i.e., unsupervised learning). STDP is one of the types of unsupervised learning methods in the brain. This concept is in contrast to supervised learning, which is learning processed as the supervisor intended. Supervised learning needs prior information about processing data, and the supervisor needs to label all the data. As the amount of data to process has increased, this labeling process has become more demanding. Natural data is continuously changing, and it is difficult to label all the input data. Thus, unsupervised learning is more appropriate to deal with natural data than supervised learning.

Unsupervised learning has a simpler mechanism than supervised learning. Training a multilayer artificial neural network (ANN), however, requires accurate data control over the entire network (i.e., input/output of the network and input/output of the layers in the network). The systematic implementation of unsupervised learning in a multilayer ANN is essential to develop neuromorphic hardware whose basic function is analogous to the biological neural network, and that can consequently process natural data. From the user-centered point of view, however, with unsupervised learning, it is difficult to determine whether the training has been completed, and the accuracy can be lower than that of supervised learning. On the other hand, users can train the ANN as they intend with supervised learning, it is easier to analyze the training results, and there are many methods to improve accuracy. However, the machine learning algorithms used to train ANNs need computations based on current synaptic weights. In addition to the computations, those computed synaptic weights have to be applied exactly and updated. Extra effort is needed to measure the resistance of a single memristor device in the memristor neural network and to record the entire hysteresis. Only then can the memristor resistance be accurately modified. These accompanied processes compromise the energy efficiency and integrity of memristor neural networks. Therefore, it is hard to realize supervised learning when the ANN consists of a memristor device.

Considering the circumstances of neuromorphic hardware implementation with a memristor ANN, a clock synchronous neuromorphic hardware system for both supervised learning and unsupervised learning was designed in this paper. The designed system was available for a multilayer memristor ANN with an unsupervised learning method. A guide-training algorithm capable of training a nonlinear memristor neural network in a supervised manner without backpropagation was devised in addition to the neuromorphic system. A memristor ANN was adopted as the synapse array for the designed neuromorphic hardware system, and the memristor ANN was trained in both unsupervised learning with the Hebbian training algorithm and supervised learning with the guide-training algorithm.

2. Materials and Methods

2.1. Clock Synchronous Neuromorphic Hardware System

The control of network input/output and neural layer input/output is the most important aspect of implementing unsupervised learning in an ANN. In supervised learning, the network input is completely processed through the network, and then network output is computed. The synaptic weights are then updated according to the backpropagation. On the contrary, unsupervised learning, such as the Hebbian or STDP algorithm, updates the synaptic weights only based on the correlations between inputs and outputs. There is a single pair of input and output in a single-layer ANN, and the correlation between them is clear. However, the situation changes when it comes to the multilayer ANN. Based on the network structure, the layers can be wide or deep, and the number of neurons contained in each layer differs. As a result, the data processing time between layers also differs. For instance, consider the circumstances in a 9-6-3 double-layer ANN (nine input neurons and six output neurons in layer 1 (L1) and six input neurons and three output neurons in layer 2 (L2)). The input to L1 is the network input (NI), and the output of L2 is the corresponding network output (NO). The output of L1 (L1out) is input to L2 (L2input). First, the network input NI1 is applied and the corresponding L1 output (L1out1) is propagated to L2, but the corresponding L2 output (NO1) is not computed yet. What happens if the second network input NI2 is applied again? In the best case, the corresponding L2 output for L2iniput1 is computed, and then the L1 output corresponding to the NI2 (L1out2) is applied to L2. However, in the worst case, the L1 output corresponding to the NI2 (L1out2) is applied to L2 before L2input1 is computed. As a result, the synaptic weights of layer 1 are updated based on the correlations between two different inputs NI1, NI2, and two different outputs L1out1 and L1out2. However, the synaptic weights of layer 2 are updated based on the correlations between two different inputs L2input1, L2input2 and a single merged output of two different inputs. This kind of timing error can result in a learning error, and there are far more possibilities in a deeper ANN than this example case.

The neuromorphic system proposed in this paper divides the data process to a single input data into four steps, and synchronizes the entire layer with the clock: allowing the input to be received by the layer, computing the output, updating the synaptic weights, propagating the output. This system can perform unsupervised learning without timing error. In accordance with the clock signal (Clock, Figure 1a), the four processing steps are performed by word line control logic (WLControl, Figure 1b), bit line control logic (BLControl, Figure 1c), output computation block (WTALogic, Figure 1e), and output propagating logic (OutputSpikeGenerator, Figure 1f). All layers of the ANN simultaneously receive the data, compute the output, update the synaptic weights based on the input and output, and propagate the output to the next layer. At this point, the propagated output from the previous layer is not instantly applied to the next layer. Rather, it is applied to the layer as the next clock signal for receiving input data. The aforementioned timing error can be improved with this neuromorphic hardware system. All training and inference simulations are performed using this designed system. The memristor ANN (Figure 1d) is applied to the neuromorphic hardware system, and both Hebbian training and guide training are performed and analyzed. The detailed methods of Hebbian training and guide training are explained in Sections 2.3 and 2.4, respectively, and the corresponding training and inference results are presented in Sections 3.1 and 3.2, respectively.

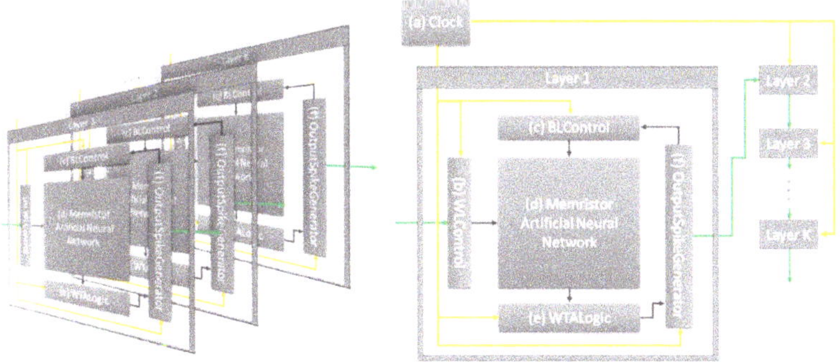

Figure 1. Clock synchronous neuromorphic hardware system. (**a**) System clock; (**b**) Word line control logic; (**c**) Bit line control logic; (**d**) memristor artificial neural network; (**e**) winner-takes-all logic; and (**f**) output spike-generating logic.

2.2. Memristor Neural Network Array

As shown in Figure 2a, the memristor device is connected between the top electrode (Word Line, WL) and the bottom electrode (Bit Line, BL). Input data were applied as voltage to the WL, and the current flowed through the memristor from the WL to the BL according to the input voltage. The winner-takes-all logic (Figure 1e) determined the neuron where the largest current flows. Based on this computation, OutputSpikeGenerator (Figure 1f) propagated output spikes to BLControl (Figure 1c) and the next layer. WLControl and BLControl apply the appropriate voltage to modify memristor conductance according to the learning algorithm.

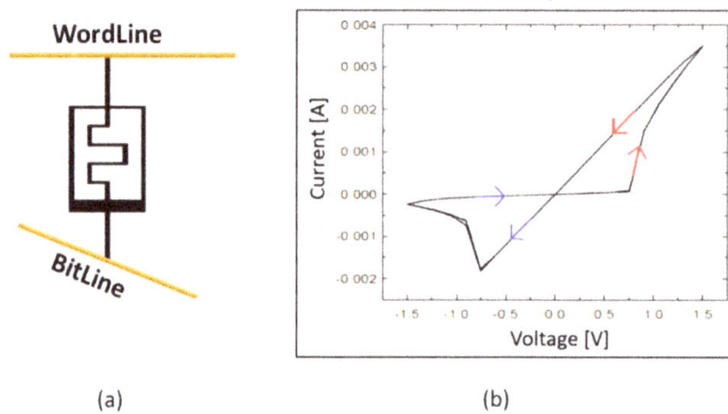

(a) (b)

Figure 2. (**a**) Memristor neural network structure; and (**b**) I–V characteristic of memristor device model used in this paper.

Memristor devices can be modeled using various parameters in the equations, and various models have been reported [29–33]. The memristor device model used in this study refers to References [31] and [32]. The simulation result in Figure 6 of reference [32] is based on the experimental data in [33]. In this study, we used a memristor neural network and a peripheral neuromorphic system instead of a single memristor device. Therefore, the modeling parameters were adjusted to the 1.5 V of system operating voltage while maintaining the device current analogous to the experimental data in [33]. The modeling parameters used are shown in Table 1. The I–V characteristic of the memristor device model used in this paper is shown in Figure 2b. To change the memristor device resistance, a voltage larger than 0.75 V had to be applied across the memristor. To increase the synaptic weight, 1.5 V was applied to the word line for 150 ns and 0 V was applied to the bit line. Conversely, to decrease the synaptic weight, 0 V was applied to the word line and 1.5 V to the bit line. For the Hebbian training, an M × N memristor ANN was implemented by adopting the single-memristor structure for the M inputs and N outputs, and there were N different classification images. For the guide training, an M × 2N memristor ANN was implemented by adopting the double-synapse memristor structure for the M inputs and N outputs, and there were N different classification images.

Table 1. Memristor device modeling parameters.

Symbol	Value	Symbol	Value
a_1	0.05	A_n	6×10^3
a_2	0.05	x_p	0.5
b	0.05	x_n	0.5
V_p	0.75 V	α_p	10
V_n	0.75 V	α_n	10
A_p	6×10^3	x_o	0.5

2.3. Hebbian Training Method

To train the memristor neural network in an unsupervised learning manner, we used the Hebbian training method shown in Table 2. The synaptic connections between input data without output increased. On the contrary, the synaptic connections between output data without input were decreased.

Otherwise, the synaptic weights remained the same. In the table, while 1 represents the existence of input or output, 0 represents the absence of input or output.

Table 2. Hebbian training method.

Input	Output	Modification
1	1	Remained
1	0	Increased
0	1	Decreased
0	0	Remained

2.4. Guide Training Method

The guide-training algorithm literally guides the memristor neural networks to make them perform a cognitive task, and it utilizes the features of both the Hebbian algorithm and a supervised learning algorithm. The Hebbian learning algorithm updates the synaptic weights only according to the correlations between the inputs and outputs. Thus, there are no mathematical formulas or computations to deduce the change in the synaptic weight. One of the significant drawbacks of unsupervised learning is that the learning results are unpredictable. Training results can differ with different initial synaptic weights even if the training data are the same. In contrast, supervised learning algorithms are based on mathematical formulas. Synaptic weights are changed according to these formulas so that the neural network can respond as the supervisor intended. However, the mathematical computations are very complex. The guide-training algorithm proposed in this paper updates synaptic weights according to the correlation information between the input, output, and intended target output determined by the supervisor. It guides the synaptic weights with this information so that the neural network can respond according to the predefined learning pattern. The guide-training algorithm does not compute derivations or integrations as the backpropagation algorithm does. It just compares the correlations between the inputs and the outputs and then determines whether the synaptic weights increase or decrease. This extremely simple learning algorithm is highly suitable for implementing and training nonlinear memristor neural networks.

A double-synapse structure was used for the guide training with two synapses for a single pair of input and output. For the M inputs and N different target classification images, an M × 2N double-synapse memristor array was constructed. M inputs were applied to the rows, and the $\{2 \times j - 1\}$th column and the $\{2 \times j\}$th column were the positive column (PCj) and the negative column (NCj) of output neuron j (Nj) for every N output neuron. The specific guide training method used in this paper is shown in Table 3. While 1 represents the existence of input or output, 0 represents the absence of input or output. K represents the type of input data, and T represents the predefined target output neuron for this input data. Users can define this learning pattern. In this study, only the input data and predefined training pattern were considered. Only the synaptic weights where input existed were updated. For instance, if the target output neuron for the K input image was T, and the i-th input existed, then the positive-column synaptic weight of the target output neuron increased. The negative-column synaptic weight of the target output neuron decreased. The positive-column synaptic weights of the other non-target output neurons decreased, and the negative-column synaptic weights of the other non-target output neurons increased.

Table 3. Guide training method.

Input Image	Predefined Output Neuron	i-th Input	$W (i, 2 \times j - 1)$ $j = T$	$W (i, 2 \times j)$ $j = T$	$W (i, 2 \times j - 1)$ $j \neq T$	$W (i, 2 \times j)$ $j \neq T$
K	T	1	Increased	Decreased	Decreased	Increased
		0	Remained	Remained	Remained	Remained

2.5. Training and Inference Dataset

For every new training trial, the memristor ANN was randomized before training. To train the 3 × 3 T, X, and V letter images (corresponding to Tref, Xref, and Vref in Figure 3a), 135 images were contained in a single training dataset: 45 images of each Tref, Xref, and Vref images. The arrangements of T, X, and V images in a single training dataset were randomized. Thus, if 30 sets of training data were used for a single learning trial, then the arrangements of T, X, and V images in all 30 datasets were different. The original image data and one-pixel flipped images (Figure 3a) of the original image data were used to perform the inference simulations.

To make the memristor ANN learn the 10 × 10 digit images (Figure 3b), 2,708 of the original digit images were used for the training. Three different levels of inference tests were conducted: noise 0% images, noise 3% images, and noise 5% images. These images are shown in Figure 3b–d, respectively. The noise 3% images consisted of images with three randomly chosen pixels flipped. For each digit, 50 different noise images were tested.

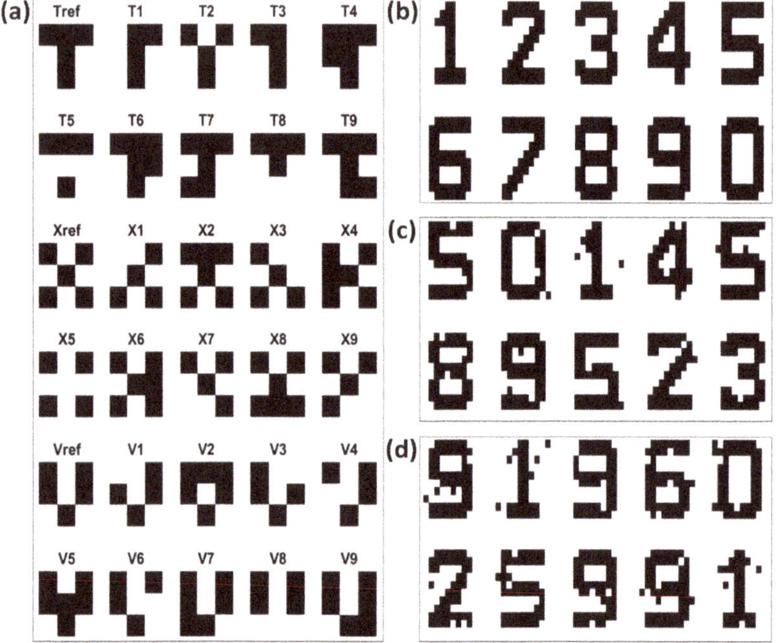

Figure 3. (**a**) 3 × 3 T, X, and V letter images. Tref, Xref, and Vref are original letter images. T1 to T9, X1 to X9, and V1 to V9 are one-pixel flipped noise images of Tref, Xref, and Vref; (**b**) 10 × 10 digit images; (**c**) 3% noise image data of 10 × 10 digit images (three randomly chosen pixels among 100 pixels are flipped); and (**d**) 5% noise image data of 10 × 10 digit images (five randomly chosen pixels among 100 pixels are flipped).

3. Results

3.1. Inference Results after Hebbian Training

Synaptic weights were trained according to the Hebbian training method shown in Table 2. Figure 4a shows the changing pattern of synaptic weights during the Hebbian training. Figure 4b shows the output responses of each output neuron during the Hebbian training. For the initial stage of training, output neuron 1 (N1) did not respond to any input image, output neuron 2 (N2) responded to both T and X images, and output neuron 3 (N3) responded to T, X, and V images. However, as the training continued, N1 trained to the T image, N2 trained to the X image, and N3 trained to the V image. The Tref, Xref, Vref, T1, T3, X1, X3, V1, and V3 images in Figure 3a were used for the inference test after the Hebbian training. Table 4 shows the initial voltages of the memristor ANN used for the training in Figure 4. The memristor ANN was randomized before every new training. The average accuracy of the inference test of Tref, T1, T3, Xref, X1, X3, Vref, V1, and V3 was 100%, 97.62%, 100%, 100%, 95.24%, 97.62%, 100%, 95.24%, and 90.48%, respectively.

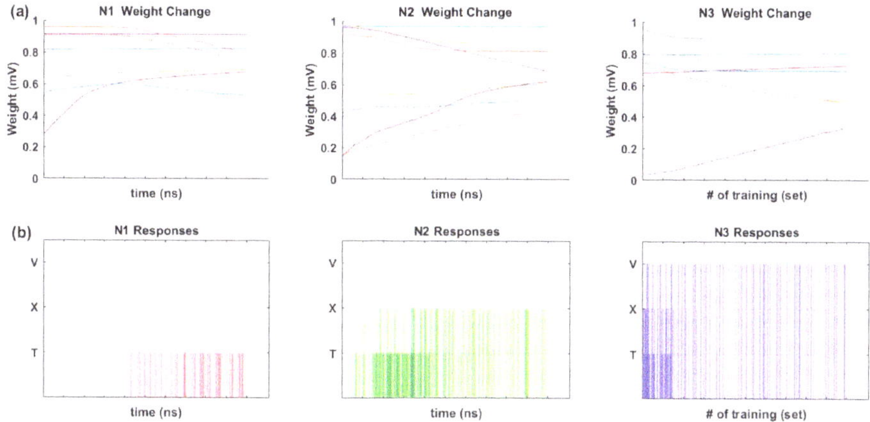

Figure 4. (**a**) Synaptic weight changes during Hebbian training; and (**b**) output neuron responses during Hebbian training.

Table 4. Initial random weight W (i, j) (mV).

i	W (i, 1)	W (i, 2)	W (i, 3)
1	814.7	964.8	792.0
2	905.7	157.6	959.4
3	126.9	970.5	655.7
4	913.3	957.1	35.7
5	632.3	485.3	849.1
6	97.5	800.2	933.9
7	278.4	141.8	678.7
8	546.8	421.7	757.7
9	957.5	915.7	743.1

3.2. Inference Results after Guide Training

3.2.1. Inference Results of 9 × 6 Memristor Neural Network

Output neuron 1 was targeted to learn the T image, output neuron 2 was targeted for the X image, and output neuron 3 was targeted for the V image. The inference test was performed after the 50 sets of guide training with this predefined training pattern. For the inference test, the 30 test images in Figure 3a were used. In the best result case, 10 different T images were responded to by output neuron 1 (N1), 10 different X images were responded to by N2, and 10 different V images were responded to by N3. The test results were the same as the predefined learning pattern, and the error rate was zero. Figure 5a shows the inference test results with error after the 50 sets of guide training. Nine different T images were responded to by N1, and the other images of letters X and V were responded to by N2 and N3, respectively. Thus, the single non-responding case of N1 to a T test image was counted as an error. The average accuracy of T, X, and V letter image classification was 92%, 99%, and 100%, respectively. The changes in the 18 synaptic weights of output neuron 1 are shown in Figure 5b.

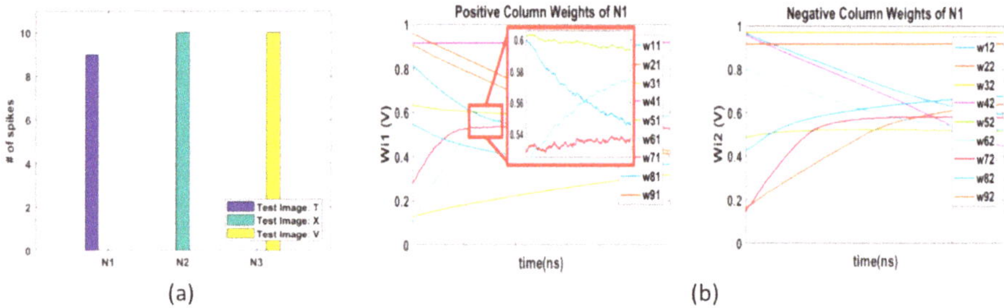

Figure 5. (**a**) 3 × 3 T, X, and V letter image classification test results. Nine different T images were responded to by N1, 10 different X images were responded to by N2, and 10 different V images were responded to by N3. The test results show that the memristor ANN was successfully trained as the predefined learning pattern; (**b**) synaptic weight changes of output neuron 1 during 50 sets of guide training. wij represents the memristor conductance between the i-th top electrode and j-th bottom electrode.

3.2.2. Inference Results of 100 × 20 Memristor Neural Network

In order to train digit images (Figure 3b), 2,708 of the original digit images were used for the training. For the 10 × 10 digit image classification, the learning pattern was predefined as follows: digit 1 was set to output neuron 1, digit 2 was set to output neuron 2, ... , digit 9 was set to output neuron 9, and digit 0 was set to output neuron 10. Thus, we expected the corresponding output of the digit 0 image to be [0, 0, 0, 0, 0, 0, 0, 0, 0, 1]. Figure 6a shows the initial random synaptic weights before the guide training, while Figure 6b shows the trained synaptic weights after the guide training. As shown in Figure 6b, the positive and negative synaptic weights of output neuron 1, Wi1 and Wi2, were successfully trained in the shape of digit 1. Other output neurons were also trained as intended. The average accuracy of the inference test of each noise image in Figure 3b–d is shown in Table 5.

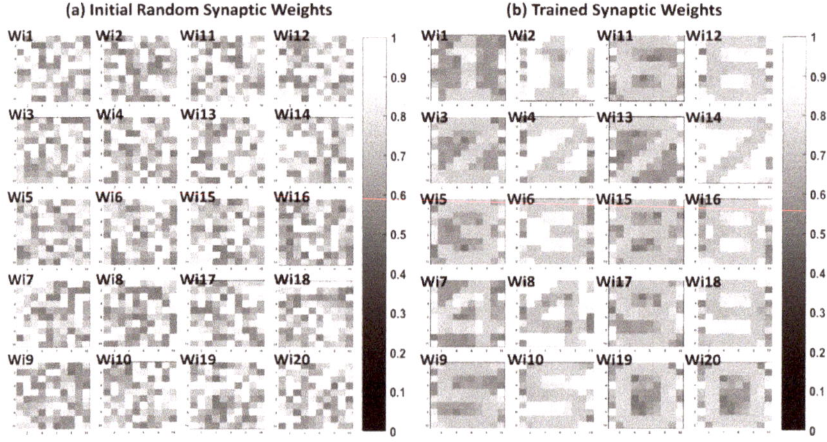

Figure 6. Synaptic weight matrix before and after guide training. A 100 × 20 memristor neural network is utilized for 10 × 10 digit image classification. Each output neuron has positive weights and negative weights. Wi1 represents the positive-column weights of output neuron 1, and Wi2 represents the negative-column weights of output neuron 1. The 100 memristor synapses of 20 columns are shown in the 10 × 10 2D images. (**a**) Initial random synaptic weights. (**b**) Trained synaptic weights after guide training. Trained synaptic weights are trained according to the predefined learning pattern.

Table 5. Average accuracy of inference test of 10 × 10 digit image classification.

Noise %	Digit 0	Digit 1	Digit 2	Digit 3	Digit 4	Digit 5	Digit 6	Digit 7	Digit 8	Digit 9
0	100%	100%	100%	100%	100%	98%	100%	96%	100%	100%
3	100%	100%	97%	96%	100%	91%	95%	100%	84%	100%
5	99%	100%	95%	93%	100%	84%	88%	86%	84%	92%

4. Discussion

In a real on-chip simulation, training has to be conducted with random, nonlinear memristor arrays. In this study, training was conducted on a random memristor array without any initialization process, considering the real-world applications. Unsupervised learning with the Hebbian training method was performed using the proposed neuromorphic hardware system with a nonlinear random memristor ANN, and it successfully classified images. In addition, a new training algorithm optimized to train memristor neural networks was developed. The guide-training algorithm only uses the correlations between the inputs and the outputs like the Hebbian learning algorithm, but the supervisor can configure the training pattern. The training of memristor neural networks poses many intrinsic problems related to the device characteristics. In contrast, the guide-training algorithm proposed in this paper is sufficiently simple to be implemented in an actual circuit and is effective enough to train a memristor neural network. With the guide training algorithm, the 3 × 3 T, X, and V letter image classification and the 10 × 10 digit image classification were successfully conducted with the nonlinear random memristor neural network. The proposed neuromorphic hardware system and guide training algorithm have the potential to train more enhanced memristor ANNs. In the 10 × 10 digit image classification, the digits with large common sections were responded to by corresponding output neurons. The flipped images of digits 5, 6, and 8 were usually responded to by N5, N6, and N8. Moreover, the flipped images of digits 2 and 7 were usually responded to by N2 and N7. Thus, the untrained synapses, which corresponded to the background images, are considered the main contributor to those unintended inference responses. Ongoing studies on the different approaches of the guide-training algorithm are being conducted to overcome these background effects.

Author Contributions: Conceptualization, H.S. and S.J.; software, S.J. and B.K.; formal analysis, S.J. and W.S.; writing—original draft preparation, S.J.; writing—review and editing, H.S.; supervision, S.K., J.P., and H.S.; project administration, W.K., J.P., and S.K.; funding acquisition, W.K. and J.P., and H.S.

Funding: This research was supported by the Basic Science Research Program through the National Research Foundation of Korea (NRF), which was funded by the Ministry of Education, Science, and Technology, grant numbers NRF-2016R1A6A3A11931998 and NRF-2017R1A2B4002540, and RP-Grant 2017 of Ewha Womans University.

References

1. Ananthanarayanan, R.; Esser, S.K.; Simon, H.D.; Modha, D.S. The cat is out of the bag: Cortical simulations with 10^9 neurons, 10^{13} synapses. In Proceedings of the Conference on High Performance Computing Networking, Storage and Analysis—SC '09, Portland, OR, USA, 14–20 November 2009; ACM Press: New York, NY, USA, 2009; pp. 1–12.

2. Merolla, P.A.; Arthur, J.V.; Alvarez-Icaza, R.; Cassidy, A.S.; Sawada, J.; Akopyan, F.; Jackson, B.L.; Imam, N.; Guo, C.; Nakamura, Y.; et al. A million spiking-neuron integrated circuit with a scalable communication network and interface. *Science* **2014**, *345*, 668–673. [CrossRef] [PubMed]

3. Misra, J.; Saha, I. Artificial neural networks in hardware: A survey of two decades of progress. *Neurocomputing* **2010**, *74*, 239–255. [CrossRef]

4. Seo, J.; Brezzo, B.; Liu, Y.; Parker, B.D.; Esser, S.K.; Montoye, R.K.; Rajendran, B.; Tierno, J.A.; Chang, L.; Modha, D.S.; et al. A 45 nm CMOS neuromorphic chip with a scalable architecture for learning in networks of spiking neurons. In Proceedings of the 2011 IEEE Custom Integrated Circuits Conference (CICC), San Jose, CA, USA, 19–21 September 2011; pp. 1–4.

5. Arthur, J.V.; Merolla, P.A.; Akopyan, F.; Alvarez, R.; Cassidy, A.; Chandra, S.; Esser, S.K.; Imam, N.; Risk, W.; Rubin, D.B.D.; et al. Building block of a programmable neuromorphic substrate: A digital neurosynaptic core. In Proceedings of the 2012 International Joint Conference on Neural Networks (IJCNN), Brisbane, Australia, 10–15 June 2012; pp. 1–8.

6. Schuman, C.D.; Potok, T.E.; Patton, R.M.; Birdwell, J.D.; Dean, M.E.; Rose, G.S.; Plank, J.S. A Survey of Neuromorphic Computing and Neural Networks in Hardware. *arXiv* **2017**, arXiv:1705.06963.

7. Walter, F.; Röhrbein, F.; Knoll, A. Neuromorphic implementations of neurobiological learning algorithms for spiking neural networks. *Neural Netw.* **2015**, *72*, 152–167. [CrossRef] [PubMed]

8. Afifi, A.; Ayatollahi, A.; Raissi, F. Implementation of biologically plausible spiking neural network models on the memristor crossbar-based CMOS/nano circuits. In Proceedings of the 2009 European Conference on Circuit Theory and Design, Antalya, Turkey, 23–27 August 2009; pp. 563–566.

9. de Garis, H.; Shuo, C.; Ruiting, L. A world survey of artificial brain projects, Part I: Large-scale brain simulations. *Neurocomputing* **2010**, *74*, 3–29. [CrossRef]

10. Mayr, C.; Noack, M.; Partzsch, J.; Schuffny, R. Replicating experimental spike and rate based neural learning in CMOS. In Proceedings of the 2010 IEEE International Symposium on Circuits and Systems, Paris, France, 30 May–2 June 2010; pp. 105–108.

11. Akopyan, F.; Sawada, J.; Cassidy, A.; Alvarez-Icaza, R.; Arthur, J.; Merolla, P.; Imam, N.; Nakamura, Y.; Datta, P.; Nam, G.-J.; et al. TrueNorth: Design and Tool Flow of a 65 mW 1 Million Neuron Programmable Neurosynaptic Chip. *IEEE Trans. Comput. Des. Integr. Circuits Syst.* **2015**, *34*, 1537–1557. [CrossRef]

12. Benjamin, B.V.; Gao, P.; McQuinn, E.; Choudhary, S.; Chandrasekaran, A.R.; Bussat, J.-M.; Alvarez-Icaza, R.; Arthur, J.V.; Merolla, P.A.; Boahen, K. Neurogrid: A Mixed-Analog-Digital Multichip System for Large-Scale Neural Simulations. *Proc. IEEE* **2014**, *102*, 699–716. [CrossRef]

13. Painkras, E.; Plana, L.A.; Garside, J.; Temple, S.; Galluppi, F.; Patterson, C.; Lester, D.R.; Brown, A.D.; Furber, S.B. SpiNNaker: A 1-W 18-Core System-on-Chip for Massively-Parallel Neural Network Simulation. *IEEE J. Solid-State Circuits* **2013**, *48*, 1943–1953. [CrossRef]

14. Rachmuth, G.; Shouval, H.Z.; Bear, M.F.; Poon, C.-S. A biophysically-based neuromorphic model of spike rate- and timing-dependent plasticity. *Proc. Natl. Acad. Sci. USA* **2011**, *108*, E1266–E1274. [CrossRef] [PubMed]

15. Kim, Y.; Zhang, Y.; Li, P. A digital neuromorphic VLSI architecture with memristor crossbar synaptic array for machine learning. In Proceedings of the 2012 IEEE International SOC Conference, Niagara Falls, NY, USA, 12–14 September 2012; pp. 328–333.

16. Cassidy, A.S.; Alvarez-Icaza, R.; Akopyan, F.; Sawada, J.; Arthur, J.V.; Merolla, P.A.; Datta, P.; Tallada, M.G.; Taba, B.; Andreopoulos, A.; et al. Real-Time Scalable Cortical Computing at 46 Giga-Synaptic OPS/Watt with ~100× Speedup in Time-to-Solution and ~100,000× Reduction in Energy-to-Solution. In Proceedings of the SC14: International Conference for High Performance Computing, Networking, Storage and Analysis, New Orleans, LA, USA, 16–21 November 2014; pp. 27–38.

17. Esser, S.K.; Merolla, P.A.; Arthur, J.V.; Cassidy, A.S.; Appuswamy, R.; Andreopoulos, A.; Berg, D.J.; McKinstry, J.L.; Melano, T.; Barch, D.R.; et al. Convolutional networks for fast, energy-efficient neuromorphic computing. *Proc. Natl. Acad. Sci. USA* **2016**, *113*, 11441–11446. [CrossRef] [PubMed]

18. Merolla, P.; Arthur, J.; Akopyan, F.; Imam, N.; Manohar, R.; Modha, D.S. A digital neurosynaptic core using embedded crossbar memory with 45pJ per spike in 45 nm. In Proceedings of the 2011 IEEE Custom Integrated Circuits Conference (CICC), San Jose, CA, USA, 19–21 September 2011; pp. 1–4.

19. Esser, S.K.; Andreopoulos, A.; Appuswamy, R.; Datta, P.; Barch, D.; Amir, A.; Arthur, J.; Cassidy, A.; Flickner, M.; Merolla, P.; et al. Cognitive computing systems: Algorithms and applications for networks of neurosynaptic cores. In Proceedings of the 2013 International Joint Conference on Neural Networks (IJCNN), Dallas, TX, USA, 4–9 August 2013; pp. 1–10.

20. Cassidy, A.S.; Merolla, P.; Arthur, J.V.; Esser, S.K.; Jackson, B.; Alvarez-Icaza, R.; Datta, P.; Sawada, J.; Wong, T.M.; Feldman, V.; et al. Cognitive computing building block: A versatile and efficient digital neuron model for neurosynaptic cores. In Proceedings of the 2013 International Joint Conference on Neural Networks (IJCNN), Dallas, TX, USA, 4–9 August 2013; pp. 1–10.

21. Amir, A.; Datta, P.; Risk, W.P.; Cassidy, A.S.; Kusnitz, J.A.; Esser, S.K.; Andreopoulos, A.; Wong, T.M.; Flickner, M.; Alvarez-Icaza, R.; et al. Cognitive computing programming paradigm: A Corelet Language for composing networks of neurosynaptic cores. In Proceedings of the 2013 International Joint Conference on Neural Networks (IJCNN), Dallas, TX, USA, 4–9 August 2013; pp. 1–10.

22. Preissl, R.; Wong, T.M.; Datta, P.; Flickner, M.; Singh, R.; Esser, S.K.; Risk, W.P.; Simon, H.D.; Modha, D.S. Compass: A scalable simulator for an architecture for cognitive computing. In Proceedings of the 2012 International Conference for High Performance Computing, Networking, Storage and Analysis, Salt Lake City, UT, USA, 10–16 November 2012; pp. 1–11.

23. Drachman, D.A. Do we have brain to spare? *Neurology* **2005**, *64*, 2004–2005. [CrossRef] [PubMed]

24. Jo, S.H.; Chang, T.; Ebong, I.; Bhadviya, B.B.; Mazumder, P.; Lu, W. Nanoscale Memristor Device as Synapse in Neuromorphic Systems. *Nano Lett.* **2010**, *10*, 1297–1301. [CrossRef] [PubMed]

25. Kim, K.-H.; Gaba, S.; Wheeler, D.; Cruz-Albrecht, J.M.; Hussain, T.; Srinivasa, N.; Lu, W. A Functional Hybrid Memristor Crossbar-Array/CMOS System for Data Storage and Neuromorphic Applications. *Nano Lett.* **2012**, *12*, 389–395. [CrossRef] [PubMed]

26. Wang, H.; Li, H.; Pino, R.E. Memristor-based synapse design and training scheme for neuromorphic computing architecture. In Proceedings of the 2012 International Joint Conference on Neural Networks (IJCNN), Brisbane, Australia, 10–15 June 2012; pp. 1–5.

27. Indiveri, G.; Linares-Barranco, B.; Legenstein, R.; Deligeorgis, G.; Prodromakis, T. Integration of nanoscale memristor synapses in neuromorphic computing architectures. *Nanotechnology* **2013**, *24*, 384010. [CrossRef] [PubMed]

28. Strukov, D.B.; Snider, G.S.; Stewart, D.R.; Williams, R.S. The missing memristor found. *Nature* **2008**, *453*, 80–83. [CrossRef] [PubMed]

29. Pershin, Y.V.; Martinez-Rincon, J.; Di Ventra, M. Memory Circuit Elements: From Systems to Applications. *J. Comput. Theor. Nanosci.* **2011**, *8*, 441–448. [CrossRef]

30. Amirsoleimani, A.; Shamsi, J.; Ahmadi, M.; Ahmadi, A.; Alirezaee, S.; Mohammadi, K.; Karami, M.A.; Yakopcic, C.; Kavehei, O.; Al-Sarawi, S. Accurate charge transport model for nanoionic memristive devices. *Microelectron. J.* **2017**, *65*, 49–57. [CrossRef]

31. Yakopcic, C.; Taha, T.M.; Subramanyam, G.; Pino, R.E. Memristor SPICE model and crossbar simulation based on devices with nanosecond switching time. In Proceedings of the 2013 International Joint Conference on Neural Networks (IJCNN), Dallas, TX, USA, 4–9 August 2013; pp. 1–7.

32. Yakopcic, C.; Taha, T.M.; Subramanyam, G.; Pino, R.E.; Rogers, S. A Memristor Device Model. *IEEE Electron Device Lett.* **2011**, *32*, 1436–1438. [CrossRef]

33. Oblea, A.S.; Timilsina, A.; Moore, D.; Campbell, K.A. Silver chalcogenide based memristor devices. In Proceedings of the 2010 International Joint Conference on Neural Networks (IJCNN), Barcelona, Spain, 18–23 July 2010; pp. 1–3.

Article

Tight Evaluation of Real-Time Task Schedulability for Processor's DVS and Nonvolatile Memory Allocation

Sunhwa A. Nam [1], Kyungwoon Cho [2] and Hyokyung Bahn [1,*]

[1]   Department of Computer Engineering, Ewha University, Seoul 03760, Korea; sunhwa.nam@gmail.com
[2]   Embedded Software Research Center, Ewha University, Seoul 03760, Korea; cezanne@ewha.ac.kr
*   Correspondence: bahn@ewha.ac.kr; Tel.: +82-2-3277-2368

Received: 1 May 2019; Accepted: 1 June 2019; Published: 3 June 2019

Abstract: A power-saving approach for real-time systems that combines processor voltage scaling and task placement in hybrid memory is presented. The proposed approach incorporates the task's memory placement problem between the DRAM (dynamic random access memory) and NVRAM (nonvolatile random access memory) into the task model of the processor's voltage scaling and adopts power-saving techniques for processor and memory selectively without violating the deadline constraints. Unlike previous work, our model tightly evaluates the worst-case execution time of a task, considering the time delay that may overlap between the processor and memory, thereby reducing the power consumption of real-time systems by 18–88%.

Keywords: real-time system; dynamic voltage scaling; task placement; low-power technique; nonvolatile memory

---

1. Introduction

As IoT (internet-of-things) technologies grow rapidly for emerging applications such as smart living and health care, reducing power consumption in battery-based IoT devices becomes an important issue. An IoT device is a type of real-time system, of which, power-saving has been widely studied in terms of the processor's dynamic voltage scaling (DVS). DVS lowers the supplied voltage of a processor when a load of tasks is less than the processor's full capacity, thereby saving power consumption without violating the deadline constraints of real-time tasks. Although the execution time will increase due to the lowered supplied voltage, it would spend less power, as the power consumption in the CMOS (complementary metal-oxide semiconductor) digital circuits is proportional to the square of the supplied voltage [1].

Meanwhile, recent research has shown that memory subsystems are reaching a significant portion of power consumption in real-time embedded systems [2]. Such tremendous power consumption results mainly from the refresh operations of DRAM (dynamic random access memory) [2,3]. As DRAM is a volatile medium, it requires continuous power recharge in order to retain its data even in idle states. This article shows that the power consumption of real-time systems can be further reduced by combining a processor's voltage scaling with hybrid memory technologies, consisting of DRAM and NVRAM.

NVRAM (nonvolatile random access memory) technologies have emerged as an attempt of saving the power consumption of DRAM, as NVRAM does not need refresh operations [3]. NVRAM is byte-addressable memory similar to DRAM but it is better than DRAM in terms of energy-consumption and scalability. Thus, NVRAM is expected to be used as a main memory medium like DRAM in the not too far future [3–5]. Unfortunately, NVRAM has two critical weaknesses that prevent the total substitution of DRAM memory. First, the number of write operations allowed for each NVRAM cell is limited. For example, the current write endurance of PRAM (phase-change memory), a kind of representative NVRAM media, is known to be about 10^7–10^8 [3,6]. The second drawback is that the access time of NVRAM is expected to be slower than that of DRAM [5,7,8].

Despite these limitations, the prospect of NVRAM is still bright. One way of coping with the slow access latency and the write endurance problem of NVRAM is to adopt DRAM along with NVRAM [5,7]. This can hide the slow performance of NVRAM and also increase the lifespan of NVRAM. Two different memory architectures that comprise DRAM and NVRAM can be considered. The first architecture, depicted in Figure 1a, uses DRAM as an upper-level memory of NVRAM, which we call, the hierarchical memory architecture. The other memory architecture, depicted in Figure 1b, presents both DRAM and NVRAM at the same main memory level, managing them together under a single physical address space [5]. We call this architecture the hybrid memory architecture. In general-purpose time-sharing systems, the hierarchical memory architecture can improve the performance of virtual memory systems, as changing the backing store from slow HDD (hard disk drive) to fast NVRAM significantly reduces the page fault handling latency. However, as we focus on real-time systems, virtual memory is difficult to use, since page fault situations cannot be predicted beforehand, making the deadline guaranteed service difficult. This implies that the size of the DRAM should be large enough not to incur unexpected page faults, which is not fit for our target system, as we focus on reducing the use of DRAM for saving power consumption. Thus, we adopt the hybrid memory architecture and determine the location of tasks between DRAM and NVRAM in order to satisfy the deadline constraints by estimating the memory access latency beforehand.

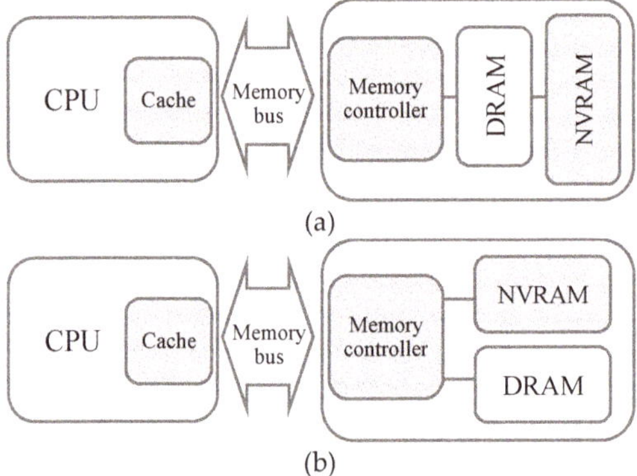

Figure 1. Architecture of the proposed system. (**a**) Hierarchical memory architecture, (**b**) hybrid memory architecture.

Although a task in NVRAM needs more time to be accessed, we can expect that an NVRAM resident task is still likely to be schedulable if it is executed under a low voltage mode of a processor. Our aim is to load tasks on NVRAM if it does not violate the deadline of real-time tasks, thereby reducing the power consumption further. To do so, we incorporate the task's memory placement problem into the processor voltage scaling and evaluate the effectiveness of the unified approach. Simulation experiments show that our technique reduces the power consumption of real-time systems by 18–88%.

2. The Proposed Policy

Let $\Gamma = \{\tau_1, \tau_2, \ldots, \tau_n\}$ be the set of n independent tasks in a real-time system, which has a processor capable of dynamic voltage scaling, and main memory consisting of DRAM and NVRAM as shown in Figure 1b. Each task τ_i is characterized by $<n_i, CPI_i, p_i, s_i>$, where n_i is the number of instructions to be executed, CPI_i is the clock cycles per instruction for τ_i, p_i is the period of τ_i, and s_i is the size of τ_i's memory reference stream during its execution. By considering common assumptions used in previous works [1], we make five assumptions for our system model.

A1. The size of the DRAM is large enough to accommodate entire task sets, but the power is turned off for the part of the DRAM where tasks are not loaded;
A2. Each task is executed independently and does not affect others;
A3. Context switch overhead (the overhead of switching a processor from one task to another) and voltage switching overhead (the overhead of switching the voltage mode of a processor from one to another) are negligible;
A4. The frequency of a processor is set to an appropriate level as the voltage supply is adjusted;
A5. We consider periodic tasks, and thus the period of a task implicitly determines the deadline of the task.

In our task model, the worst-case execution time (WCET) of a task can be determined based on the number of instructions to be executed in the processor and the memory access latency of the task. As modern embedded processors have an on-chip cache, main memory is accessed only upon a cache miss. Thus, memory delay caused by NVRAM also occurs only when a requested block is not in the on-chip cache. Once a block is loaded on the cache, then accessing a part of data within the block does not incur memory accesses. Let c be the cache block size and s_i be the total size of memory reference stream in task τ_i. Then, in the worst case, the number of memory accesses can be represented as s_i/c.

In our task model, WCET of a task is decided by the slower time component of executing instructions and accessing memory with the given voltage mode and the memory type. Specifically, WCET t_i of a task τ_i with the processor's voltage level v_i and the memory type m_i is defined as:

$$t_i = \max\{t_{i,\mathrm{cpu}}\,(v_i,\,n_i),\,t_{i,\mathrm{mem}}\,(m_i,\,s_i)\} \tag{1}$$

where $t_{i,\mathrm{cpu}}\,(v_i,\,n_i)$ is the execution time of n_i instructions in the processor with the voltage level of v_i and $t_{i,\mathrm{mem}}\,(m_i,\,s_i)$ is the memory access time of task τ_i with the memory type m_i and the size of reference stream s_i, which can be subsequently defined as follows:

$$t_{i,\mathrm{cpu}}\,(v_i,\,n_i) = (n_i/v_i) \times CPI_i \times LC \tag{2}$$

$$t_{i,mem}\,(m_i,\,s_i) = (s_i/c) \times LT(m_i) \tag{3}$$

where CPI_i represents clock cycles per instruction for the task τ_i, LC is the cycle time, c is the cache block size, and $LT(m_i)$ is the memory access latency of the memory type m_i. Note that the voltage level v_i is set to 1 for the default voltage mode, and becomes less than 1 as the processor is set to a low voltage mode.

The schedulability of a real-time task set Γ is tested by the utilization U of a processor as follows:

$$U = \sum_{i=1}^{n} \frac{t_i}{p_i} \leq 1 \tag{4}$$

We use the earliest deadline first (EDF) scheduling algorithm as it is known to perform scheduling without deadline misses, provided that there exist any feasible schedules on that task set [1]. Now, let us take a look at an example task set consisting of three tasks τ_1, τ_2, and τ_3, whose worst-case execution times t_1, t_2, and t_3 are 2, 1, and 1, respectively, under the default setting (i.e., normal voltage mode and DRAM only placement), and their periods are 8, 10, and 14, respectively. The schedulability of the task set is tested by calculating the utilization of the tasks τ_1, τ_2, and τ_3, and adding up them i.e., $U = 2/8 + 1/10 + 1/14 = 0.421$. As the total utilization is less than 1, the task set is schedulable. Figure 2a shows the scheduling result for the task set with the EDF. Although the task set is schedulable, idle intervals reach up to 50% of the total possible working time of the processor. This inefficiency can be relieved by lowering the processor's voltage for some idle intervals. For example, if two low voltage levels of 0.5 and 0.25 are applied for tasks τ_2 and τ_3, respectively, t_2 and t_3 will be 2 and 4, respectively. Accordingly, the utilization of the processor is increased to $U = 2/8 + 2/10 + 4/14 = 0.736$, which is still less than 1 and thus schedulable. Also, if we locate τ_3 in NVRAM whose access latency is twice that of the DRAM, one may think that t_3 will be 8, and thus it is not schedulable as $U = 2/8 + 2/10 + 8/14 = 1.021 > 1$. However, we tightly model WCET (worst case execution time) considering the overlapped time delay between the processor and memory, as shown in Equation (1), and thus t_3 is still 4. Therefore, in our model, the utilization of the processor by applying both DVS and NVRAM becomes less than 1, still being schedulable. Figure 2b shows the scheduling result with our model when the aforementioned voltage scaling and memory mapping is adopted. As we see, idle intervals are decreased significantly when compared with the result in Figure 2a.

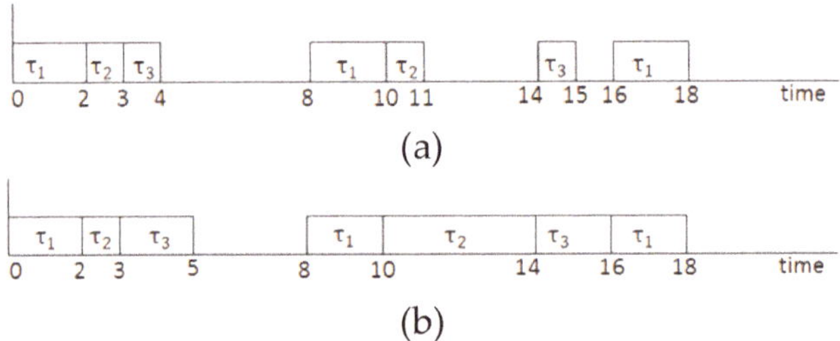

Figure 2. Comparison of the scheduled task set. (**a**) Scheduling result by original earliest deadline first (EDF), (**b**) scheduling result by the proposed approach.

As we deal with hard real-time systems, we assume that task scheduling is performed beforehand (i.e., off-line scheduling) and the scheduling does not change during the execution of the tasks. That is, the schedulability test is performed with the given resources (the voltage modes and memory types) at the design phase and the system resources are determined based on the schedulability test results, not to miss the deadlines of all tasks. This is a typical procedure for real-time task scheduling, and we extend it for memory placement. Note that as traditional real-time systems do not use virtual memory swapping due to the unpredictable page fault handling I/O (input/output) latency, the full address space of a task is loaded on the physical memory once it starts its execution. Thus, we also assume that the memory footprint of a task is determined at the scheduling phase, which is set to the maximum value for satisfying the deadline constraints in the worst case.

Algorithm 1 depicts the pseudo-code of our task setting and scheduling, of which the objective function is the maximization of *power_saving* with the constraint of U less than 1, implying that there are no deadline misses in the task set. Our problem can be modeled similar to the 0/1 knapsack problem. Thus, we solve the problem based on dynamic programming, which is one of the most efficient techniques to solve the 0/1 knapsack problem [9]. The algorithm tries to lower the power mode of each task i ($1 \le i \le n$) without exceeding the given utilization of each step. The state of each task with the utilization 1 will be our final solution. One can refer to the approximation algorithm of 0/1 knapsack problem for more details [9]. To decide the increment of utilization, we performed empirical analysis and found that the increment of 0.1 was appropriate in our case as the results were not sensitive when it became less than that value.

Algorithm 1 Task setting and scheduling

Input: n, number of tasks; $power\_saving_{(i, U)}$, maximum power savings for tasks 1, 2, ..., i with utilization less than U; $power\_saving_i$, power savings obtained by adopting task i to low power mode; U_i, increased utilization by adopting task i to low power mode
Output: Φ, a schedule of all tasks

for i is 1 to n **do**
    $power\_saving_{(i, 0)} \leftarrow 0$;
end for
    for U is 0.0 to 1.0 by 0.1 **do**
    $power\_saving_{(0, U)} \leftarrow 0$;
end for
for i is 1 to n **do**
    for U is 0.0 to 1.0 by 0.1 **do**
    if $U_i > U$
        $power\_saving_{(i, U)} \leftarrow power\_saving_{(i-1, U)}$;
    else
        $power\_saving_{(i, U)} \leftarrow \max\{power\_saving_{(i-1, U)},$
            $power\_saving_{(i-1, U - Ui)} + power\_saving_i \}$;
        end if
    end for
    end for
set processor and memory states based on $power\_saving_{(n, 1)}$;
schedule task set via EDF;

---

3. Performance Evaluations

We compare our technique, called DVS-HM (dynamic voltage scaling with hybrid memory), with DVS-DRAM, HM (hybrid memory), and DRAM, which operate as follows.

- DVS-DRAM: This algorithm uses the processor's dynamic voltage scaling, similar to DVS-HM, but does not use NVRAM and all tasks reside in DRAM;
- HM: This algorithm does not use the processor's dynamic voltage scaling, but uses hybrid memory consisting of DRAM and NVRAM, and places tasks in NVRAM if it is still schedulable;
- DRAM: This is a baseline condition that does not adopt either the processor's dynamic voltage scaling or hybrid memory technologies. That is, the processor is executed with its full voltage mode and all tasks reside in the DRAM.

The sizes of the DRAM and NVRAM are equally set to accommodate the entire task set. Table 1 shows the access latency and the power consumption of the DRAM and PRAM (phase-change random access memory), which is a type of NVRAM we experimented with. In theoretical aspects, there is no limitation in the level of the processor's operating modes. However, as DVS-supported processors usually allow a very limited number of operating modes for practical reasons, we also allow four voltage levels of 1, 0.5, 0.25, and 0.125.

Table 1. DRAM (dynamic random access memory) and NVRAM (nonvolatile random access memory) characteristics.

Characteristics	DRAM	PRAM
Read latency	50 (ns)	100 (ns)
Write latency	50 (ns)	350 (ns)

Table 1. *Cont.*

Characteristics	DRAM	PRAM
Read energy	0.1 (nJ/bit)	0.2 (nJ/bit)
Write energy	0.1 (nJ/bit)	1.0 (nJ/bit)
Idle power	1 (W/GB)	0.1 (W/GB)

Power consumption in the memory system can be divided into active and idle power consumption. Idle power consumption includes the leakage power and refresh power. The leakage power is power consumed even when the memory is idle and the leakage power of NVRAM is negligible compared to that of DRAM. DRAM memory cells store data in small capacitors that lose their charge over time and must be recharged. This process is called refresh. Regardless of the read and write operations, DRAM consumes considerable refresh power to sustain refresh cycles to retain its data. However, this is not required in NVRAM because of its non-volatile characteristics. Active power consumption, on the other hand, refers to the power dissipated when data is being read and written. In our experiments, power consumptions of processor and memory are separately evaluated and then accumulated. The total power consumption $Power_{total}$ is evaluated as:

$$Power_{total} = Power_{cpu} + Power_{mem} \tag{5}$$

where:

$$Power_{cpu} = \sum_{cpu\_mode} \{Unit\_Power_{cpu\_mode} \times Cycles_{cpu\_mode}\} \tag{6}$$

and:

$$Power_{mem} = \sum_{mem\_type} \{Unit\_Active\_Power_{mem\_type} \times Active\_Cycles_{mem\_type} + Unit\_Idle\_Power_{mem\_type} \times Idle\_Cycles_{mem\_type}\}. \tag{7}$$

$Unit\_Power_{cpu\_mode}$ is the unit power consumption per cycle for the given CPU mode and $Cycles_{cpu\_mode}$ is the number of CPU cycles with the given CPU mode. $Unit\_Active\_Power_{mem\_type}$ is the active power per cycle for accessing the given memory type, $Active\_Cycles_{mem\_type}$ is the number of memory cycles for accessing the given memory type. $Unit\_Idle\_Power_{mem\_type}$ is the static power per cycle, including both the leakage power and refresh power for the given memory type, and $Idle\_Cycles_{mem\_type}$ is the number of memory cycles for the idle period for the given memory type.

We performed experiments under both synthetic and realistic workload conditions. In the synthetic workload, we created 10 task sets varying the load of tasks for a given processor capacity, similar to previous studies [10]. In the case of the realistic workload, we used two workloads, the robotic highway safety marker (RSM) workload [11] and the IoT workload [12]. Tables 2 and 3 list the workload configurations for the RSM and IoT workloads that we experimented with [11,12].

Table 2. Robotic highway safety marker (RSM) task set parameters. WCET = worst case execution time; PID = process id.

Task	Period	WCET
Serial	7.8125 ms	100 µs
Length	7.8125 ms	1 ms
Way Point	23.4375 ms	2.5 ms
Encoder	23.4375 ms	350 µs
PID	23.4375 ms	1.06 ms
Motor	23.4375 ms	250 µs

Table 3. Internet-of-things (IoT) task set parameters. WCET = worst case execution time; GUI = graphical user interface.

Task	Period	WCET
Sense Temperature	100 ms	10 µs
Send data to server	1 min	6 ms
Sense Vibration	10 ms	600 µs
Compress and send	1 s	7.5 ms
Get info. & calc.	10 ms	1 ms
Control machine	10 ms	1 ms
Update GUI	1 s	20 ms

3.1. Experiments with Synthetic Workloads

Figure 3 shows the power consumption in processor and memory for the four schemes when the synthetic workload is used. As shown in Figure 3a, DVS-DRAM and DVS-HM, which adopt voltage scaling, similarly saved a substantial amount of processor's power consumption. HM and DRAM, which do not use DVS, showed a relatively higher power consumption than DVS-DRAM and DVS-HM, although the gap became small in some cases. In particular, DVS was less effective as a task set's load approached the full capacity of a processor. This was because the chance of utilizing idle periods of a processor by DVS becomes difficult in such cases. Note that the load of a task set became heavy as the task set number increases in our cases.

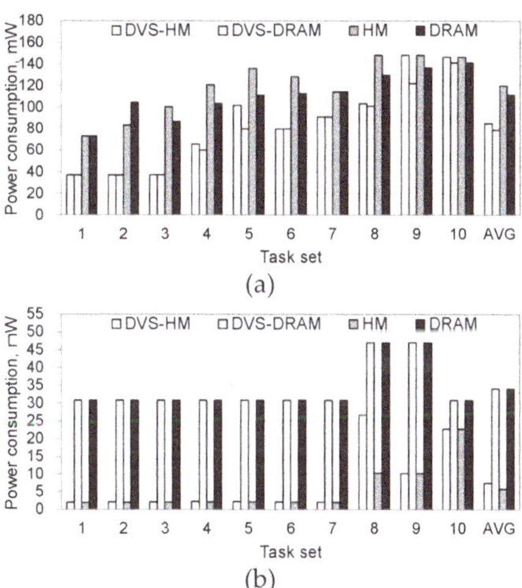

Figure 3. Processor and memory power consumption under synthetic workloads. (**a**) Power consumption in processor, (**b**) power consumption in memory. DVS-HM = dynamic voltage scaling with hybrid memory; DVS-DRAM = dynamic voltage scaling with dynamic random access memory; HM = hybrid memory; DRAM = dynamic random access memory.

Figure 3b shows the power consumption in the memory. The DVS-HM and HM, which use NVRAM along with DRAM, consumed less energy than the DVS-DRAM and DRAM, which only use DRAM. This is because the idle power of NVRAM is close to zero, and thus the reduced size of DRAM—by adopting NVRAM—saved the refresh power of the DRAM. However, as the latency of NVRAM is longer than that of DRAM, executing a task in NVRAM may increase the execution time in the processor, possibly increasing the processor's power consumption. However, as shown in Figure 3a, such a phenomenon happened only in HM and it disappeared in DVS-HM, which uses voltage scaling along with hybrid memory placement. When comparing the DVS-DRAM and DVS-HM, we can see that adopting NVRAM does not increase the processor's power consumption if DVS is used. This is because power-saving can be maximized by executing a processor in a low voltage mode when the task is located in NVRAM.

Figure 4a shows the total energy consumption by adding up the consumed energy in processor and memory. DVS-HM saved the energy consumption of the DRAM, DVS-DRAM, and HM by 36%, 18%, and 28%, respectively. Figure 4b,c separately show the active and idle power consumptions. Although the DVS-HM performed worse than the DVS-DRAM, in terms of active power consumption, it performed the best in idle power consumption, leading to the minimized total power consumption. Figure 5 shows the processor's utilization. As we see, DVS-HM showed the highest utilization in all cases and was close to 100% in some cases.

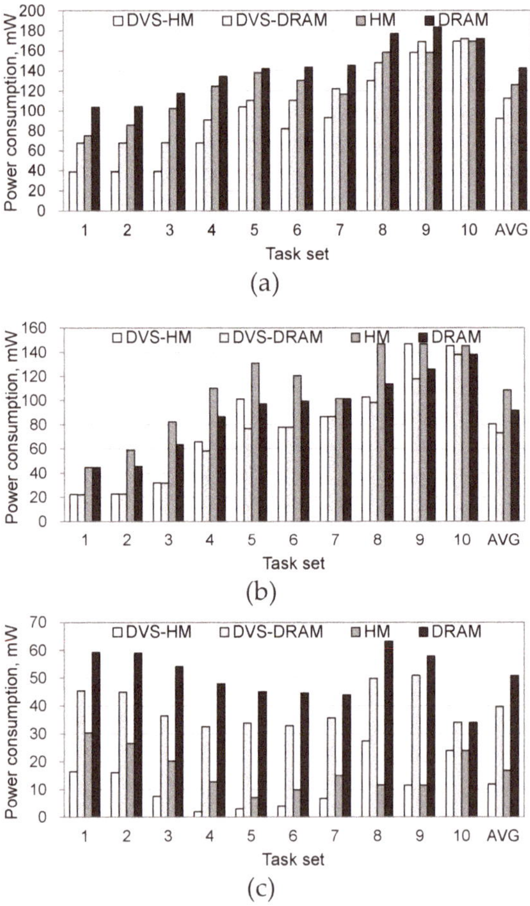

Figure 4. Power consumptions under synthetic workloads. (**a**) Total power consumptions, (**b**) active power consumptions, (**c**) idle power consumptions.

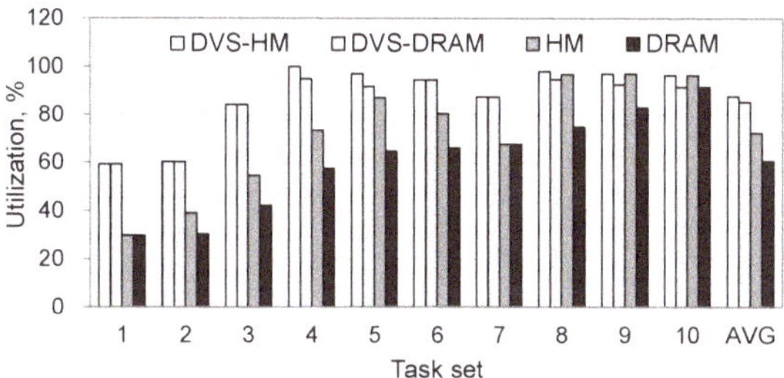

Figure 5. Processor utilizations under synthetic workloads.

3.2. Experiments with Realistic Workloads

To see the effectiveness of the proposed algorithm in more realistic situations, we performed additional experiments under two realistic workload conditions, a robotic highway safety marker (RSM) workload [11] and an IoT workload [12]. Similar to the synthetic workload cases, we show that the proposed algorithm is effective in increasing the processor's utilization and decreasing the power consumption. Figure 6 shows the power consumptions in processor and memory separately when the RSM and IoT workloads are used. For both workloads, power consumption in the processor was significantly reduced when the DVS was used. Specifically, DVS-HM and DVS-DRAM saved the processor's power consumption by 86–88% in comparison with HM and DRAM, as shown in Figure 6a.

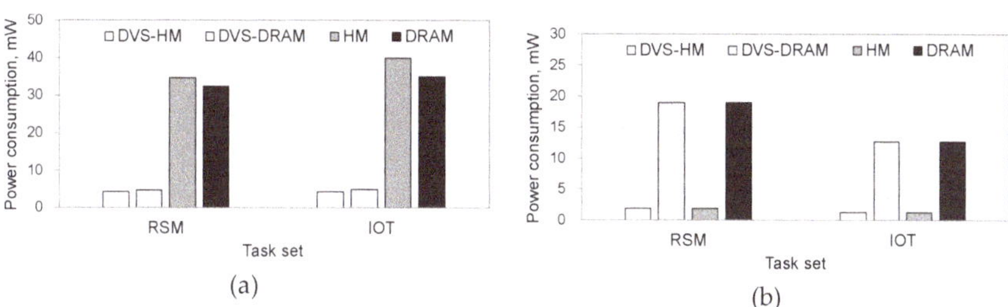

Figure 6. Processor and memory power consumptions under realistic workloads. (**a**) Power consumption in processor, (**b**) power consumption in memory.

When we compared the power consumption in memory, algorithms that use NVRAM along with DRAM significantly reduced the power consumption, as shown in Figure 6b. Specifically, the DVS-HM and HM consumed 89–90% less power than the DVS-DRAM and DRAM, which only use DRAM. This is because the idle power of NVRAM is very small.

Figure 7 shows the total power consumption when realistic workloads are used. As shown in the figure, the trends of the graphs are consistent with the synthetic workload cases. Specifically, DVS-HM saved the power consumption of DRAM, DVS-DRAM, and HM by 88%, 74%, and 83%, respectively, in the RSM workload and 88%, 68%, and 87%, respectively, in the IoT workload. Figure 8 shows the processor's utilization for realistic workloads. As can be seen from the figure, DVS-HM exhibited the highest utilization by adopting low-power resource configurations in both the processor and memory. The HM also showed a high utilization similar to DVS-HM, but this was not due to the low voltage setting of the processor, as HM does not use DVS. In fact, the high utilization of the HM was caused by the stalls in executing the instructions while accessing the slow NVRAM memory. Due to this reason, the HM presented a significantly larger power consumption in the processor, although its utilization became high, which was different from the DVS-HM cases.

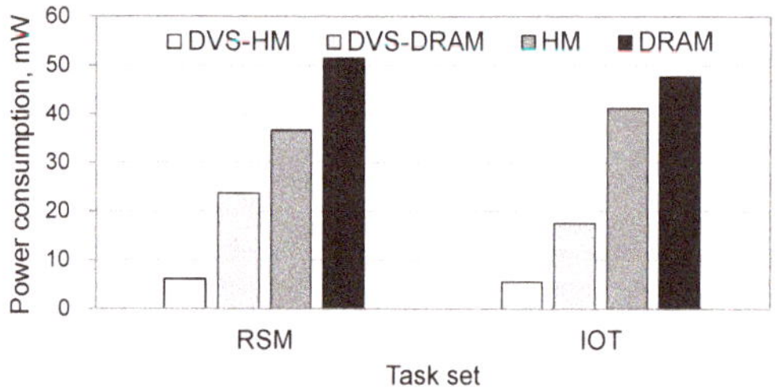

Figure 7. Total power consumption under realistic workloads.

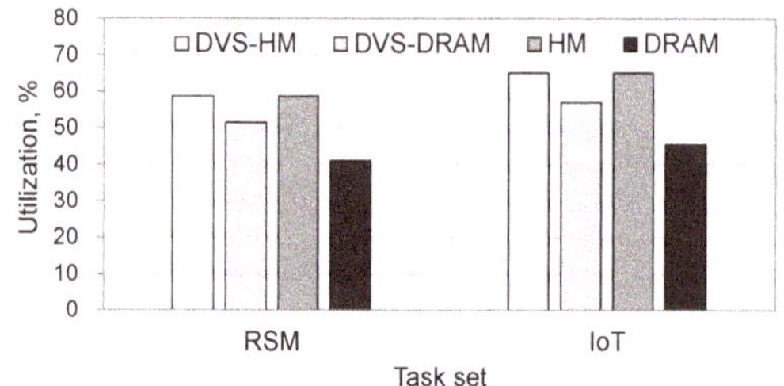

Figure 8. Processor utilizations under realistic workloads.

4. Related Works

4.1. Hybrid Memory Technologies

Recently, hybrid memory technologies consisting of DRAM and NVRAM have been catching interest. As NVRAM is byte-accessible, similar to DRAM, but consumes less energy and provides higher scalability than DRAM, it is anticipated to be adopted in the main memory hierarchy of future computer systems. Mogul et al. suggest an efficient memory management policy for DRAM and PRAM hybrid memory [4]. Their policy tries to place read-only pages in PRAM, while writable pages in DRAM, thereby reducing the slow PRAM writes [4]. Dhiman et al. propose a hybrid memory architecture consisting of PRAM and DRAM, which dynamically moves data between PRAM and DRAM in order to balance the write count of PRAM [5]. Qureshi et al. propose a hierarchical memory architecture consisting of DRAM and PRAM [7]. Specifically, they use DRAM as the write buffer of PRAM in order to prolong the lifespan of PRAM and hide the slow write performances of PRAM. Lee et al. propose the CLOCK-DWF (clock with dirty bits and write frequency) policy for hybrid memory architecture, consisting of DRAM and PRAM [6]. They allocate read-intensive pages to PRAM and write-intensive pages to DRAM by online characterization of memory access patterns. Zhou et al. propose a hierarchical memory architecture consisting of DRAM and PRAM [8]. In particular, they propose a page replacement policy that tries to reduce both the cache misses and the write-backs from DRAM. Narayan et al. propose a page allocation approach for hybrid memory architectures at the memory object level [13]. They characterize memory objects and allocate them to their best-fit memory module to improve performance and energy efficiency. Kannan et al. propose heterogeneous memory management in virtualized systems [14]. They designed a heterogeneity-aware guest operating system (OS), which allows for placing data in the appropriate memory, which avoids page migrations. They also present migration policies for performance-critical pages and memory sharing policies for guest machines.

4.2. Low-Power Techniques for Real-time Scheduling

Many studies have been performed on DVS in order to reduce power consumption in real-time systems [15–18]. Pillai and Shin propose a mechanism of selecting the lowest operating frequency that will meet deadlines for a given task set [19]. They propose three algorithms for DVS: Static DVS, cycle-conserving DVS, and look-ahead DVS. Static DVS selects the voltage of a processor statically, whereas cycle-conserving DVS uses reclaimed cycles for lowering the voltage when the actual execution time of a task is shorter than the worst-case execution time. Look-ahead DVS lowers the voltage further by determining future computation requirements and deferring the execution of the task in accordance. Lee et al. use the slack time to lower the processor's voltage [1]. Specifically, initial voltages can be dynamically switched upon reclaiming unused clock cycles when a task completes before its deadline. Lin et al. point out that there is a memory mapping problem, as heterogeneous memory types are used [10]. They use dynamic programming and greedy approximation for solving the problem. Zhang et al. propose task placement in hybrid memory to save energy consumption [20]. In their scheme, tasks are located one by one in the NVRAM and the schedulability is checked. This procedure is repeated until the locations of all tasks are determined. Ghor and Aggoune propose a slack-based method to find the least voltage schedule for real-time tasks [16]. They stretch the execution time of tasks through off-line computing and schedule tasks as late as possible without missing their deadlines.

5. Conclusions

This article presented a new real-time task scheduling approach that unifies the processor's voltage scaling and task placement in hybrid memory. Our approach incorporates the task placement in hybrid memory into the task model of voltage scaling in order to maximize the power-saving of real-time systems. The experimental results showed that the proposed technique reduces the power consumption of real-time systems by 18–88%. In the future, we will perform measurement studies in real systems in order to assess the effectiveness of the proposed approach in more realistic situations.

Author Contributions: S.N. designed the architecture and algorithm, K.C. performed the experiments. H.B. supervised the work and provided expertise.

Funding: This research was funded by the ICT R & D program of MSIP/IITP (2019-0-00074, developing system software technologies for emerging new memory that adaptively learn workload characteristics) and also by the Basic Science Research Program through the NRF grant funded by Korea Government (MSIP) (No. 2019R1A2C1009275).

Conflicts of Interest: The authors declare no conflict of interest.

References

1. Lee, Y.H.; Doh, Y.; Krishna, C.M. EDF scheduling using two-mode voltage clock scaling for hard real-time systems. In *Proceedings of the 2001 International Conference on Compilers, Architecture, and Synthesis for Embedded Systems*; ACM: New York, NY, USA, 2001; pp. 211–228.
2. Liu, S.; Pattabiraman, K.; Moscibroda, T.; Zorn, B.G. Flikker: Saving DRAM refresh-power through critical data partitioning. *ACM SIGPLAN Not.* **2012**, *47*, 213–224. [CrossRef]
3. Eilert, S.; Leinwander, M.; Crisenza, G. Phase Change Memory: A New Memory Technology to Enable New Memory Usage Models. 2011. Available online: https://www.ecnmag.com/article/2010/01/phase-change-memory-new-memory-technology-enable-new-memory-usage-models (accessed on 3 June 2019).
4. Mogul, J.C.; Argollo, E.; Shah, M.; Faraboschi, P. Operating system support for NVM+DRAM hybrid main memory. In *12th USENIX Workshop on Hot Topics in Operating Systems (HotOS)*; USENIX: Monte Verita, Switzerland, 2009.
5. Dhiman, G.; Ayoub, R.; Rosing, T. PDRAM: A hybrid PRAM and DRAM main memory system. In *2009 46th ACM/IEEE Design Automation Conference*; IEEE: Piscataway, NJ, USA, 2009.
6. Lee, S.; Bahn, H.; Noh, S.H. CLOCK-DWF: A Write-History-Aware Page Replacement Algorithm for Hybrid PCM and DRAM Memory Architectures. *IEEE Trans. Comput.* **2013**, *63*, 2187–2200. [CrossRef]
7. Qureshi, M.K.; Srinivasan, V.; Rivers, J.A. Scalable high performance main memory system using phase-change memory technology. In *ACM SIGARCH Computer Architecture News*; ACM: New York, NY, USA, 2009.

8. Zhou, P.; Zhao, B.; Yang, J.; Zhang, Y. A durable and energy efficient main memory using phase change memory technology. In *ACM SIGARCH Computer Architecture News*; ACM: New York, NY, USA, 2009.

9. Ibarra, O.H.; Kim, C.E. Fast approximation algorithms for the knapsack and sum of subset problems. *J. ACM* **1975**, *22*, 463–468. [CrossRef]

10. Lin, Y.; Guan, N.; Deng, Q. Allocation and scheduling of real-time tasks with volatile/non-volatile hybrid memory systems. In *2015 IEEE Non-Volatile Memory System and Applications Symposium (NVMSA)*; IEEE: Piscataway, NJ, USA, 2015; pp. 1–6.

11. Qadi, A.; Goddard, S.; Farritor, S. A dynamic voltage scaling algorithm for sporadic tasks. In *RTSS 2003. 24th IEEE Real-Time Systems Symposium*; IEEE: Piscataway, NJ, USA, 2003.

12. Wang, Z.; Liu, Y.; Sun, Y.; Li, Y.; Zhang, D.; Yang, H. An energy-efficient heterogeneous dual-core processor for Internet of Things. In *2015 IEEE International Symposium on Circuits and Systems (ISCAS)*; IEEE: Piscataway, NJ, USA, 2015.

13. Narayan, A.; Zhang, T.; Aga, S.; Narayanasamy, S.; Coskun, A. MOCA: Memory object classification and allocation in heterogeneous memory systems. In *2018 IEEE International Parallel and Distributed Processing Symposium (IPDPS)*; IEEE: Piscataway, NJ, USA, 2018.

14. Kannan, S.; Gavrilovska, A.; Gupta, V.; Schwan, K. HeteroOS—OS design for heterogeneous memory management in datacenter. In *2017 ACM/IEEE 44th Annual International Symposium on Computer Architecture (ISCA)*; IEEE: Piscataway, NJ, USA, 2017.

15. Choi, K.; Lee, W.; Soma, R.; Pedram, M. Dynamic voltage and frequency scaling under a precise energy model considering variable and fixed components of the system power dissipation. In *Proceedings of the 2004 IEEE/ACM International Conference on Computer-Aided Design*; IEEE Computer Society: Washington, DC, USA, 2004.

16. Ghor, H.E.; Aggoune, E.H.M. Energy saving EDF scheduling for wireless sensors on variable voltage processors. *J. Adv. Comput. Sci. Appl.* **2014**, *5*, 158–167.

17. David, H.; Fallin, C.; Gorbatov, E.; Hanebutte, U.R.; Mutlu, O. Memory power management via dynamic voltage/frequency scaling. In *Proceedings of the 8th ACM International Conference on Autonomic Computing*; ACM: New York, NY, USA, 2011.

18. Chetto, H.; Chetto, M. Some results of the earliest deadline scheduling algorithm. *IEEE Trans. Software Eng.* **1989**, *10*, 1261–1269. [CrossRef]

19. Pillai, P.; Shin, K.G. Real-time dynamic voltage scaling for low-power embedded operating systems. In *ACM SIGOPS Operating Systems Review*; ACM: New York, NY, USA, 2001.

20. Zhang, Z.; Jia, Z.; Liu, P.; Ju, L. Energy efficient real-time scheduling for embedded systems with hybrid main memory. *J. Signal Proc. Syst.* **2016**, *84*, 69–89. [CrossRef]

Article

Resistance Switching Statistics and Mechanisms of Pt Dispersed Silicon Oxide-Based Memristors

Xiaojuan Lian [1,*], Xinyi Shen [1], Liqun Lu [1], Nan He [1], Xiang Wan [1], Subhranu Samanta [2] and Yi Tong [1,*]

[1] The Department of Microelectronics, Nanjing University of Posts and Telecommunications, Nanjing 210023, China; 1218023032@njupt.edu.cn (X.S.); b17020715@njupt.edu.cn (L.L.); 1018020830@njupt.edu.cn (N.H.); wanxiang@njupt.edu.cn (X.W.)
[2] The Department of Electrical and Computer Engineering, National University of Singapore, Singapore 117576, Singapore; subhranu.samanta@gmail.com
* Correspondence: xjlian@njupt.edu.cn (X.L.); tongyi@njupt.edu.cn (Y.T.); Tel.: +86-025-85866321 (X.L.)

Received: 30 April 2019; Accepted: 29 May 2019; Published: 1 June 2019

Abstract: Silicon oxide-based memristors have been extensively studied due to their compatibility with the dominant silicon complementary metal–oxide–semiconductor (CMOS) fabrication technology. However, the variability of resistance switching (RS) parameters is one of the major challenges for commercialization applications. Owing to the filamentary nature of most RS devices, the variability of RS parameters can be reduced by doping in the RS region, where conductive filaments (CFs) can grow along the locations of impurities. In this work, we have successfully obtained RS characteristics in Pt dispersed silicon oxide-based memristors. The RS variabilities and mechanisms have been analyzed by screening the statistical data into different resistance ranges, and the distributions are shown to be compatible with a Weibull distribution. Additionally, a quantum points contact (QPC) model has been validated to account for the conductive mechanism and further sheds light on the evolution of the CFs during RS processes.

Keywords: silicon oxide-based memristors; resistance switching mechanism; variability; conductive filament; Weibull distribution; quantum point contact

---

1. Introduction

Memristors are nonvolatile resistance switching (RS) devices which can keep their internal resistance depending on the applied voltage and current status [1–6]. Currently, memristors have attracted considerable attention due to their great potentials for next generation scalable nonvolatile memory applications and neuromorphic computing [7–24]. Among numerous RS materials, silicon oxide-based memristors have been intensively investigated, owing to their compatibility with the dominant silicon complementary metal–oxide–semiconductor (CMOS) fabrication technology [25–35]. However, the variability of RS parameters is a major challenge for the progression of silicon oxide-based memristors from research to application.

In this work, we fabricated Pt dispersed silicon oxide-based memristors and successfully obtained their RS characteristics. In order to investigate the variability of RS parameters, the statistics of RS parameters have been analyzed by screening the statistical data into different resistance ranges in both the Reset and Set processes. Additionally, a quantum point contact model has been validated to account for the conductive mechanism and further shed light on the evolution of the conductive filaments (CFs) during RS processes.

2. Materials and Methods

The studied Pt/Pt:SiO$_x$/Ta memristors (the inset of Figure 1a) were fabricated on a Si wafer. Metallic Ta and Pt layers were deposited by DC sputter deposition at ambient temperature. The RS layers of the Pt:SiO$_x$ films were deposited by radiofrequency (RF) magnetron co-sputtering in pure Ar, using SiO$_2$ and Pt targets as dielectric and metal sources, respectively. The as-grown Pt dispersed SiO$_2$ thin films were composed of a SiO$_2$ matrix with 2–3 nm-sized Pt nanoclusters. Pt concentrations were of about 20–45 atomic%, which were controlled by the RF power of the Pt sputtering target [36,37]. The sandwich structure of the Pt/Pt:SiO$_x$/Ta memristors consisted of (from bottom to top) a 10 nm Ta bottom electrode, a 7 nm silicon dioxide blanket layer, and a 16 nm Pt disc (the diameter is about 50 µm) top electrode.

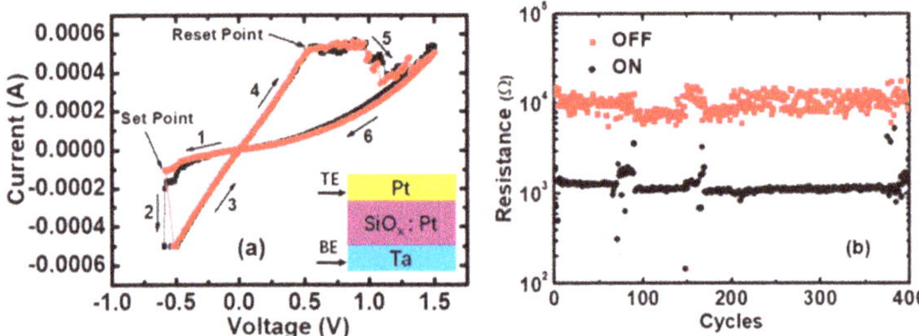

Figure 1. The Current–Voltage (I–V) characteristics in Pt/Pt:SiO$_x$/Ta memristors. (**a**) The I–V curves for the Set and Reset transitions. A current compliance limit of 0.5 mA is given in the Set process to avoid the breakdown; (**b**) The ON and OFF resistance states in 400 cycles, extracted at low voltage (0.1 V).

The Current–Voltage (I–V) switching curves and resistance measurements were performed by using an Agilent B1500 semiconductor parameter analyzer. After the electroforming operation, long lasting repetitive cycling experiments were performed using voltage ramp stress for both the Set and Reset processes, and a current compliance limit of 0.5 mA was given in the Set process to avoid the breakdown. The Pt/Pt:SiO$_x$/Ta memristors show a bipolar switching behavior, i.e., Set to the low-resistance state (LRS) under negative voltages and Reset to the high-resistance state (HRS) under positive voltages, as shown in Figure 1a. Figure 1b presents the ON and OFF resistance states of 400 cycles, and the average RS range is approximately from 1 to 10 kΩ.

3. Results

3.1. Statistical Distributions

To investigate the variability of RS parameters in both the Set and Reset processes, the statistics of RS parameters versus the initial resistances has been done, and are shown in Figure 2. Figure 2a,b shows the Reset voltage and Reset current (V_{RESET} and I_{RESET}) versus the ON-state resistance (R_{ON}), which is calculated at a low voltage (0.1 V). According to the statistics results, we can see that V_{RESET} is nearly independent of R_{ON}, whereas I_{RESET} is inversely proportional to R_{ON}. This observation is compatible with the thermal-activated dissolution model [38]. In this model, the Reset event happens only when the temperature of the CFs reaches a critical value. Figure 2c,d shows the Set voltage and Set current (V_{SET} and I_{SET}) versus the OFF-state resistance (R_{OFF}), also calculated at 0.1 V. From these two figures, it can be seen that V_{SET} is proportional to R_{OFF}, whereas I_{SET} is nearly independent of R_{OFF}. Through the statistics of RS parameters, we can know that the variations of R_{ON} and R_{OFF} have a strong impact on the uniform distributions of RS parameters. We could improve the performance of memristors by controlling the sizes of the CFs before the Reset and Set transitions.

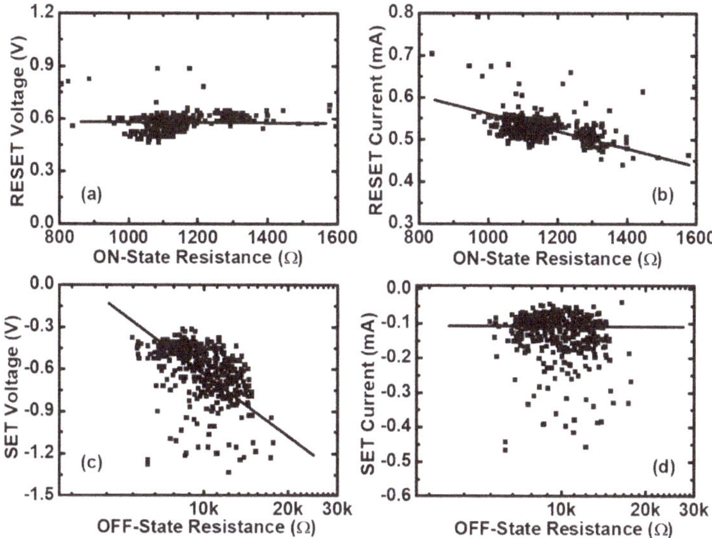

Figure 2. The statistics of resistance switching (RS) parameters in Pt/Pt:SiO$_x$/Ta memristors. (**a**) The Reset voltages and (**b**) the Reset currents versus the ON-state resistances for the measured 400 cycling data of the same device. (**c**) The Set voltages and (**d**) the Set currents versus the OFF-state resistances for the measured 400 cycling data of the same device.

Next, the nature of the variation of RS parameters was explored using a data screening method. The cumulative distributions of V_{RESET} and I_{RESET} in different ON-state resistance ranges are shown in Figure 3a,b, respectively, and the cumulative distributions of V_{SET} and I_{SET} in different OFF-state resistance ranges are shown in Figure 4a,b, respectively. In these four cases, the cumulative distributions are almost straight lines, which are compatible with the Weibull distribution. Therefore, we can use the Weibull distribution function to fit the experimental data of RS parameters in different resistance ranges to obtain the Weibull parameters. The Weibull distribution is defined as:

$$F = 1 - \exp\left[-(x/x_{63\%})^{\beta}\right] \tag{1}$$

or

$$W \equiv Ln(-Ln(1-F)) = \beta Ln(x/x_{63\%}) \tag{2}$$

where β is the Weibull slope or shape factor, which represents the statistical dispersion. $x_{63\%}$ is the scale factor parameter, which is the value of F $\approx$ 63%. After fitting of the experimental data by the Weibull distribution, we can obtain the Weibull parameters of V_{RESET} and I_{RESET}, as shown in Figure 3c,d. The scale factor of V_{RESET} ($V_{RESET63\%}$) is independent of R_{ON}, and the scale factor of I_{RESET} ($I_{RESET63\%}$) is inversely proportional to R_{ON}, which is consistent with the scatter plots of Figure 2a,b. The Weibull slope of V_{RESET} and I_{RESET} is nearly independent of the ON-state resistances, which means that there are no microstructure variations of the CFs before the Reset point [38,39]. Similarly, the Weibull parameters of V_{SET} and I_{SET} can be obtained by fitting the experimental data using the Weibull distribution function, as shown in Figure 4c,d, respectively. The scale factor of V_{SET} ($V_{SET63\%}$) is proportional to R_{OFF}, and the scale factor of I_{SET} ($I_{SET63\%}$) is independent of R_{OFF}, which is consistent with the scatter plots of Figure 2c,d. The Weibull slopes of V_{SET} and I_{SET} are nearly independent of the OFF-state resistances, which means that there are no obvious microstructure variations of the CFs before the Set point [40].

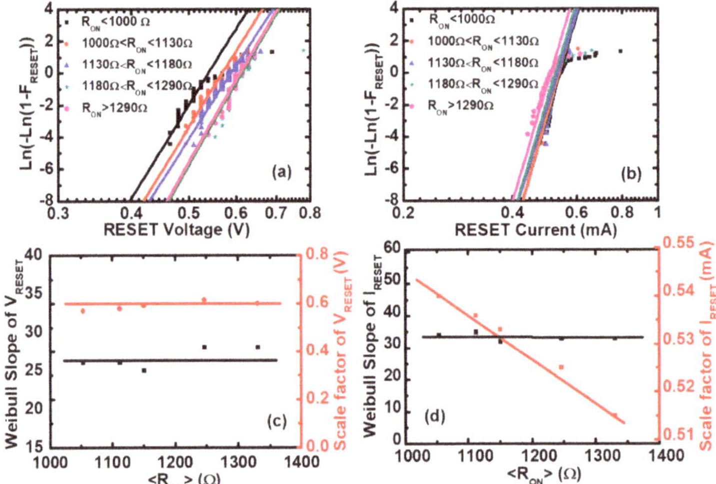

Figure 3. The Weibull distributions of the Reset voltage and the Reset current in Pt/Pt:SiO$_x$/Ta devices. Experimental distributions (symbols) and the fitting to Weibull distribution (lines) of (**a**) the Reset voltage and (**b**) the Reset current as functions of the ON-state resistance. Weibull slopes and scale factors of (**c**) the Reset voltage and (**d**) the Reset current versus <R_{ON}>, where <R_{ON}> is the average value of the ON-state resistance (R_{ON}) in each screening range. It can be seen that the Weibull slopes of the Reset voltage and the Reset current are independent of <R_{ON}>, and the scale factor of the Reset voltage is constant, whereas the Reset current is inversely proportional to <R_{ON}>.

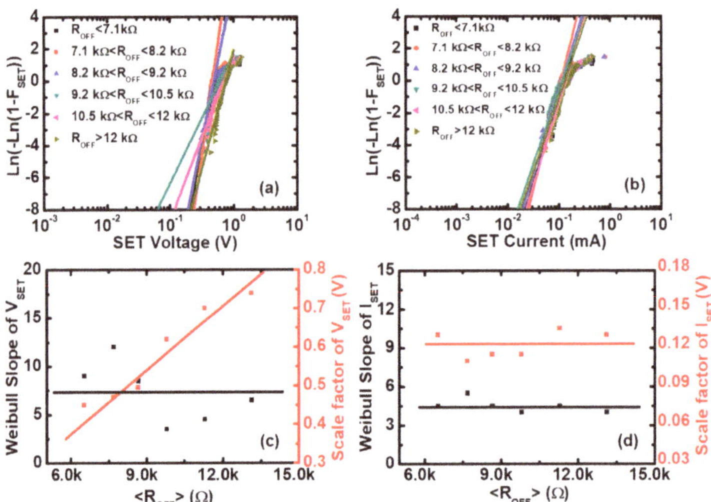

Figure 4. The Weibull distributions of the Set voltage and the Set current in Pt/Pt:SiO$_x$/Ta devices. Experimental distributions (symbols) and the fitting to Weibull distribution (lines) of (**a**) the Set voltage and (**b**) the Set current as functions of the OFF-state resistance. Weibull slopes and scale factors of (**c**) the Set voltage and (**d**) the Set current versus <R_{OFF}>, where <R_{OFF}> is the average value of the OFF-state resistance (R_{OFF}) in each screening range. It can be seen that the Weibull slopes of the Set voltage and the Set current are independent of <R_{OFF}>, and the scale factor of the Set voltage is proportional to <R_{OFF}>, whereas the Set current is constant.

3.2. Quantum Point Contact Model

Many different conduction models have been proposed for the HRS, including Schottky emission [41–44], trap-assisted tunneling [45–47], Poole–Frenkel conduction [43,48], space-charge limited current [49–52], thermally activated hopping [53,54], and the Quantum Point Contact model (QPC) [55–61], among others. Specifically, the QPC model can provide a smooth transition from tunneling in the HRS to Ohmic conduction in the LRS for several kinds of RS devices [58–61]. To analyze the conductive mechanisms of RS processes for Pt/Pt:SiO$_x$/Ta memristors, the QPC model has been introduced here to fit the I–V curves in both the Reset and Set processes.

The QPC model is based on the Landauer transmission approach to calculate conduction along narrow microscopic constrictions [57,58]. According to the Landauer's approach, the current flowing through a CF with N paths can be calculated as [62]:

$$I(V) = \frac{2e}{h} N \int_{-\infty}^{\infty} T(E)\{f(E - \beta eV) - f(E + (1 - \beta)eV)\}\, dE \tag{3}$$

where f is the Fermi–Dirac distribution function, E is the energy, $T(E)$ is the transmission probability, β is the averaged asymmetry parameter (with the constraint $0 < \beta \leq 1$), and V is the applied voltage assumed to drop at the cathode and anode interfaces with a fraction of β and $(1 - \beta)$, respectively. Assuming an inverted parabolic potential barrier, we can obtain an expression for the tunneling probability [63–65], $T(E) = \{1 + exp[-\alpha(E - \Phi)]\}^{-1}$, where Φ is the barrier height, $\alpha = t_B \pi^2 h^{-1} \sqrt{2m^*/\Phi}$ is related to the inverse of the potential barrier curvature, m^* is the effective electron mass, and t_B is the barrier width at the equilibrium Fermi energy, assumed to be equal to t_{gap}. Inserting the tunneling probability into Equation (3), we can obtain:

$$I = \frac{2e}{h} N \left\{ eV + \frac{1}{\alpha} Ln \left[\frac{1 + exp\{\alpha[\Phi - \beta eV]\}}{1 + exp\{\alpha[\Phi + (1 - \beta)eV]\}} \right] \right\} \tag{4}$$

There are four parameters in Equation (4). In order to simplify the fitting process, here we fixed $\Phi = 0.5$ eV and $\beta = 1$ by considering the asymmetry structure of the devices. Then, we extracted the number of CF paths N and the average t_{gap} from the fitting experimental data of 400 cycles by using Equation (4) and the least squares estimation (LSE) method. The I–V fitting results are excellent in both log and linear scales, as shown in Figure 5a,b. Furthermore, Figure 5c,d shows the exacted QPC parameters versus the CF resistance. It can be seen that the average t_{gap} is approximately 0.1 nm the in LRS (ON-state) and 0.25 nm in the HRS (OFF-state), and the average number of CF paths is about 30 in the LRS and five in the HRS.

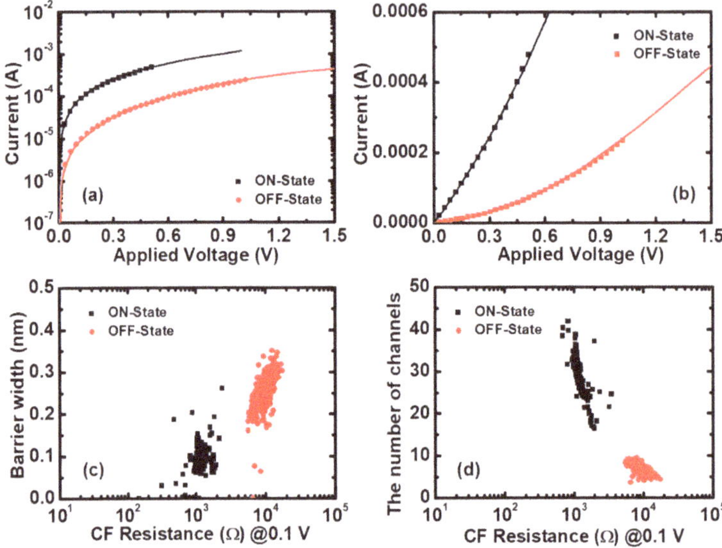

Figure 5. The quantum points contact (QPC) model applied to Pt/Pt:SiO$_x$/Ta memristors. The I–V fitting results together with experimental data of ON and OFF states (**a**) in log scale and (**b**) linear scale. (**c**) The barrier thickness and (**d**) the number of CF paths versus the initial resistance, respectively. The averaged values are: $< t_{gap} > = 0.1$ nm, $< N > = 30$ in the ON-state; and $< t_{gap} > = 0.25$ nm, $< N > = 5$ in the OFF-state.

4. Discussion

According to the screening of the statistical data into different resistance ranges, the distributions of RS parameters were shown to be compatible with a Weibull distribution. After using the Weibull distribution function to fit the experimental data of RS parameters into different resistance ranges, we can obtain that $V_{RESET63\%}$ is independent of R_{ON} and $I_{RESET63\%}$ is inversely proportional to R_{ON}, whereas $V_{SET63\%}$ is proportional to R_{OFF} and $I_{SET63\%}$ is independent of R_{OFF}, which are consistent with the experimental results. Besides, the Weibull slopes of V_{RESET}, I_{RESET}, V_{SET}, and I_{SET} are nearly independent of the initial resistances, which means that there are no microstructure variations of the CFs before the Reset and Set points. Furthermore, the QPC model has been validated to account for the conductive mechanism and further show the evolution of the CFs during RS processes. From the LRS to HRS, the number of CF paths would decrease, while the barrier gap would increase.

Combining the fitting results of the QPC model with the statistics of RS parameters, we now try to propose the conductive mechanisms of RS processes. During the ON switching, the RS process is mainly driven by an applied electric field, and the CFs are more likely to grow along the locations of Pt nanostructures. Cation migration and metallic CF formation in RS layers can be identified as a candidate RS mechanism due to the abrupt increase of the current in I–V curves (Figure 1a) [66,67]. During the OFF switching, cations are driven out of the CFs and thus introduce a gap between the CFs and the top Pt electrode. Therefore, the number of CF paths would decrease, while the barrier gap would increase from the LRS to the HRS. The Reset event happens only when the temperature of the CFs reaches a critical value, according to the thermal-activated dissolution model. In addition, according to the statistics, we can know that the variations of the RS parameters can be significantly reduced and the performance of memristors could be improved by controlling the sizes of the CFs before the Reset and Set transitions. That is to say, the variability of RS parameters can be reduced by doping in RS regions, where CFs can be induced to grow along the locations of impurities, or by inserting a two-dimensional material with engineered nanopores, which can modify the RS characteristics of memristors.

Author Contributions: Conceptualization, X.L. and Y.T.; methodology, X.L.; software, X.L.; formal analysis, X.L.; investigation, X.L.; resources, X.L.; data curation, X.L.; writing—original draft preparation, X.L.; writing—review and editing, X.L., X.S., L.L., N.H., X.W., S.S. and Y.T.; visualization, X.L.; supervision, X.L.; project administration, X.L. and Y.T.; funding acquisition, X.L. and Y.T.

Funding: This research was funded in part by the National Natural Science Foundation of China (grant number 61804079), the University Natural Science Foundation of Jiangsu Province (grant number 18KJD510005), the Senior Talent Foundation of Jiangsu Province (grant number SZDG2018007), and the Science Research Funds for Nanjing University of Posts and Telecommunications (grant number NY218110, NY217116).

Acknowledgments: Xiaojuan Lian also thanks the research groups of Feng Miao and J. Joshua Yang for helping with the device's preparation.

Conflicts of Interest: The authors declare no conflict of interest.

References

1. Chua, L.O. Memristor-the missing circuit element. *IEEE Trans. Circuit Theory* **1971**, *18*, 507–519. [CrossRef]
2. Chua, L.O.; Kang, S.M. Memristive devices and systems. *Proc. IEEE* **1976**, *64*, 209–223. [CrossRef]
3. Waser, R.; Aono, M. Nanoionics-Based Resistive Switching Memories. *Nat. Mater.* **2007**, *6*, 833–840. [CrossRef] [PubMed]
4. Strukov, D.B.; Snider, G.S.; Stewart, D.R.; Williams, R.S. The missing memristor found. *Nature* **2008**, *453*, 80–83. [CrossRef] [PubMed]
5. Sawa, A. Resistive switching in transition metal oxides. *Mater. Today* **2008**, *11*, 28–36. [CrossRef]
6. Chua, L.O. Resistance switching memories are memristors. *Appl. Phys. A Mater. Sci. Process.* **2011**, *102*, 765–783. [CrossRef]
7. Yang, J.J.; Pickett, M.D.; Li, X.; Ohlberg, D.A.A.; Stewart, D.R.; Williams, R.S. Memristive switching mechanism for metal/oxide/metal nanodevices. *Nat. Nanotechnol.* **2008**, *3*, 429–433. [CrossRef]
8. Wong, H.-S.P.; Lee, H.-Y.; Yu, S.; Chen, Y.-S.; Wu, Y.; Chen, P.-S.; Lee, B.; Chen, F.T.; Tsai, M.-J. Metal–Oxide RRAM. *Proc. IEEE* **2012**, *100*, 1951–1970. [CrossRef]
9. Xia, Q.F.; Robinett, W.; Cumbie, M.W.; Banerjee, N.; Cardinali, T.J.; Yang, J.J.; Wu, W.; Li, X.M.; Tong, W.M.; Strukov, D.B.; et al. Memristor-CMOS hybrid integrated circuits for reconfigurable logic. *Nano Lett.* **2009**, *9*, 3640–3645. [CrossRef]
10. Jo, S.H.; Chang, T.; Ebong, I.; Bhadviya, B.B.; Mazumder, P.; Lu, W. Nanoscale memristor device as synapse in neuromorphic systems. *Nano Lett.* **2010**, *10*, 1297–1301. [CrossRef] [PubMed]
11. Lee, M.-J.; Lee, C.B.; Lee, D.; Lee, S.R.; Chang, M.; Hur, J.H.; Kim, Y.-B.; Kim, C.-J.; Seo, D.H.; Seo, S.; et al. A fast, high-endurance and scalable non-volatile memory device made from asymmetric Ta_2O_{5-x}/TaO_{2-x} bilayer structures. *Nat. Mater.* **2011**, *10*, 625–630. [CrossRef]
12. Chen, W.; Fang, R.; Balaban, M.B.; Yu, W.; Velo, Y.G.; Barnaby, H.J.; Kozicki, M.N. A CMOS-compatible electronic synapse device based on $Cu/SiO_2/W$ programmable metallization cells. *Nanotechnology* **2016**, *27*, 255202. [CrossRef] [PubMed]
13. Burgt, Y.; Lubberman, E.; Fuller, E.J.; Keene, S.T.; Faria, G.C.; Agarwal, S.; Marinella, M.J.; Talin, A.A.; Salleo, A. A non-volatile organic electrochemical device as a low-voltage artificial synapse for neuromorphic computing. *Nat. Mater.* **2017**, *16*, 414–418. [CrossRef]
14. Hu, M.; Graves, C.E.; Li, C.; Li, Y.; Ge, N.; Montgomery, E.; Davila, N.; Jiang, H.; Williams, R.S.; Yang, J.; et al. Memristor-Based Analog Computation and Neural Network Classification with a Dot Product Engine. *Adv. Mater.* **2018**, *30*, 1705914. [CrossRef] [PubMed]
15. Li, C.; Belkin, D.; Li, Y.; Yan, P.; Hu, M. Efficient and self-adaptive in-situ learning in multilayer memristor neural networks. *Nat. Commun.* **2018**, *9*, 2385. [CrossRef] [PubMed]
16. Gavrilov, D.; Strukov, D.; Likharev, K.K. Capacity, Fidelity, and Noise Tolerance of Associative Spatial-Temporal Memories Based on Memristive Neuromorphic Networks. *Front. Neurosci.* **2018**, *12*, 195. [CrossRef] [PubMed]
17. Bayat, F.M.; Prezioso, M.; Chakrabarti, B.; Nili, H.; Kataeva, I.; Strukov, D. Implementation of multilayer perceptron network with highly uniform passive memristive crossbar circuits. *Nat. Commun.* **2017**, *9*, 2331. [CrossRef]
18. Li, D.; Wu, B.; Zhu, X.; Wang, J.; Ryu, B.; Lu, W.D.; Lu, W.; Liang, X. MoS_2 Memristors Exhibiting Variable Switching Characteristics towards Bio-Realistic Synaptic Emulation. *ACS Nano* **2018**, *9*, 9240–9252. [CrossRef]

19. Zhu, X.; Lu, W.D. Optogenetics-Inspired Tunable Synaptic Functions in Memristors. *ACS Nano* **2018**, *12*, 1242–1249. [CrossRef]

20. Shi, Y.; Liang, X.; Yuan, B.; Chen, V.; Li, H.; Hui, F.; Yu, Z.; Yuan, F.; Pop, E.; Wong, H.-S.P.; et al. Electronic synapses made of layered two-dimensional materials. *Nat. Electron.* **2018**, *1*, 458–465. [CrossRef]

21. Zhao, Y.; Jiang, J.; Nanosci, J. Electronic synapses made of layered two-dimensional materials. *Nanotechnol* **2018**, *18*, 8003.

22. Park, M.H.; Lee, Y.H.; Mikolajick, T.; Schroeder, U.; Hwang, C.S. Review and Perspective on Ferroelectric HfO$_2$-based Thin Films for Memory Applications. *MRS Commun.* **2018**, *8*, 795–808. [CrossRef]

23. Li, Y.; Han, K.; Kang, Y.; Kong, E.Y.J.; Gong, X. Extraction of Polarization-dependent Damping Constant for Dynamic Evaluation of Ferroelectric Films and Devices. *IEEE Electron. Device Lett.* **2018**, *39*, 1211–1214. [CrossRef]

24. Li, Y.; Kang, Y.; Gong, X. Evaluation of Negative Capacitance Ferroelectric MOSFET for Analog Circuit Applications. *IEEE Trans. Electron. Devices* **2017**, *64*, 4317–4321. [CrossRef]

25. Bricalli, A.; Ambrosi, E.; Laudato, M.; Maestro, M.; Rodriguez, R.; Lelmini, D. Resistive Switching Device Technology Based on Silicon Oxide for Improved ON-OFF Ratio—Part II: Select Devices. *IEEE Trans. Electron. Devices* **2018**, *65*, 122–128. [CrossRef]

26. Liu, C.; Lin, C.; Liu, S.; Bai, C.; Zhang, Y. Improvement of switching uniformity in Cu/SiO$_2$/Pt resistive memory achieved by voltage prestress. *Jpn. J. Appl. Phys.* **2015**, *54*, 031801. [CrossRef]

27. Liu, C.; Huang, J.; Lai, C.; Lin, C. Influence of embedding Cu nano-particles into a Cu/SiO$_2$/Pt structure on its resistive switching. *Nanoscale Res. Lett.* **2013**, *8*, 156. [CrossRef]

28. Nandakumar, S.R.; Minvielle, M.; Nagar, S.; Dubourdieu, C.; Rajendran, B. A 250 mV Cu/SiO$_2$/W Memristor with Half-Integer Quantum Conductance States. *Nano Lett.* **2016**, *16*, 1602–1608. [CrossRef]

29. Mehonic, A.; Shluger, A.L.; Gao, D.; Valov, I.; Miranda, E.; Ielmini, D.; Bricalli, A.; Ambrosi, E.; Li, C.; Yang, J.J.; et al. Silicon Oxide (SiO$_x$): A Promising Material for Resistance Switching? *Adv. Mater.* **2018**, *30*, 1801187. [CrossRef]

30. Ng, W.H.; Mehonic, A.; Buckwell, M.; Montesi, L.; Kenyon, A.J. High-Performance Resistance Switching Memory Devices Using Spin-On Silicon Oxide. *IEEE Trans. Nanotechnol.* **2018**, *17*, 884–888. [CrossRef]

31. Munde, M.S.; Mehonic, A.; Ng, W.H.; Buckwell, M.; Montesi, L.; Bosman, M.; Shluger, A.L.; Kenyon, A.J. Intrinsic Resistance Switching in Amorphous Silicon Suboxides: The Role of Columnar Microstructure. *Sci. Rep.* **2017**, *7*, 9274. [CrossRef] [PubMed]

32. Mehonic, A.; Buckwell, M.; Montesi, L.; Munde, M.S.; Gao, D.; Hudziak, S.; Chater, R.J.; Fearn, S.; McPhail, D.; Bosman, M.; et al. Nanoscale Transformations in Metastable, Amorphous, Silicon-Rich Silica. *Adv. Mater.* **2016**, *28*, 7486–7493. [CrossRef] [PubMed]

33. Tappertzhofen, S.; Valov, I.; Tsuruoka, T.; Hasegawa, T.; Waser, R.; Aono, M. Generic Relevance of Counter Charges for Cation-Based Nanoscale Resistive Switching Memories. *ACS Nano* **2013**, *7*, 6396–6402. [CrossRef] [PubMed]

34. Tsuruoka, T.; Terabe, K.; Hasegawa, T.; Valov, I.; Waser, R.; Aono, M. Effects of Moisture on the Switching Characteristics of Oxide-Based, Gapless-Type Atomic Switches. *Adv. Funct. Mater.* **2012**, *22*, 70–77. [CrossRef]

35. Yang, Y.; Gao, P.; Li, L.; Pan, X.; Tappertzhofen, S.; Choi, S.; Waser, R.; Valov, I.; Lu, W.D. Electrochemical dynamics of nanoscale metallic inclusions in dielectrics. *Nat. Commun.* **2014**, *5*, 4232. [CrossRef] [PubMed]

36. Choi, B.J.; Torrezan, A.C.; Norris, K.J.; Miao, F.; Strachan, J.P.; Zhang, M.X.; Ohlberg, D.A.; Kobayashi, N.P.; Yang, J.J.; Williams, R.S. Electrical Performance and Scalability of Pt Dispersed SiO$_2$ Nanometallic Resistance Switch. *Nano Lett.* **2013**, *13*, 3213–3217. [CrossRef]

37. Choi, B.J.; Ge, N.; Yang, J.J.; Zhang, M.-X.; Williams, R.S.; Norris, K.J.; Kobayashi, N.P. New materials for memristive switching. *IEEE Int. Symp. Circuits Syst. (ISCAS)* **2014**, *10*, 2808–2811.

38. Lian, X.; Wang, M.; Yan, P.; Yang, J.J.; Miao, F. Reset switching statistics of TaO$_x$-based Memristor. *J. Electroceram.* **2017**, *39*, 132–136. [CrossRef]

39. Long, S.; Lian, X.; Cagli, T.; Ye, C.; Perniola, L.; Miranda, E.; Liu, M.; Sune, J. Cycle-to-Cycle Intrinsic RESET Statistics in HfO$_2$-Based Unipolar RRAM. *IEEE Electron. Device Lett.* **2013**, *34*, 623–625. [CrossRef]

40. Lian, X.; Miao, F.; Wan, X.; Guo, Y.-F.; Tong, Y.; Electroceram, J. Set transition statistics of different switching regimes of TaO$_x$ memristor. *J. Electroceram.* **2019**. [CrossRef]

41. Michalas, L.; Stathopoulos, S.; Khiat, A.; Prodromakis, T. Conduction mechanisms at distinct resistive levels of Pt/TiO$_{2-x}$/Pt memristors. *Appl. Phys. Lett.* **2018**, *113*, 143503. [CrossRef]

42. Chen, K.-H.; Tsai, T.-M.; Cheng, C.-M.; Huang, S.-J.; Chang, K.-C.; Liang, S.-P.; Young, T.-F. Schottky Emission Distance and Barrier Height Properties of Bipolar Switching Gd:SiOx RRAM Devices under Different Oxygen Concentration Environments. *Materials* **2017**, *11*, 43. [CrossRef] [PubMed]

43. Gul, F. Carrier transport mechanism and bipolar resistive switching behavior of a nano-scale thin film TiO_2 memristor. *Ceram. Int.* **2018**, *44*, 11417–11423. [CrossRef]

44. Gismatulin, A.A.; Kruchinin, V.N.; Gritsenko, V.A.; Prosvirin, I.P.; Yen, T.-J.; Chin, A. Charge transport mechanism of high-resistive state in RRAM based on SiO_x. *Appl. Phys. Lett.* **2019**, *114*, 033503. [CrossRef]

45. Bersuker, G.; Gilmer, D.C.; Veksler, D.; Kirsch, P.; Vandelli, L.; Padovani, A.; Larcher, L.; McKenna, K.; Shluger, A.; Iglesias, V.; et al. Metal oxide resistive memory switching mechanism based on conductive filament properties. *J. Appl. Phys.* **2011**, *110*, 124518. [CrossRef]

46. Yan, X.; Zhou, Z.; Ding, B.; Zhao, J.; Zhang, Y. Superior resistive switching memory and biological synapse properties based on a simple $TiN/SiO_2/p$-Si tunneling junction structure. *J. Mater. Chem. C* **2017**, *5*, 2259–2267. [CrossRef]

47. Yu, S.; Guan, X.; Wong, H.-S.P. Conduction mechanism of TiN/HfOx/Pt resistive switching memory: A trap-assisted-tunneling model. *Appl. Phys. Lett.* **2011**, *99*, 063507. [CrossRef]

48. Chang, K.-C.; Tsai, T.-M.; Chang, T.-C.; Wu, H.-H.; Chen, J.-H.; Syu, Y.-E.; Chang, G.-W.; Chu, T.-J.; Liu, G.-R.; Su, Y.-T.; et al. Characteristics and Mechanisms of Silicon-Oxide-Based Resistance Random Access Memory. *IEEE Electron. Device Lett.* **2013**, *34*, 399–401. [CrossRef]

49. Kim, K.M.; Choi, B.J.; Shin, Y.C.; Choi, S.; Hwang, C.S. Anode-interface localized filamentary mechanism in resistive switching of TiO_2 thin films. *Appl. Phys. Lett.* **2007**, *91*, 012907. [CrossRef]

50. Hsieh, W.-K.; Lam, K.-T.; Chang, S.-J. Characteristics of tantalum-doped silicon oxide-based resistive random access memory. *Mater. Sci. Semicond. Process.* **2014**, *27*, 293–296. [CrossRef]

51. Sun, C.; Lu, S.M.; Jin, F.; Mo, W.Q.; Song, J.L.; Dong, K.F. The Resistive Switching Characteristics of $TiN/HfO_2/Ag$ RRAM Devices with Bidirectional Current Compliance. *J. Electron. Mater.* **2019**, *48*, 2992–2999. [CrossRef]

52. Liu, M.; Abid, Z.; Wang, W.; He, X.; Liu, Q.; Guan, W. Multilevel resistive switching with ionic and metallic filaments. *Appl. Phys. Lett.* **2009**, *94*, 233106. [CrossRef]

53. Kunchur, M.N.; Liang, M.; Gurevich, A. Thermally activated dynamics of spontaneous perpendicular vortices tuned by parallel magnetic fields in thin superconducting films. *Phys. Rev. B* **2012**, *86*, 024521. [CrossRef]

54. Graves, C.E.; Dávila, N.; Merced-Grafals, E.J.; Lam, S.-T.; Strachan, J.P.; Williams, R.S. Temperature and field-dependent transport measurements in continuously tunable tantalum oxide memristors expose the dominant state variable. *Appl. Phys. Lett.* **2017**, *110*, 123501. [CrossRef]

55. Miranda, E.; Mehonic, A.; Ng, W.H.; Kenyon, A.J. Simulation of Cycle-to-Cycle Instabilities in SiOx- Based ReRAM Devices Using a Self-Correlated Process with Long-Term Variation. *IEEE Electron. Device Lett.* **2018**, *40*, 28–31. [CrossRef]

56. Holloway, G.W.; Ivanov, O.; Gavrilov, R.; Bluschke, A.G.; Hold, B.K.; Baugh, J. Electrical Breakdown in Thin Si Oxide Modeled by a Quantum Point Contact Network. *IEEE Trans. Electron. Devices* **2016**, *63*, 3005–3010. [CrossRef]

57. Sune, J.; Miranda, E.; Nafria, M.; Aymerich, X. Point contact conduction at the oxide breakdown of MOS devices. In Proceedings of the International Electron Devices Meeting 1998, Technical Digest (Cat. No.98CH36217), San Francisco, CA, USA, 6–9 December 1998.

58. Sune, J.; Miranda, E. Understanding soft and hard breakdown statistics, prevalence ratios and energy dissipation during breakdown runaway. In Proceedings of the International Electron Devices Meeting, Technical Digest (Cat. No.01CH37224), Washington, DC, USA, 2–5 December 2001.

59. Lian, X.; Cartoixà, X.; Miranda, E.; Perniola, L.; Rurali, R.; Long, S.; Liu, M.; Suñé, J. Multi-scale quantum point contact model for filamentary conduction in resistive random access memories devices. *J. Appl. Phys.* **2014**, *115*, 244507. [CrossRef]

60. Zhang, M.; Long, S.; Wang, G.; Xu, X.; Li, Y.; Liu, Q.; Lv, H.; Lian, X.; Miranda, E.; Suñé, J.; et al. Set statistics in conductive bridge random access memory device with $Cu/HfO_2/Pt$ structure. *Appl. Phys. Lett.* **2014**, *105*, 193501. [CrossRef]

61. Lian, X.; Wang, M.; Rao, M.; Yan, P.; Yang, J.J.; Miao, F. Characteristics and transport mechanisms of triple switching regimes of TaO_x memristor. *Appl. Phys. Lett.* **2017**, *110*, 173504. [CrossRef]

62. Datta, S. *Electronic Transport in Mesoscopic Systems*; University of Cambridge: Cambridgeshire, UK, 1997.

63. Buttiker, M. Quantized transmission of a saddle-point constriction. *Phys. Rev. B* **1990**, *41*, 7906. [CrossRef]
64. Bogachek, E.N.; Scherbakov, A.G.; Landman, U. Nonlinear magnetoconductance of nanowires. *Phys. Rev. B* **1997**, *56*, 14917. [CrossRef]
65. Miranda, E.; Sune, J. From post-breakdown conduction to resistive switching effect in thin dielectric films. In Proceedings of the 2012 IEEE International Reliability Physics Symposium (IRPS), Anaheim, CA, USA, 15–19 April 2012; p. 367.
66. Lübben, M.; Valov, I. Active Electrode Redox Reactions and Device Behavior in ECM Type Resistive Switching Memories. *Adv. Electron. Mater.* **2019**, 1800933. [CrossRef]
67. Valov, I.; Luebben, M.; Wedig, A.; Waser, R. Mobile Ions, Transport and Redox Processes in Memristive Devices. *ECS Trans.* **2016**, *75*, 27–39. [CrossRef]

Article

New Logic-In-Memory Paradigms: An Architectural and Technological Perspective

Giulia Santoro *, Giovanna Turvani and Mariagrazia Graziano

Dipartimento di Elettronica e Telecomunicazioni, Politecnico di Torino; Corso Castelfidardo 39,
10129 Torino, Italy; giovanna.turvani@polito.it (G.T.); mariagrazia.graziano@polito.it (M.G.)
* Correspondence: giulia.santoro@polito.it

Received: 30 April 2019; Accepted: 22 May 2019; Published: 31 May 2019

Abstract: Processing systems are in continuous evolution thanks to the constant technological advancement and architectural progress. Over the years, computing systems have become more and more powerful, providing support for applications, such as Machine Learning, that require high computational power. However, the growing complexity of modern computing units and applications has had a strong impact on power consumption. In addition, the memory plays a key role on the overall power consumption of the system, especially when considering data-intensive applications. These applications, in fact, require a lot of data movement between the memory and the computing unit. The consequence is twofold: Memory accesses are expensive in terms of energy and a lot of time is wasted in accessing the memory, rather than processing, because of the performance gap that exists between memories and processing units. This gap is known as the memory wall or the von Neumann bottleneck and is due to the different rate of progress between complementary metal–oxide semiconductor (CMOS) technology and memories. However, CMOS scaling is also reaching a limit where it would not be possible to make further progress. This work addresses all these problems from an architectural and technological point of view by: (1) Proposing a novel Configurable Logic-in-Memory Architecture that exploits the in-memory computing paradigm to reduce the memory wall problem while also providing high performance thanks to its flexibility and parallelism; (2) exploring a non-CMOS technology as possible candidate technology for the Logic-in-Memory paradigm.

Keywords: in-memory computing; logic-in-memory; non-von Neumann architecture; configurable logic-in-memory architecture; memory wall; convolutional neural networks; emerging technologies; perpendicular Nano Magnetic Logic (pNML)

1. Introduction

The von Neumann paradigm is the foundation of all modern computing systems. This paradigm is based on the exchange of data between a Central Processing Unit (CPU) and a memory. In particular, the CPU executes instructions on data that it retrieves from the memory, and writes back results in the memory. This data exchange mechanism is exacerbated when dealing with applications that require the manipulation of large data quantities (i.e., data-intensive applications). While through the years CPUs have become more and more powerful thanks to complementary metal–oxide semiconductor (CMOS) technology scaling, memories have not improved at the same rate, with the bandwidth being the main limitation. As a consequence, memories are not able to provide data as fast as CPUs are able to compute them. This problem is known as *von Neumann bottleneck* or *memory wall* and it limits the performance of systems based on the von Neumann architectural model as a lot of time is spent in retrieving data from the memory rather than computing them. This side effect is particularly visible when dealing with memory bound algorithms. Another critical consequence of the physical separation between the processing unit and the memory is related to the energy spent in performing memory accesses. In fact, especially for data-intensive applications, the large quantity of memory accesses required has a big impact on the overall power consumption. The very well known Moore's law, according to which the number of transistors in an integrated circuit doubles every two years, has been obeyed for decades, but the growth rate predicted by Moore is now slowing down because of the limitations that technological scaling is facing. In fact, as foretold in the 2013 International Technology Roadmap for Semiconductors (ITRS) [1], CMOS scaling is reaching a boundary where further progresses will be impeded by physical, technological and economical limitations.

The drawbacks related to the von Neumann computing model and to the CMOS technology scaling are the main factors that drive this research. On the one side, the in-memory computational paradigm is explored as an alternative to the von Neumann one. The aim is to go beyond the conventional separation between computation and storage by integrating simple logic directly inside the memory cell. We refer to this approach as Logic-in-Memory (LiM). Its key benefits are mainly: (1) Bringing the computation directly inside the memory allows one to exploit the full internal bandwidth, mitigating the memory wall problem; (2) data are computed directly inside the memory without the need to move them between the computing and the storage units, drastically reducing the amount of memory accesses and the associated energy consumption and latency. On the other side, from a technological point of view, a non-CMOS technology, namely perpendicular Nano Magnetic Logic (pNML), is considered as a possible alternative to CMOS for implementing in-memory computing architectures as it intrinsically provides non volatility and computing capabilities in the same device.

The rest of this paper is organized as follows: Section 2 presents a taxonomy of the main in-memory computing approaches, based on how the memory is used for data computation; following the proposed taxonomy, we classify the main works found in literature. In Section 3 we present the main concepts and ideas behind the Configurable Logic-in-Memory Architecture (CLiMA) that is presented here for the first time. Section 4 describes an adaptation of CLiMA for quantized Convolutional Neural Networks that in Section 5 is compared to a non in-memory architecture and Section 6 describes the adaptation for pNML.

2. State of the Art

The state of the art on in-memory computing is vast. The works found in literature differentiate from each other mainly for the role that the memory has in computing data. Starting from this observation, a taxonomy for classifying previous works was defined. According to this taxonomy the in-memory computing approaches can be divided in four main categories, as represented in Figure 1.

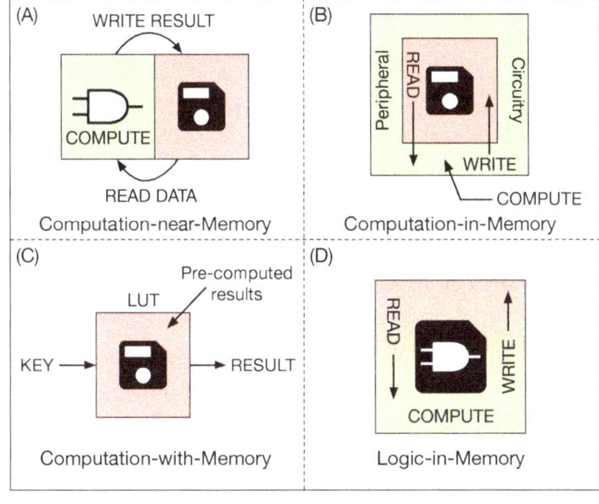

Figure 1. Depending on how the memory is used for computing data, four main in-memory computing approaches can be defined. (**A**) Computation-near-Memory (CnM): 3D-integration technologies allow one to bring computation and storage closer together by reducing the length of the interconnections. Logic and storage are still two separate entities. (**B**) Computation-in-Memory (CiM): The standard memory structure is not modified, while data computation is performed in the peripheral circuitry. (**C**) Computation-with-Memory (CwM): Memory is used as a Look Up Table to retrieve pre-computed results. (**D**) Logic-in-Memory (LiM): Data computation is performed directly inside the memory by adding simple logic in each memory cell.

The four main approaches are described in the following.

(A) Computation-near-Memory (CnM, Figure 1A): Thanks to the 3D Stacked Integrated Circuit technology (3D-SIC) [2], computation and storage are brought closer together, from which the name CnM, by stacking the two units one on top of the other. This technique has a two-fold advantage: Reducing the length of the interconnections and widening the memory bandwidth. However, this approach cannot be considered as true in-memory computing, since computation and storage are still two separate entities, but more as an evolution of conventional architectures based on the von Neumann model. Works belonging to this category are [3–8].

(B) Computation-in-Memory (CiM, Figure 1B): The structure of the memory array is not modified, while its intrinsic analog functionality is exploited to perform computation. In particular, in-memory computation is achieved by reading data from the memory which is then sensed by sense amplifiers (SAs). SAs are specifically modified in order to support the computation of a few simple logic operations (AND, OR, . . .). The result is then written back in the memory array. Decoders are also adapted in order to read more than one data from the array and execute row-wise (between data on different rows) or column-wise (between data on different columns) operations. Works belonging to this class are [9–14] and they all use a resistive non-volatile memory technology (RRAM). The approach followed in [15] is the same but here authors use a commodity volatile memory (DRAM, Dynamic Random Access Memory).

(C) Computation-with-Memory (CwM, Figure 1C): This approach uses memory as a Content Addressable Memory (CAM) to retrieve pre-computed results by means of a Look Up Table (LUT). The working principle of this kind of computation is that any Boolean function involving two or more inputs can be encoded in a memory by storing its truth table. In particular, input combinations are stored in a LUT, while results are stored in a CAM. Then the LUT is accessed through an input combination and an address is retrieved. These addresses are used to access the CAM and obtain the final result. Works that follows this approach are [16–20].

(D) Logic-in-Memory (LiM, Figure 1D): In this case logic is directly integrated inside the memory cell. Differently from the other three approaches, here data are computed locally without the need to move them outside the array (towards a close computing unit as in a CnM approach or towards the peripheral circuitry as in a CiM approach). Internal readings are performed in order to execute operations on data stored in different cells, by exploiting inter-cells connections. Internal writings are executed to locally save the result of the operation. There are a few works belonging to this category, such as [21–24].

3. Configurable Logic-In-Memory Architecture (CLiMA): Main Ideas

Our approach to in-memory computing, while mainly targeting the Logic-in-Memory concept, is not limited to it and also exploits the other approaches when required.

The novelties that we introduce with respect to existing works are manifold:

- The idea of an architecture that exploits various approaches to in-memory computing in order to adapt to different requirements and applications (Section 3);
- Configurability, hence flexibility, at different levels:

    - The basic block of CLiMA is a 1-bit Configurable LiM (CLiM) cell that can be programmed to perform different logic and arithmetic operations (Section 4.4);
    - More 1-bit CLiM cells can be grouped together to from a multi-bit CLiM cell that supports more complex operations such as bit-wise logic operations, multi-bit addition/subtraction, multiplication, shifts (Sections 3 and 4.4);

- A data flow for Convolutional Neural Networks workload and an inter-cells connection fabric specifically optimized to minimize memory accesses outside CLiMA, to maximize data-reuse inside the CLiM array and to support high parallelism (Sections 4.3–4.5);
- A pNML-based design of the 1-bit and multi-bit CLiM cells and a small version of the CLiM array (Section 6).

We demonstrate the effectiveness of our approach by comparing CLiMA to a non in-memory Deep Learning Accelerator, showing promising results in terms of performance and a significant reduction of external memory accesses, which are the main limitations of the von Neumann bottleneck. The innovations presented in this work will be thoroughly explained and highlighted in the following sections.

3.1. Overview

Figure 2 depicts the conceptual structure, in its most generic form, of the proposed in-memory computing architecture called CLiMA, Configurable Logic-in-Memory Architecture.

The key point in the definition of CLiMA is the flexibility. In fact, the idea is to conceive an architecture that well adapts to various applications that can benefit from in-memory computing in general and this means providing flexibility on different levels. In fact, applications differ for:

- Type of operations (logic, arithmetic);
- Complexity of operations (e.g., a logic function with respect to division);
- Data movement.

These parameters have an influence on the hardware requirements of the architecture. Depending on the type of operations and on their complexity, some of them can be executed directly in memory while others cannot. For this reason, as shown in Figure 2, CLiMA is conceived as a heterogeneous architecture composed of an in-memory (LiM and/or CiM) computing unit, the CLiM arrays, and a near-memory (CnM) computing unit. Operations that can be executed in-memory are dispatched to CLiM arrays, while the ones that cannot be executed in memory are assigned to the CnM unit. Each CLiM array is composed of different CLiM cells and, eventually, some extra-array (extra-row or extra-column) logic. A CLiM cell is thought as composed of a storage cell enhanced with simple logic that can be configured to perform different types of operations, from which the name Configurable Logic-in-Memory (CLiM) cell. The extra-array logic might be needed for further data processing outside the array and it can be considered as the CiM unit of CLiMA. The flexibility of CLiMA derives from its configurability (possibility of executing operations that differ for type and complexity) and from the presence of various degrees of in-memory computation (CnM, CiM, LiM).

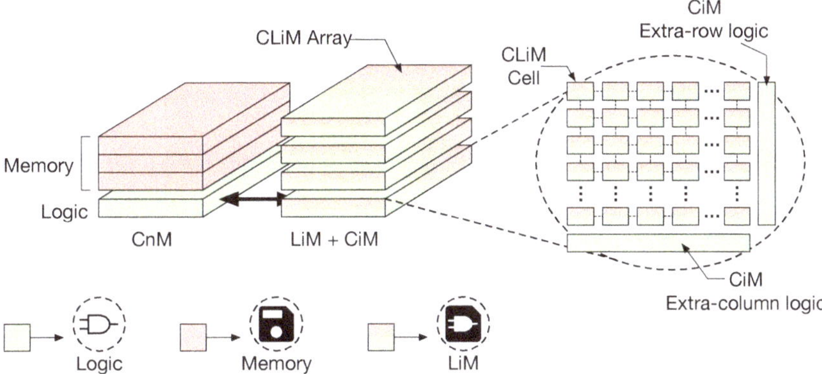

Figure 2. Conceptual structure of Configurable Logic-in-Memory Architecture (CLiMA): It can be seen as an heterogeneous unit that exploits configurability and different degrees of in-memory computation (CnM, CiM, LiM) to guarantee flexibility.

3.2. Type of Operations and Data Movement in CLiM Array

A more detailed view of CLiM array is shown in Figure 3.

The array is composed of CLiM cells whose reading/writing operations are controlled by bit lines (BL) and word lines (WL) as in a standard memory. Each CLiM cell is a logic-enhanced memory cell where data can be computed locally. In the example depicted in Figure 3, each CLiM cell is composed of a storage cell (MEM), a configurable logic block (CONFIG LOGIC) that can be configured to support different logic functions, and a full adder.

In addition to the local data computation inside each cell, CLiM cells are interconnected between them in order to support other kinds of operations inside the array (Figure 4):

- Intra-row computation between cells in the same row (black dashed arrow in Figure 4);
- Intra-column computation between cells in the same column (black solid arrow in Figure 4);
- Inter-row computation between two rows, an instance being an operation between a data stored in row 0 and one stored in row 1;
- Inter-column computation between two columns, an instance being an operation between a data stored in column 0 and one stored in column 1.

Intra-row connections can be exploited to implement in-memory addition. In fact, as shown in Figure 3, full adders belonging to different cells can be connected together to propagate the carry and build a Ripple Carry Adder (RCA, highlighted by the red box). Similarly, inter-row connections can be used to build an Array Multiplier (AM) by connecting two RCAs. In this way, it is possible to implement complex arithmetic functions completely in memory. The disadvantage is that RCAs and AMs are not fast arithmetic circuits, hence, applications that have a large number of additions and/or multiplications might be slowed down (especially for what concerns multiplications, since an AM is much slower than a RCA). A solution to this problem could be to delegate these operations to a fast non in-memory unit when the considered application is characterized by a very large number of arithmetic operations.

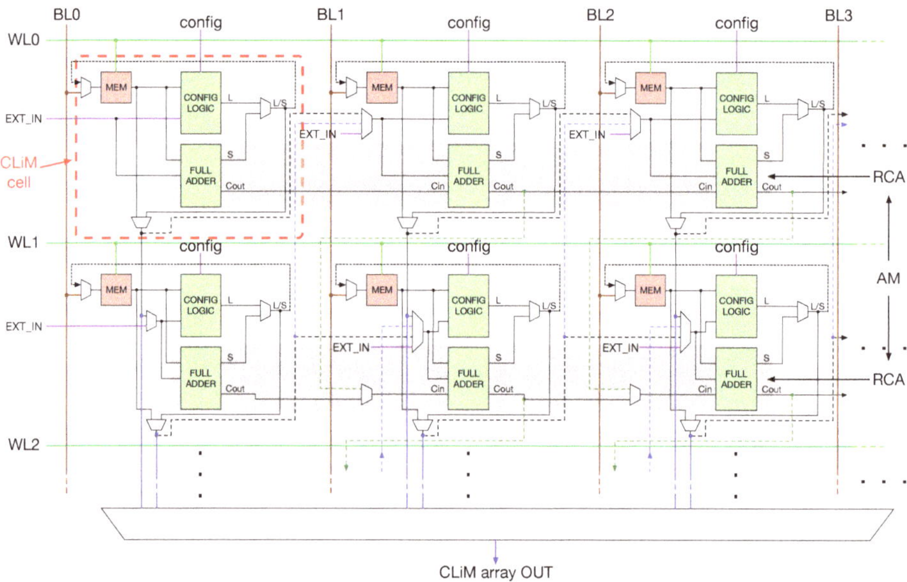

Figure 3. Detailed internal structure of the Configurable Logic-in-Memory (CLiM) array. Each CLiM cell can be represented as a logic-enhanced memory cell where data can be computed locally. By exploiting inter-cells connections it is possible to build more complex in-memory functions (e.g., a Ripple Carry Adder (RCA) or and Array Multiplier (AM)).

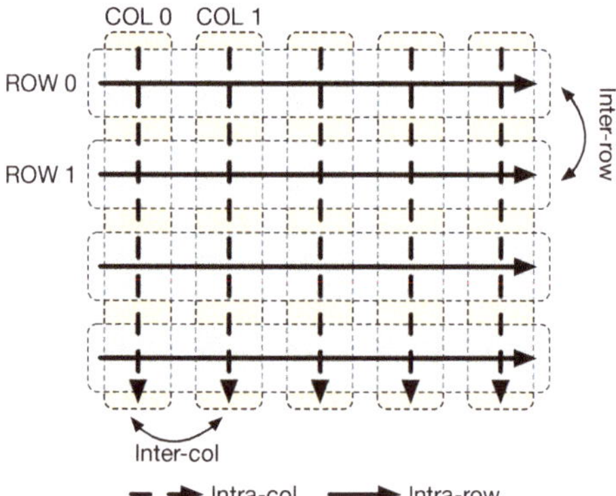

Figure 4. Possible types of data computation inside CLiM array.

4. CLiMA for Quantized Convolutional Neural Networks

On the basis of the ideas and concepts presented in Section 3, here a version of CLiMA is presented for quantized Convolutional Neural Networks. The reasons why CNNs have been chosen as target application are manifold:

- CNNs are an extremely popular application nowadays because they are a powerful method for solving many complex problems such as image recognition and classification, language processing, etc.;

- CNNs are data-intensive, hence, memory accesses represent the bottleneck;
- CNNs are computational-intensive, hence, they require hardware acceleration.

CLiMA is the ideal candidate for CNNs as it enables in-memory computation, drastically reducing the number of required memory accesses, and a high degree of parallelism, providing acceleration for time consuming applications like CNNs.

4.1. Convolutional Neural Networks (CNNs)

Convolutional Neural Networks (CNNs) [25–27] are a family of Artificial Neural Networks used for pattern recognition and classification. A CNN, as depicted in Figure 5, is composed of many 3D layers that are responsible for feature extraction and classification.

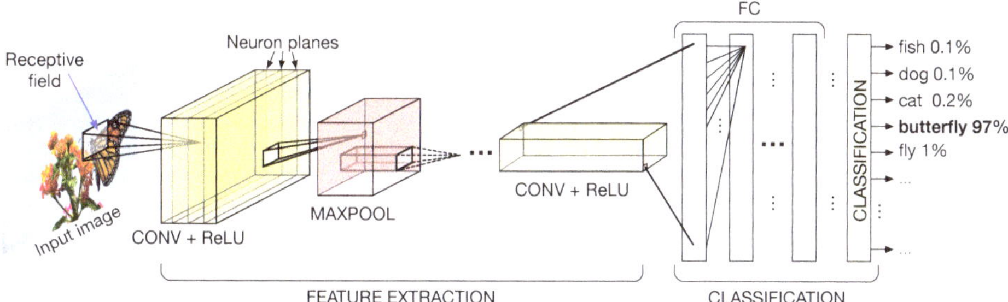

Figure 5. Convolutional Neural Networks (CNNs) are composed of different 3D layers. Each layer extracts different features from the input image.

Layers are three-dimensional as they are composed of a number of neuron planes, where each neuron analyzes a small portion of the input image, called the receptive field, extracting some key features. The feature extraction process is carried out by filtering the image with a kernel of weights (a filter), that is shared over a plane of neurons. The extraction of features by using the kernels of weights is called convolution, from which the name of the network. The output produced by the convolution operation is called the output feature map (i.e., the filtered image) and it is the input of the subsequent layer. Convolutional layers (CONV) are responsible for the extraction of features. Other type of layers are used to down sample feature maps (e.g., maxpooling) or to introduce linear rectification (e.g., Rectifying Linear Unit (ReLU)). Fully connected (FC) layers are responsible for the actual classification.

Figure 6 shows in more detail how the convolution operation works.

The input image is usually composed of different input channels (C_{in}) with dimensions $R \times C$. The kernels used to extract features have the same number of channels C_{in} as the input image and dimensions $K \times K$, which can vary in each layer. Kernels are slid on the input feature map by a quantity called stride (S). The number of kernels (F) determines the number of channels (C_{out}) of the resulting output feature map, which has dimensions $O \times P$. The dimensions of the output feature map depend on the input image dimensions, the kernel dimensions and the stride, according to Equation (1).

$$O = \frac{R - K}{S} + 1; \; P = \frac{C - K}{S} + 1. \tag{1}$$

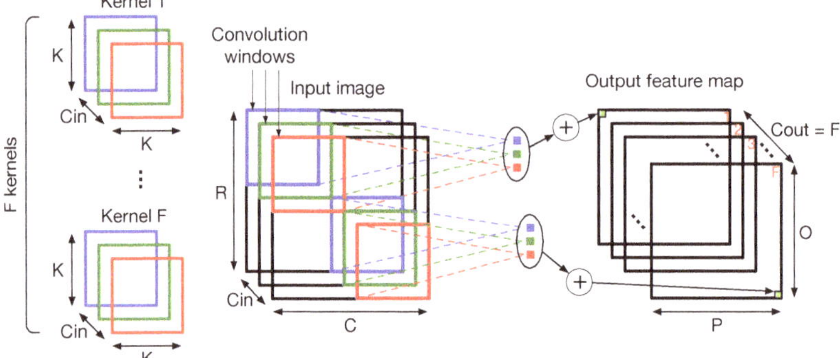

Figure 6. High-dimensional convolution operation.

CNNs are characterized by a complex structure and, over the years, network architectures have become more and more complex. The consequences of this growth are the need for very high-performance systems able to sustain such large throughput, and the increase of memory requirements because of the large number of parameters.

4.2. ShiftCNN: A Quantized CNN

Since an in-memory implementation can support only simple operations and limited precision, quantized CNNs are the perfect fit for in-memory computing architectures, since memory and computational requirements are greatly reduced in exchange for a small loss in prediction accuracy. In [28] authors propose to use power-of-two weights to eliminate the need for multiplications, which are instead transformed in simple shift operations. Moreover, according to their quantization algorithm, all weights are values of the type 2^{-n}, hence, shift operations are all arithmetic right shifts. ShiftCNN has been chosen as target application for CLiMA.

4.3. CNN Data Flow Mapping Scheme for CLiMA

In this section we present a CNN data flow mapping scheme specifically optimized for CLiMA. Differently from the commonly used unrolling technique, this mapping scheme avoids data redundancy while guaranteeing parallel computation.

The convolution operation, as highlighted in Figure 7, consists in applying a kernel of weights over the input feature map.

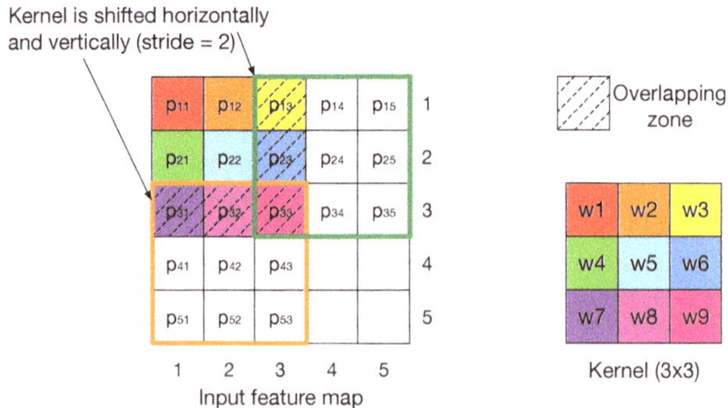

Figure 7. The kernel of weights is slid over the entire input image by a quantity called stride. The sub-region of the input image on which the kernel is applied is called convolution window. Convolution widows partially overlap.

As explained in Section 4.1, the kernel is slid horizontally and vertically by a quantity called stride. In the example in Figure 7 the stride is equal to 2. The sub-region of the input feature map on which the kernel is applied is called the convolution window. It can be seen that convolution windows partially overlap so, in order to allow parallel computation, they are unrolled and overlapping regions are replicated causing data redundancy. The impact of unrolling convolution widows is exacerbated as the size of the kernel increases and the stride decreases, since the overlapping region gets larger. The graph in Figure 8 shows how the number of input features vary when applying unrolling, for each convolutional layer of two popular CNNs, AlexNet [29] and ResNet-18 [30].

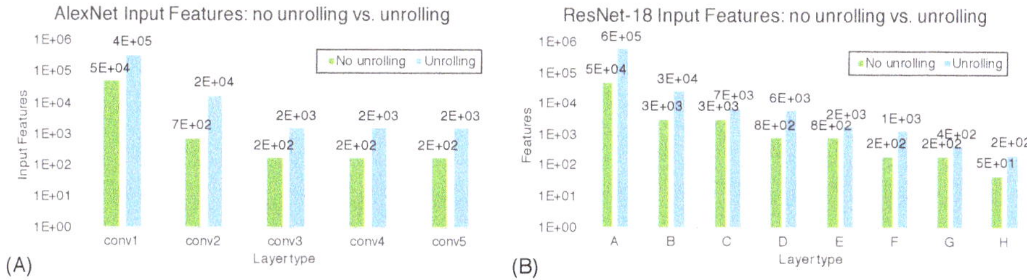

Figure 8. Data redundancy caused by unrolling in (**A**) AlexNet and (**B**) ResNet-18. Green columns represent the number of input features when applying no unrolling, blue columns represent the number of input features when applying unrolling. Input features are shown for each convolutional layer.

It can be seen that the data redundancy is not at all negligible as the number of unrolled input features (blue columns) increases of one order of magnitude with respect to the original number of features (green columns). For an architecture such as CLiMA, data redundancy is not acceptable since the storage space must be used in the most efficient way possible. For this reason, a different data flow mapping scheme is proposed. When executing convolution, not all convolution windows overlap, hence, those that do not overlap can be executed in parallel. As shown in Figure 9, the convolution operation can be divided in different steps in which only non-overlapping convolution windows are executed in parallel.

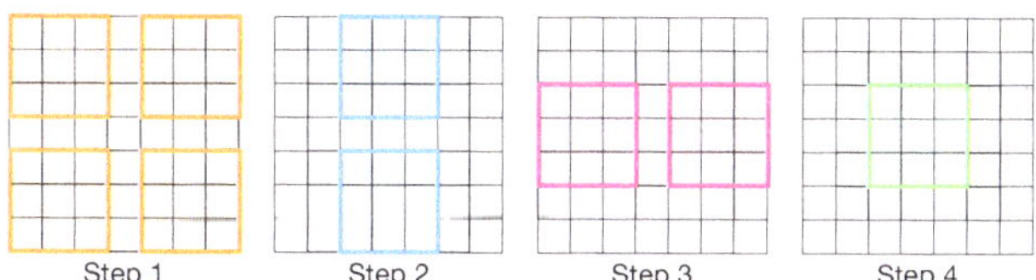

Figure 9. The convolution operation is divided in different steps. In each step only non-overlapping convolution windows are executed in parallel.

The number of steps to complete the convolution between a kernel of weights and an input feature map depends on the size of the input feature map, the kernel and on the stride. In the example in Figure 9, four steps are required to complete the convolution. The number of steps can be computed according to the following equation:

$$\#steps = \frac{tot\_conv\_windows}{parallel\_conv\_windows}. \tag{2}$$

In Equation (2), *tot_conv_windows* is the total number of convolution windows while *parallel_conv_windows* is the number of non-overlapping convolution windows that can be executed in parallel. This number can be calculated as:

$$parallel\_conv\_windows = \left(\frac{C}{K + (S - 1)} \right)^2, \, K > 1. \tag{3}$$

Equation (3) is valid for kernels with dimensions larger than one ($K > 1$). When the kernel has size 1×1 the number of non-overlapping convolution windows is equal to the number of total windows. It is clear that the advantage of this parallel non-overlapping data flow scheme is to avoid data redundancy while still guaranteeing parallel computation. This scheme can be naturally mapped on CLiMA by assigning a pixel of the input feature map to each cell of the array. Weights are instead properly distributed and shifted over the array (Section 4.5).

4.4. CLiM Array Structure

Figure 10 depicts the architecture of CLiMA for quantized CNNs.

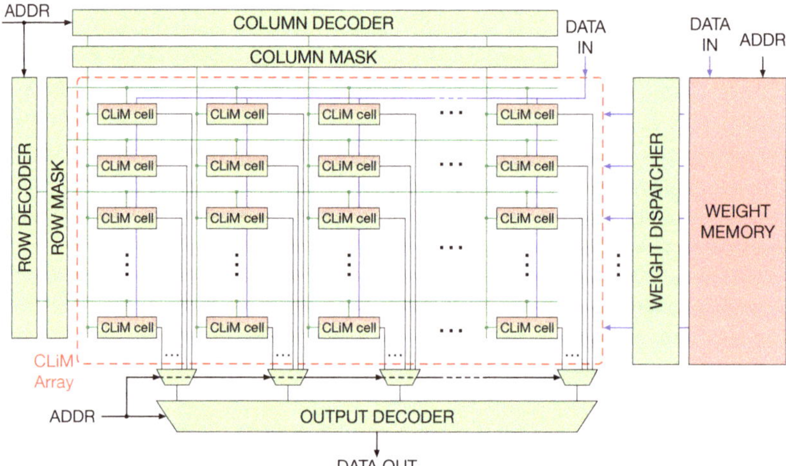

Figure 10. Architecture of CLiMA for quantized CNNs.

The main core of CLiMA is the array of CLiM cells. Each CLiM cell has both storage and computation capabilities. Modified row and column decoders are used to control the data flow inside the array. Weights are read from a weight memory, external to the array, and dispatched through a weight dispatching mechanism. More details on decoders and the weight dispatcher will be given in Section 4.5. The internal structure of the CLiM cell is shown in Figure 11.

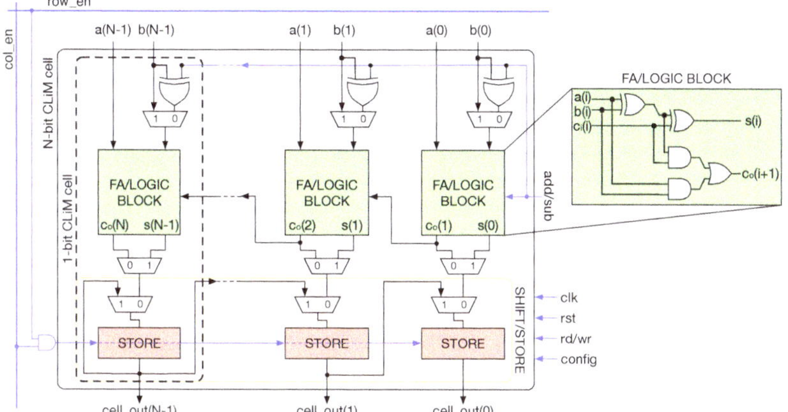

Figure 11. Internal structure of the CLiM cell. Many 1-bit CLiM cells are properly interconnected, exploiting inter-cell connections, to build a more complex N-bit CLiM cell.

It can be seen that many 1-bit CLiM cells are properly interconnected, by exploiting inter-cell connections, to create a more complex N-bit CLiM cell. Each 1-bit cell is composed of a configurable computational block, a storage cell and other simple logic. The computational block is a Full Adder (FA) that can also be used to perform logic operations by fixing one or more of the FA inputs to logic 0 or 1, as shown in Table 1.

Table 1. Logic operations that can be performed with a Full Adder by fixing one or more of the inputs. In this case A, B and C_{in} are the three inputs while S and C_{out} are the output (sum and output carry, respectively).

Fixed Input	S	C_{out}
$A = 0$	$B \oplus C_{in}$	$B \cdot C_{in}$
$A = 1$	$\overline{B \oplus C_{in}}$	$B + C_{in}$
$A = 0 \,\&\, B = 1$	$\overline{C_{in}}$	C_{in}
$A = 1 \,\&\, B = 0$	$\overline{C_{in}}$	C_{in}

In order to support multi-bit addition, the output carry (C_{out}) of the FA inside a 1-bit CLiM cell is connected to the input carry (C_{in}) of the adjacent 1-bit cell. By exploiting inter-cell connections it is possible to build an in-memory Ripple Carry Adder (RCA). In addition, storage cells are interconnected in a chain-like manner in order to implement a multi-bit storage block that can also work as a shift register. Only right shifts are supported in the case represented in Figure 11 since, as explained in Section 4.2, ShiftCNN requires only those. Nonetheless, with very simple modifications left shifts can also be handled. Moreover, for the sake of clarity, Figure 11 does not show the presence of redundant storage blocks (one for each 1-bit cell, in addition to the one that is also used as the shift register). The redundant storage block is used to retain partial results that will be reused for further elaboration.

The architecture depicted in Figure 10 does not show the interconnections between CLiM cells. These interconnections have been specifically designed to support CNN-like data flow inside the array. A detailed scheme of the interconnection fabric inside the CLiM array is shown in Figure 12.

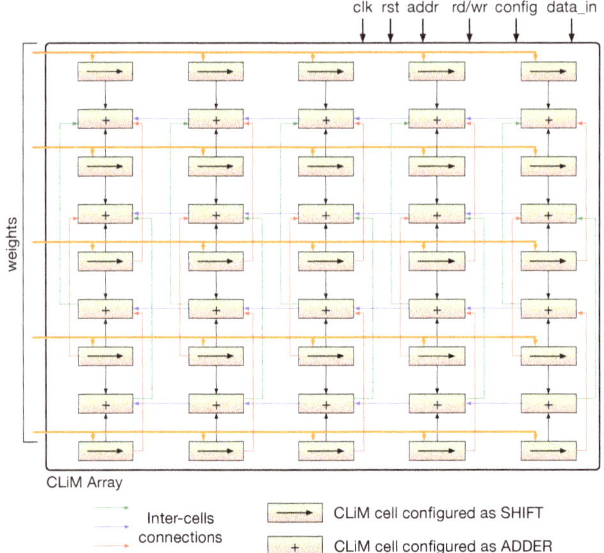

Figure 12. Interconnection fabric inside the CLiM array.

Furthermore, rows of CLiM cells are alternatively configured as shift registers (even rows) and adders (odd rows). The idea is to store pixels of the input feature map inside shift cells where they are also locally shifted according to the value of the correspondent weight. Then the shifted pixels are accumulated in the cells configured as adders. Figure 13 clarifies how convolution is managed inside the array.

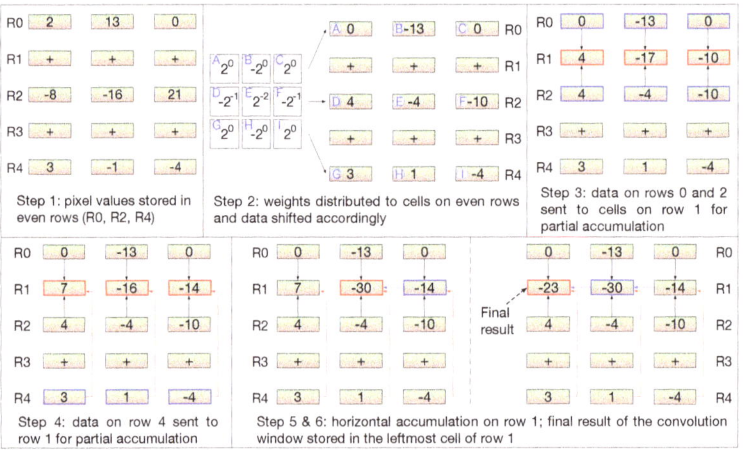

Figure 13. Management of convolution computation inside the CLiM array.

In particular, the computation of a 3×3 convolution window is shown as example. The interconnection fabric has been designed to be flexible, hence, it can support any kernel size.

4.5. Weight Dispatching Mechanism

In order to support the parallel non-overlapping data flow scheme shown in Figure 9, weights must be properly dispatched to the cells inside the CLiM array. In order to do so, the combined action of the weight dispatcher and row/column decoders is exploited. Row/column decoders are modified in order to activate multiple adjacent rows/columns. A starting and an ending address are provided to decoders that will consequently activate all rows/columns comprises between the starting and the ending address. Since, as it can be notice from Figure 9, parallel convolution windows might not be adjacent, row/column masks are used to disable those rows or columns comprised between the starting and ending address which must remain inactive. The weight dispatcher is used to properly shuffle weights over the array.

As highlighted in Figure 14A, the window shifting process is obtained by controlling which cells are active and which are not, step after step. At the same time, weights are properly shuffled, as shown in Figure 14B, so that they are distributed to the correct cells.

The weight dispatching mechanism has been optimized for 3×3 kernels since they are the most common ones. Nonetheless, other kernel sizes can be also supported.

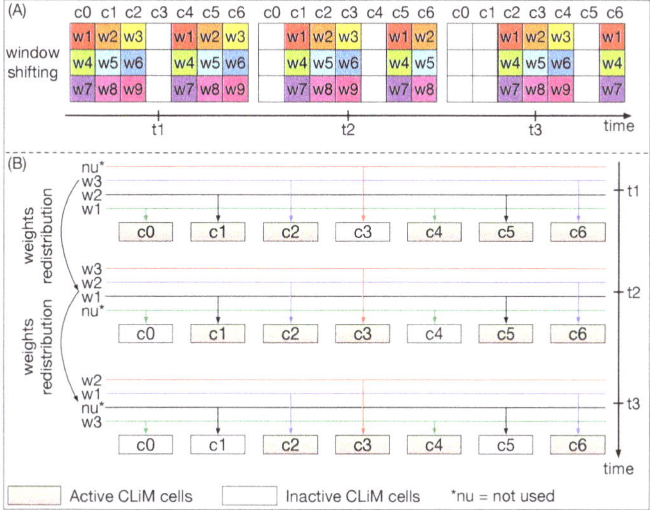

Figure 14. (A) Convolution windows are shifted over the array by properly activating/inactivating rows and columns. (B) The weight dispatcher properly distributes weights inside the CLiM array in order to reproduce the convolution window shifting process.

4.6. Data Reuse Possibilities

One of the main reasons for exploiting a Logic-in-Memory architecture such as CLiMA for Convolutional Neural Networks is the possibility of reusing data already stored and computed inside the array for further processing, without any need to move it outside.

The possibilities for data reuse in CLiMA are summarized in Figure 15 and explained in the following.

- Filters are reused across input feature maps according to the sliding window process (Figure 15A).
- Input feature maps are reused by different filters (Figure 15A).
- Partial results are reused for further processing (cross-channel accumulation) to obtain the final output feature maps (Figure 15B).

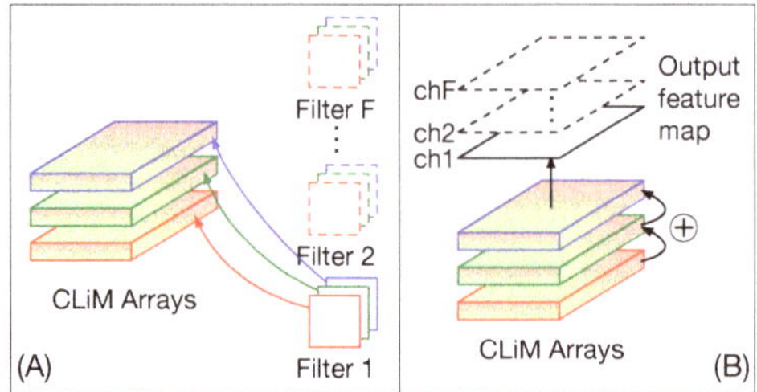

Figure 15. Data reuse in CLiMA. (**A**) Filters are reused across input feature maps according to the sliding window process. Input feature maps are also reused by different filters. (**B**) Partial results are reused for further processing to obtain the final output feature maps.

5. Results and Discussion

Before bounding it to any technology, CLiMA was modelled by using a fully parametric VHDL (VHSIC Hardware Description Language) code that was validated by means of extensive simulations and by comparing the obtained results to those obtained from an analogous model developed in MATLAB. Moreover, in order to prove the effectiveness of the CLiMA computational model, it has been compared to a conventional (non in-memory) Deep Learning Processor presented in [31,32].

An analytic computational model of CLiMA was defined. This model takes into account the following parameters:

- Convolutional layer parameters including input feature map dimensions (R, C), kernel dimensions (K), stride (S) and output feature map dimensions (O, P);
- The number of parallel non overlapping convolution windows;
- The number of execution cycles needed to complete a convolution window.

The total number of convolution widows in a layer depends on the size of the output feature map, that is given by the following equation:

$$O = P = \frac{R - K}{S} + 1. \tag{4}$$

We are assuming that input and output feature maps and kernels are square, hence, they have the same width and height $(R = C, O = P)$. The total number of convolution widows, CW_{tot}, is then equal to:

$$CW_{tot} = O \cdot P = O^2 = P^2. \tag{5}$$

The number of non overlapping convolution windows, CW_{non-ov}, is given by the following expression:

$$CW_{non-ov} = \left(\frac{R}{K + (S - 1)} \right)^2. \tag{6}$$

According to the data flow mapping scheme presented in Section 4.3, a certain number of steps is needed to complete a convolution operation. This number, C_{steps}, is equal to the upper bound of the ratio between the total number of convolution windows CW_{tot} and the number of parallel non-overlapping ones CW_{non-ov}:

$$C_{steps} = \left\lceil \frac{CW_{tot}}{CW_{non-ov}} \right\rceil. \tag{7}$$

The number of execution cycles, C_{cycles}, needed to complete a full convolution operation on a layer is given by the product between the number of cycles to execute a single convolution window, CW_{cycles}, and C_{steps}:

$$C_{cycles} = CW_{cycles} \cdot C_{steps}. \tag{8}$$

CW_{cycles} depends on the size of the convolution window that, in turn, depends on the size of the kernel. Moreover, by taking into account how a convolution window is mapped and executed inside CLiMA, the term CW_{cycles} can be calculated as following:

$$CW_{cycles} = 8 + 1 + \left(\frac{K - 1}{2} \right) + (K - 1). \tag{9}$$

In Equation (9) the following factors are taken into account:

- The number of cycles to execute shift operations; in CLiMA data are shifted 1 bit at a time. Since weights are 8-bit long, in the worst case scenario eight cycles are needed to complete the operation;
- The number of cycles to execute accumulations:

  - One cycle for partial accumulation of data couples (Figure 13, step 3); this term does not depend on the size of the kernel because these accumulations can always be done in parallel;
  - $(K - 1)/2$ cycles for partial accumulation of non-adjacent data (Figure 13, step 4); this term depends on the size of the kernel, in fact, as the convolution window dimension changes the number of non-adjacent data to accumulate changes as well;
  - $K - 1$ cycles to perform final horizontal accumulations (Figure 13, steps 5 and 6); similarly to the previous term, also this one depends on the size of the kernel.

Equations (7) and (9) can be substituted in Equation (8) to obtain the total number of cycles required to execute a full convolution operation of a layer.

This simple but effective computational model was used to extract results and carry out comparisons between CLiMA and the Deep Learning Processor, by considering AlexNet and ResNet-18. The Deep Learning Processor is composed of a number of Processing Elements (PEs) that are capable of performing different types of operations including Multiply-Accumulate (MAC) ones. PEs work in parallel and each of them has a throughput of 1 MAC per cycle. Assuming that each PE executes a convolution window, it takes $K \times K$ cycles to complete a single convolution window. For what concerns CLiMA, the assumption is that a certain number of non-overlapping convolution windows is executed in parallel inside the array. In order to perform comparisons, four different scenarios were considered. The difference between these scenarios is the parallelism that, for the Deep Learning Processor, is referred to the number of parallel PEs, while for CLiMA, it is referred to the number of parallel non-overlapping windows. Figures 16 and 17 report the average number of clock cycles needed to perform a complete convolution in different parallelism scenarios for AlexNet and ResNet, respectively.

The average number of clock cycles is simply calculated by averaging the number of clock cycles needed to complete the convolution of each layer in the considered CNN. In both graphs, the parallelism level is reported on the x axis, while the average number of clock cycles is shown in the y axis. It can be clearly seen that, for both the CNNs and for all the parallelism scenarios, CLiMA outperforms the Deep Learning Accelerator. In the AlexNet case, the average cycles are reduced by 78% percent in the worst parallelism scenarios (only 10 PEs or non-overlapping convolution windows). The percentage reduction slightly decreases as the parallelism increases, reaching -70% in the best parallelism scenario (60 PEs or non-overlapping convolution windows). For what concerns ResNet, the trend shown in Figure 17 is similar to the AlexNet one, except that the difference between the average cycles of CLiMA with respect to the Deep Learning Accelerator is smaller. In fact, it ranges from -49% in the worst parallelism scenario to -45% in the best.

For both the CNNs, CLiMA provides a reduction in terms of average cycles needed to complete the convolution in all the layers of the network that is higher when the parallelism level is smaller, as compared to the Deep Learning Accelerator, further proving the effectiveness of the CLiMA computational model. The reduction difference between AlexNet and ResNet-18 depends on the characteristics of the two networks (i.e., layers and kernels dimensions, number of channels etc.).

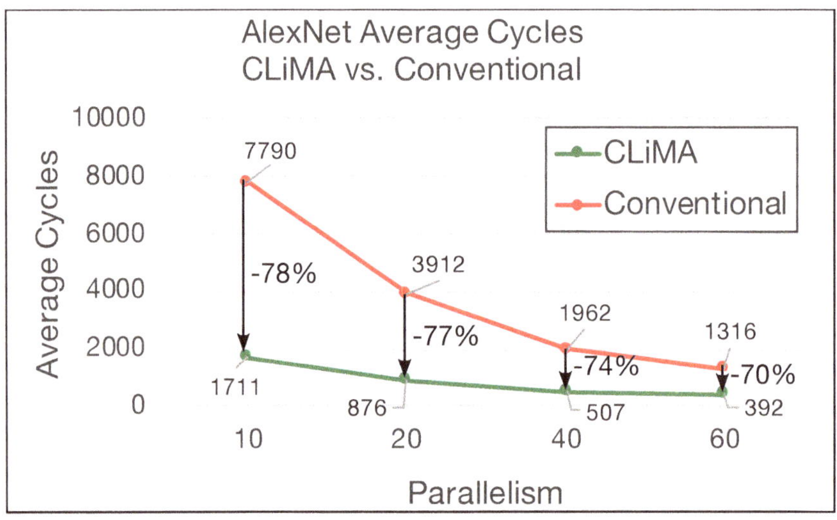

Figure 16. Average cycles needed to execute AlexNet in different scenarios: CLiMA vs. Conventional.

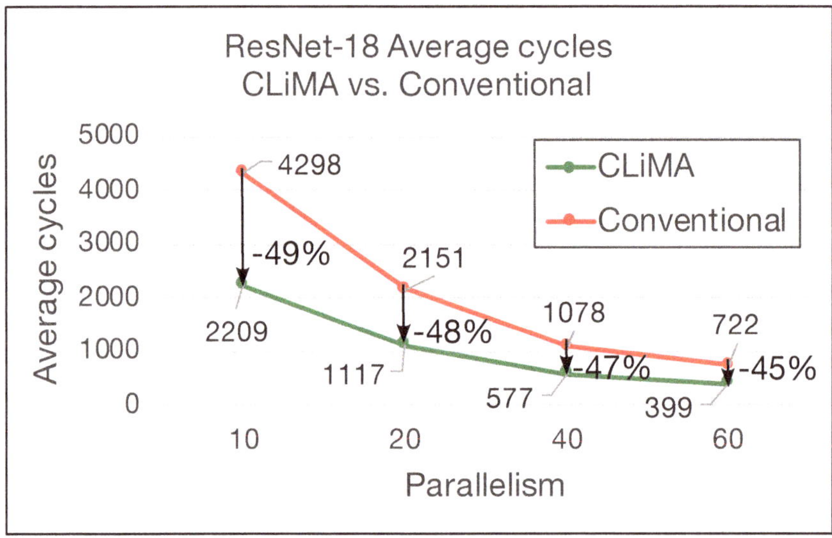

Figure 17. Average cycles needed to execute ResNet-18 in different scenarios: CLiMA vs. Conventional.

The VHDL code used to describe CLiMA was synthesized in order to get an estimation of the maximum working frequency at which the architecture can run. The technology used for the synthesis is the same used for the Deep Learning Accelerator and it is a commercial 28 nm FDSOI (Fully Depleted Silicon-on-Insulator). For both architectures a parallelism of 10 has been chosen and the maximum reachable working frequency, in both cases, is approximately 1.8 GHz. The working frequency was used to compute the execution time required by CLiMA and the Deep Learning Processor to run ALexNet and ResNet-18 when the parallelism is 10. Results are reported in Table 2.

Table 2. Performance estimation of CLiMA with respect to the Deep Learning Accelerator for AlexNet and ResNet-18 when the parallelism is 10. For both architectures the working frequency is 1.8 GHz.

CNN Type	Architecture	Average Cycles	T_{exec} (µs)
AlexNet	CLiMA	1711	0.95
	DL Acc.	7790	43.2
ResNet-18	CLiMA	2209	1.2
	DL Acc.	42,939	24

When comparing the two architectures, since the working frequency is the same, whereas the number of average cycles required by CLiMA is much lower than what the Deep Learning Accelerator requires, the resulting execution time needed to complete the convolution of Alexnet and ResNet-18 is, respectively, 45× and 20× lower for CLiMA with respect to the Deep Learning Accelerator.

The main figure of comparison between the two architectures is related to the number of memory accesses. In fact, we want to demonstrate that not only is the CLiMA computational paradigm effective in terms of execution acceleration thanks to its intrinsic massive parallelism, but it is also effective in reducing the data exchange between the processing unit and the memory. When considering CLiMA, as shown in Figure 10, we can identify the computing core that is the CLiM array and an external memory that is the weight memory. This memory is accessed to retrieve the weights that are reused over all the convolution windows inside a feature map, therefore, requiring only $K \times K$ read operations. We are assuming that the input features are already stored inside each CLiM cell of the array, neglecting the write operations required to load them for the first time as this is an initialization operation that cannot be avoided. Once the convolution operation is completed, the final results, which are then reused for cross-channel accumulation, are already stored inside the CLiM array, hence, no further external write or read operation is needed.

When considering, instead, the Deep Learning Accelerator, both input features and weights are continuously read from an input buffer and passed to the execution unit that performs MAC operations and then writes results into an output buffer. Therefore, the number of read/write operations to input/output buffers, when considering all convolution windows in a layer, is:

- $2 \times (K \times K) \times tot\_conv\_windows$ read accesses to the input buffer to retrieve input features and weights;
- $O \times P$ write accesses to the output buffer to store the convolution results.

As for CLiMA, we are not considering that input features and weights must be loaded from an external memory into the input buffer because it is an unavoidable operation.

Figures 18 and 19 show the comparison in terms of memory accesses between CLiMA and the Deep Learning Accelerator for AlexNet and ResNet-18, respectively. It can be clearly noticed that the in-memory computational model and the data reuse possibilities offered by CLiMA make it possible to drastically reduce the number of memory accesses with respect to a non-in-memory conventional paradigm, such as the one used in the Deep Learning Processor.

In general, comparing CLiMA to other architectures (either in-memory or conventional ones) is not easy because of architectural, technological and computational differences. As a result, the risk is that the comparison might be unfair. In addition, most of the time, papers lack of details about how the proposed architectures manage the computation or how there are no common comparison figures. This makes comparisons even more difficult and, for this reason, CLiMA was only compared to a conventional architecture (the Deep Learning Processor) about which we had sufficient details to be able to extract some useful data.

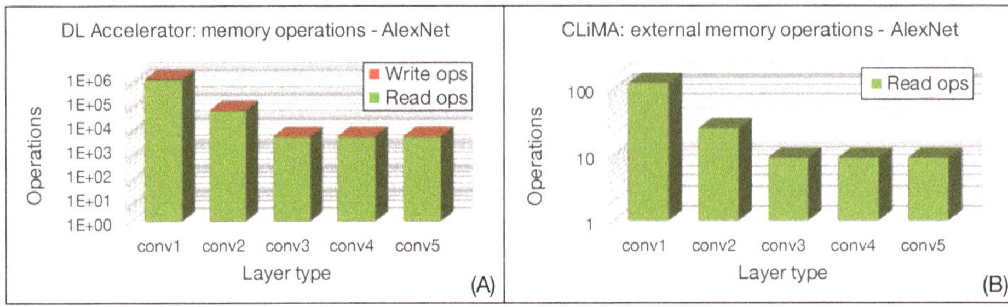

Figure 18. Memory access evaluation for AlexNet in (**A**) Deep Learning Accelerator and (**B**) CLiMA.

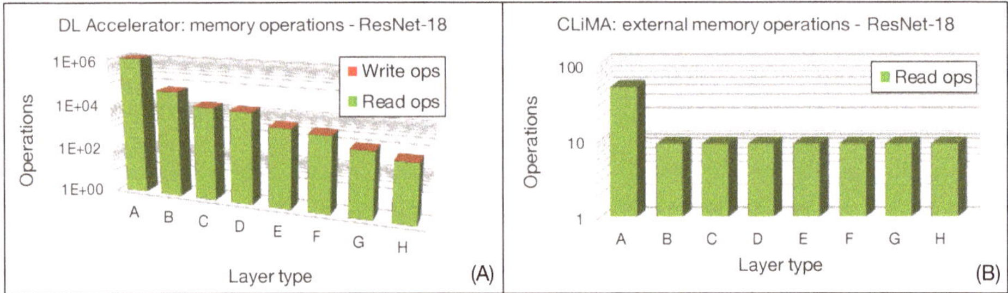

Figure 19. Memory access evaluation for ResNet-18 in (**A**) Deep Learning Accelerator and (**B**) CLiMA.

6. Beyond CMOS: A pNML Implementation

Perpendicular Nano Magnetic Logic (pNML) [33] is considered one of the most promising alternative technologies to CMOS [34] and it is perfect for in-memory computation as it intrinsically provides both non-volatility and computing capabilities [35,36]. In addition, pNML offers 3D integrability and low power consumption, all characteristics that make this technology ideal for overcoming the issues related to von Neumann architectures and CMOS scaling.

6.1. pNML Basics

pNML is based on the nanomagnet, a small (∼tens of nanometers) single domain multi-layer Co/Pt device that has bi-stable behavior. This means that, because of the perpendicular (from which the name perpendicular NML) magnetization anisotropy, it can be only in two stable magnetization states that depend on the direction of the magnetization. These states can be used to encode logic '0' and logic '1', as shown in Figure 20A.

Signal propagation in pNML depends on the magneto-static field-coupling interactions between nanomagnets [37]. In order to propagate the information in a specific direction, the magnetic properties of a small region of the nanomagnet are modified through Focused Ion Beam (FIB) irradiation [38]. The irradiated region is called the Artificial Nucleation Center (ANC). As shown in Figure 20B, neighboring pNML cells couple in a parallel or anti-parallel way, depending on their relative position, favoring signal propagation in a direction that depends on the position of the ANC. The ANC is the point where the nucleation of a domain wall starts and eventually propagates inside the magnetic device (Figure 20C). ANCs can also be obtained by changing the shape and thickness of the nanomagnet [39] (Figure 20E). The propagation of information inside pNML circuits is obtained thanks to an external magnetic field (sinusoidal as shown in Figure 20D) that is applied globally to the circuit [40]. This external magnetic field has the same function of the clock signal in CMOS circuits. Thanks to the combined action of ANCs and the clocking field, information propagation can be correctly controlled in pNML circuits. The elementary pNML blocks with which any logic circuit can be built are the inverter (Figure 20E), the notch (Figure 20F) and the minority voter

(Figure 20G and 3D version in Figure 20H). The notch works as a barrier, blocking the signal propagation unless a short depinning magnetic field is applied [41]. This block can be used to create memory elements [42,43]. Moreover, pNML technology allows one to build 3D structures by stacking different layers of nanomagnets [44–47]. Previous works such as [42,48–52] already explore the potentialities of NanoMagnetic Logic architectures (3D and non), but none of them propose a complete Logic-in-Memory design, which is instead presented in the following.

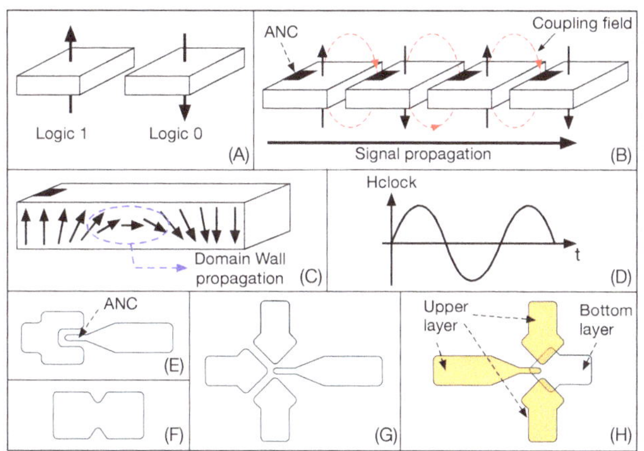

Figure 20. pNML basics. (**A**) The magnetization direction encodes logic '0' and '1'. (**B**) The Artificial Nucleation Center (ANC) guarantees correct signal propagation in a perpendicular Nano Magnetic Logic (pNML) chain of magnets. (**C**) Domain wall propagation inside the nanomagnet causes the switch of the magnetization direction. (**D**) Global out-of-plane magnetic field used as clocking mechanism. (**E**) Inverter. (**F**) Notch. (**G**) Minority voter. (**H**) 3D minority voter.

6.2. pNML-Based CLiM Array

Figure 21 depicts a small pNML-based version of the CLiM array described in Section 4.4.

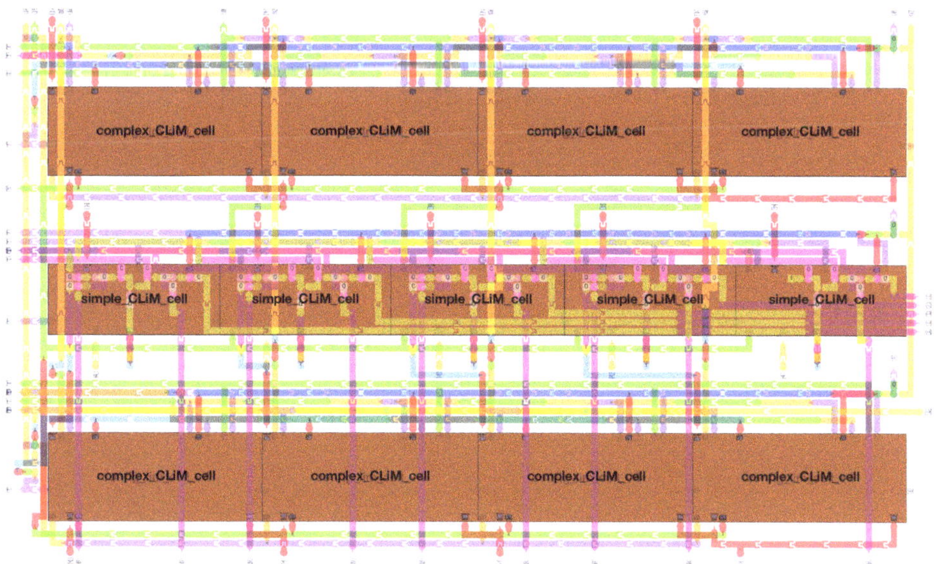

Figure 21. Small pNML-based version of the CLiM array.

The design of the array was done by using MagCAD (https://topolinano.polito.it) [53], a CAD for emerging technologies developed at the VLSI Laboratory (research group in the Department of Electronics and Telecommunication Engineering of Politecnico di Torino). MagCAD allows one to design pNML structures thanks to an intuitive and simple GUI (Graphical User Interface) in which elementary blocks can be combined together to form more complex structures and 3D designs. Starting from the designed structure, MagCAD allows the extraction of the VHDL description of the circuit, that is based on a compact VHDL model [35] of pNML devices. The generated VHDL can be used to simulate (using a common HDL simulator) and verify the functionality of the circuit [54–56]. The complexity of the pNML-based CLiM array depends on the complexity of the interconnections between CLiM cells, as it can be noticed from Section 4.4. This strongly limits the size of the design that can be implemented by hand, without any support for the routing. The design in Figure 21 uses nine layers of nanomagnets. There are two types of cells used for the pNML array, one called complex (Figure 22) and the other the simple (Figure 23) CLiM cell. Both are based on the structure shown in Figure 11, the only difference between them being that the simple CLiM cell does not support shift operations and does not have the redundant storage block. The simple CLiM cell can be used in the odd rows of the array that perform only accumulations.

Both the cells have four layers of magnets. Based on the dimensions of the nanomagnets, that in these designs are 30 × 50 nm, the area occupied by the complex cell is 22.5 µm, while the simple cell occupies 14.4 µm. The area of the CLiM array is 582 µm and interconnections occupy a big portion of it because of their complexity.

Even though in the designs here presented we have used relatively large nanomagnets (30 × 50 nm), pNML can be easily scaled to improve compactness. The designs could be also improved in order to reduce the impact of interconnections on the overall area occupation, however, as already said, the lack of an automatic and optimized routing tool makes it challenging. Nonetheless, the non-volatile nature of the technology and the total absence of current flow and leakage sources makes it an ideal beyond-CMOS technology for in-memory computing.

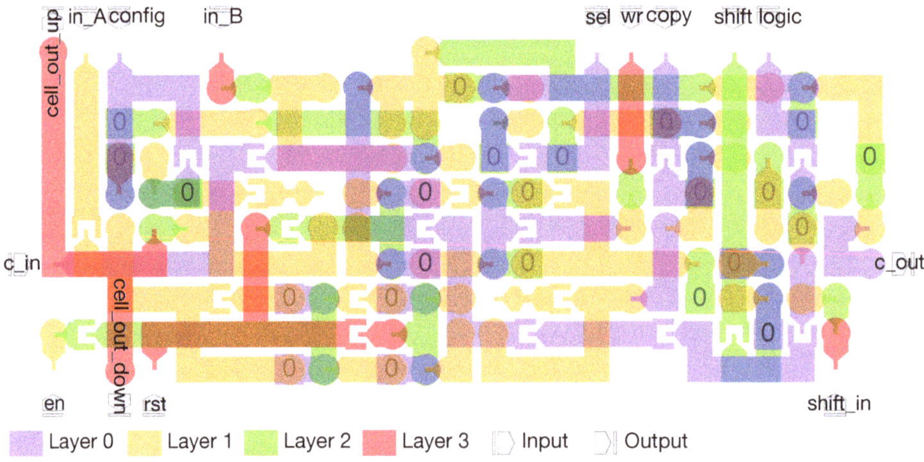

Figure 22. Complex pNML cell.

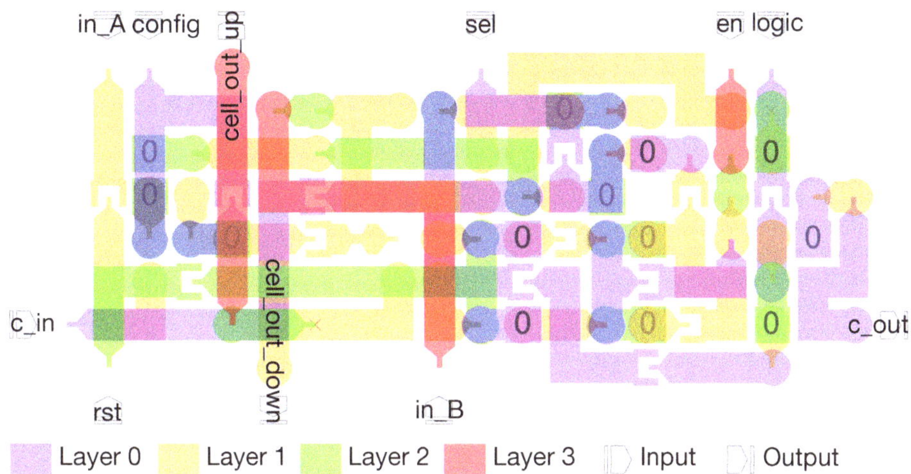

Figure 23. Simple pNML cell.

7. Conclusions

The Configurable Logic-in-Memory architecture that we have presented has strong points and issues that are worth being analyzed. Regarding its advantages, CLiMA provides:

- In-memory computation: Data are processed directly inside the memory, drastically reducing the need of data movement and favoring their reusing for further computation;
- Parallelism: The array is intrinsically highly parallel and perfect for accelerating compute and data intensive applications;
- Flexibility: The configurability of the cells and the possibility of exploiting inter-cells connections to build complex in-memory functions make CLiMA adaptable to different applications.

Regarding its limitations, mainly two can be identified:

- Not all data-flows can be supported in an array-like structure because moving data from any source to any destination is not easy and would require a very complex (but flexible) network of interconnections;
- The control of data movement between cells is complex and must be managed carefully in order to avoid cells receiving/sending wrong data from/to wrong cells.

To conclude, the Logic-in-Memory paradigm seems to be a promising alternative to the von Neumann one. We have defined a novel Configurable Logic-in-Memory Architecture that relies on in-memory computation, flexibility and parallelism to tackle the memory bottleneck problem while also providing high performance.

Author Contributions: Conceptualization, G.S.; methodology, G.S., M.G. and G.T.; investigation, G.S.; data curation, G.S.; writing—original draft preparation, G.S.; writing—review and editing, G.S., G.T. and M.G.; supervision, M.G.

Funding: This research received no external funding.

References

1. 2013 International Technology Roadmap for Semiconductors (ITRS). Available online: http://www.itrs2. net/2013-itrs.html (accessed on 31 May 2013).
2. 2009 International Technology Roadmap for Semiconductors (ITRS). Available online: https://www. semiconductors.org/wp-content/uploads/2018/09/Interconnect.pdf (accessed on 31 May 2009).

3. Kim, D.H.; Athikulwongse, K.; Healy, M.B.; Hossain, M.M.; Jung, M.; Khorosh, I.; Kumar, G.; Lee, Y.; Lewis, D.L.; Lin, T.; et al. Design and Analysis of 3D-MAPS (3D Massively Parallel Processor with Stacked Memory). *IEEE Trans. Comput.* **2015**, *64*, 112–125. [CrossRef]

4. Zhu, Q.; Akin, B.; Sumbul, H.E.; Sadi, F.; Hoe, J.C.; Pileggi, L.; Franchetti, F. A 3D-Stacked Logic-in-Memory Accelerator for Application-Specific Data Intensive Computing. In Proceedings of the 2013 IEEE International 3D Systems Integration Conference (3DIC), San Francisco, CA, USA, 2–4 October 2013; pp. 1–7. [CrossRef]

5. Ahn, J.; Hong, S.; Yoo, S.; Mutlu, O.; Choi, K. A Scalable Processing-in-Memory Accelerator for Parallel Graph Processing. In Proceedings of the 2015 ACM/IEEE 42nd Annual International Symposium on Computer Architecture (ISCA), Portland, OR, USA, 13–17 June 2015; pp. 105–117. [CrossRef]

6. Zhang, D.; Jayasena, N.; Lyashevsky, A.; Greathouse, J.L.; Xu, L.; Ignatowski, M. TOP-PIM: Throughput-oriented Programmable Processing in Memory. In Proceedings of the 23rd International Symposium on High-performance Parallel and Distributed Computing Vancouver, BC, Canada, 23–27 June 2014; pp. 85–98. [CrossRef]

7. Xie, C.; Song, S.L.; Wang, J.; Zhang, W.; Fu, X. Processing-in-Memory Enabled Graphics Processors for 3D Rendering. In Proceedings of the 2017 IEEE International Symposium on High Performance Computer Architecture (HPCA), Austin, TX, USA, 4–8 February 2017; pp. 637–648. [CrossRef]

8. Tang, Y.; Wang, Y.; Li, H.; Li, X. ApproxPIM: Exploiting realistic 3D-stacked DRAM for energy-efficient processing in-memory. In Proceedings of the 2017 22nd Asia and South Pacific Design Automation Conference (ASP-DAC), Chiba, Japan, 16–19 January 2017; pp. 396–401. [CrossRef]

9. Angizi, S.; He, Z.; Fan, D. PIMA-Logic: A Novel Processing-in-Memory Architecture for Highly Flexible and Energy-Efficient Logic Computation. In Proceedings of the 2018 55th ACM/ESDA/IEEE Design Automation Conference (DAC), San Francisco, CA, USA, 24–29 June 2018; pp. 1–6. [CrossRef]

10. Chi, P.; Li, S.; Xu, C.; Zhang, T.; Zhao, J.; Liu, Y.; Wang, Y.; Xie, Y. PRIME: A Novel Processing-in-Memory Architecture for Neural Network Computation in ReRAM-Based Main Memory. In Proceedings of the 2016 ACM/IEEE 43rd Annual International Symposium on Computer Architecture (ISCA), Seoul, Korea, 18–22 June 2016; pp. 27–39. [CrossRef]

11. Han, L.; Shen, Z.; Shao, Z.; Huang, H.H.; Li, T. A novel ReRAM-based processing-in-memory architecture for graph computing. In Proceedings of the 2017 IEEE 6th Non-Volatile Memory Systems and Applications Symposium (NVMSA), Taiwan, China, 16–18 August 2017; pp. 1–6. [CrossRef]

12. Gaillardon, P.; Amarú, L.; Siemon, A.; Linn, E.; Waser, R.; Chattopadhyay, A.; De Micheli, G. The Programmable Logic-in-Memory (PLiM) computer. In Proceedings of the 2016 Design, Automation Test in Europe Conference Exhibition (DATE), Dresden, Germany, 14–18 March 2016; pp. 427–432.

13. Li, S.; Xu, C.; Zou, Q.; Zhao, J.; Lu, Y.; Xie, Y. Pinatubo: A processing-in-memory architecture for bulk bitwise operations in emerging non-volatile memories. In Proceedings of the 2016 53nd ACM/EDAC/IEEE Design Automation Conference (DAC), Austin, TX, USA, 5–9 June 2016; pp. 1–6. [CrossRef]

14. Papandroulidakis, G.; Vourkas, I.; Abusleme, A.; Sirakoulis, G.C.; Rubio, A. Crossbar-Based Memristive Logic-in-Memory Architecture. *IEEE Trans. Nanotechnol.* **2017**, *16*, 491–501. [CrossRef]

15. Seshadri, V.; Lee, D.; Mullins, T.; Hassan, H.; Boroumand, A.; Kim, J.; Kozuch, M.A.; Mutlu, O.; Gibbons, P.B.; Mowry, T.C. Ambit: In-memory Accelerator for Bulk Bitwise Operations Using Commodity DRAM Technology. In Proceedings of the 50th Annual IEEE/ACM International Symposium on Microarchitecture, Cambridge, MA, USA, 14–18 October 2017; pp. 273–287. [CrossRef]

16. Huangfu, W.; Li, S.; Hu, X.; Xie, Y. RADAR: A 3D-ReRAM based DNA Alignment Accelerator Architecture. In Proceedings of the 2018 55th ACM/ESDA/IEEE Design Automation Conference (DAC), San Francisco, CA, USA, 24–28 June 2018; pp. 1–6. [CrossRef]

17. Kaplan, R.; Yavits, L.; Ginosar, R.; Weiser, U. A Resistive CAM Processing-in-Storage Architecture for DNA Sequence Alignment. *IEEE Micro* **2017**, *37*, 20–28. [CrossRef]

18. Yavits, L.; Kvatinsky, S.; Morad, A.; Ginosar, R. Resistive Associative Processor. *IEEE Comput. Archit. Lett.* **2015**, *14*, 148–151. [CrossRef]

19. Imani, M.; Rosing, T. CAP: Configurable resistive associative processor for near-data computing. In Proceedings of the 2017 18th International Symposium on Quality Electronic Design (ISQED), Santa Clara, CA, USA, 14–15 March 2017; pp. 346–352. [CrossRef]

20. Imani, M.; Gupta, S.; Arredondo, A.; Rosing, T. Efficient query processing in crossbar memory. In Proceedings of the 2017 IEEE/ACM International Symposium on Low Power Electronics and Design (ISLPED), Taiwan, China, 24–26 July 2017; pp. 1–6. [CrossRef]
21. Matsunaga, S.; Hayakawa, J.; Ikeda, S.; Miura, K.; Hasegawa, H.; Endoh, T.; Ohno, H.; Hanyu, T. Fabrication of a Nonvolatile Full Adder Based on Logic-in-Memory Architecture Using Magnetic Tunnel Junctions. *Appl. Phys. Express* **2008**, *1*, 091301. [CrossRef]
22. Jarollahi, H.; Onizawa, N.; Gripon, V.; Sakimura, N.; Sugibayashi, T.; Endoh, T.; Ohno, H.; Hanyu, T.; Gross, W.J. A Nonvolatile Associative Memory-Based Context-Driven Search Engine Using 90 nm CMOS/MTJ-Hybrid Logic-in-Memory Architecture. *IEEE J. Emerg. Sel. Top. Circuits Syst.* **2014**, *4*, 460–474. [CrossRef]
23. Yang, K.; Karam, R.; Bhunia, S. Interleaved logic-in-memory architecture for energy-efficient fine-grained data processing. In Proceedings of the 2017 IEEE 60th International Midwest Symposium on Circuits and Systems (MWSCAS), Boston, MA, USA, 6–9 August 2017; pp. 409–412. [CrossRef]
24. Cofano, M.; Vacca, M.; Santoro, G.; Causapruno, G.; Turvani, G.; Graziano, M. Exploiting the Logic-In-Memory paradigm for speeding-up data-intensive algorithms. *Integration* **2019**. [CrossRef]
25. LeCun, Y.; Bengio, Y. *The Handbook of Brain Theory and Neural Networks*; Chapter Convolutional Networks for Images, Speech, and Time Series; MIT Press: Cambridge, MA, USA, 1998; pp. 255–258.
26. LeCun, Y.; Bengio, Y.; Hinton, G. Deep learning. *Nature* **2015**, *521*, 436–444. [CrossRef] [PubMed]
27. LeCun, Y.; Kavukcuoglu, K.; Farabet, C. Convolutional networks and applications in vision. In Proceedings of the 2010 IEEE International Symposium on Circuits and Systems, Paris, France, 30 May–2 June 2010; pp. 253–256.
28. Gudovskiy, D.A.; Rigazio, L. ShiftCNN: Generalized Low-Precision Architecture for Inference of Convolutional Neural Networks. *arXiv* **2017**, arXiv:1706.02393.
29. Krizhevsky, A.; Sutskever, I.; Hinton, G.E. ImageNet Classification with Deep Convolutional Neural Networks. In Proceedings of the 25th International Conference on Neural Information Processing Systems—Volume 1; Curran Associates Inc.: Lake Tahoe, NV, USA, 2012; pp. 1097–1105.
30. He, K.; Zhang, X.; Ren, S.; Sun, J. Deep Residual Learning for Image Recognition. *arXiv* **2015**, arXiv:1512.03385.
31. Santoro, G.; Casu, M.R.; Peluso, V.; Calimera, A.; Alioto, M. Energy-performance design exploration of a low-power microprogrammed deep-learning accelerator. In Proceedings of the 2018 Design, Automation Test in Europe Conference Exhibition (DATE), Dresden, Germany, 19–23 March 2018; pp. 1151–1154. [CrossRef]
32. Santoro, G.; Casu, M.R.; Peluso, V.; Calimera, A.; Alioto, M. Design-Space Exploration of Pareto-Optimal Architectures for Deep Learning with DVFS. In Proceedings of the 2018 IEEE International Symposium on Circuits and Systems (ISCAS), Florence, Italy, 27–30 May 2018; pp. 1–5. [CrossRef]
33. Becherer, M.; Csaba, G.; Porod, W.; Emling, R.; Lugli, P.; Schmitt-Landsiedel, D. Magnetic Ordering of Focused-Ion-Beam Structured Cobalt-Platinum Dots for Field-Coupled Computing. *IEEE Trans. Nanotechnol.* **2008**, *7*, 316–320. [CrossRef]
34. Nikonov, D.E.; Young, I.A. Benchmarking of Beyond-CMOS Exploratory Devices for Logic Integrated Circuits. *IEEE J. Explor. Solid-State Comput. Devices Circuits* **2015**, *1*, 3–11. [CrossRef]
35. Cairo, F.; Turvani, G.; Riente, F.; Vacca, M.; Gamm, S.B.V.; Becherer, M.; Graziano, M.; Zamboni, M. Out-of-plane NML modeling and architectural exploration. In Proceedings of the 2015 IEEE 15th International Conference on Nanotechnology (IEEE-NANO), Rome, Italy, 27–30 July 2015; pp. 1037–1040. [CrossRef]
36. Causapruno, G.; Riente, F.; Turvani, G.; Vacca, M.; Roch, M.R.; Zamboni, M.; Graziano, M. Reconfigurable Systolic Array: From Architecture to Physical Design for NML. *IEEE Trans. Very Large Scale Integr. (VLSI) Systems* **2016**. [CrossRef]
37. Chiolerio, A.; Allia, P.; Graziano, M. Magnetic dipolar coupling and collective effects for binary information codification in cost-effective logic devices. *J. Magn. Magn. Mater.* **2012**, *324*, 3006–3012. [CrossRef]
38. Breitkreutz, S.; Kiermaier, J.; Ju, X.; Csaba, G.; Schmitt-Landsiedel, D.; Becherer, M. Nanomagnetic Logic: Demonstration of directed signal flow for field-coupled computing devices. In Proceedings of the European Solid-State Device Research Conference (ESSDERC), Helsinki, Finland, 12–16 September 2011; pp. 323–326. [CrossRef]

39. Kimling, J.; Gerhardt, T.; Kobs, A.; Vogel, A.; Wintz, S.; Im, M.Y.; Fischer, P.; Peter Oepen, H.; Merkt, U.; Meier, G. Tuning of the nucleation field in nanowires with perpendicular magnetic anisotropy. *J. Appl. Phys.* **2013**, *113*, 163902. [CrossRef]

40. Becherer, M.; Kiermaier, J.; Breitkreutz, S.; Eichwald, I.; Žiemys, G.; Csaba, G.; Schmitt-Landsiedel, D. Towards on-chip clocking of perpendicular Nanomagnetic Logic. *Solid-State Electron.* **2014**, *102*, 46–51. [CrossRef]

41. Goertz, J.J.W.; Ziemys, G.; Eichwald, I.; Becherer, M.; Swagten, H.J.M.; Breitkreutz-v. Gamm, S. Domain wall depinning from notches using combined in- and out-of-plane magnetic fields. *AIP Adv.* **2016**, *6*, 056407. [CrossRef]

42. Ferrara, A.; Garlando, U.; Gnoli, L.; Santoro, G.; Zamboni, M. 3D design of a pNML random access memory. In Proceedings of the 2017 13th Conference on Ph.D. Research in Microelectronics and Electronics (PRIME), Giardini Naxos, Italy, 12–15 June 2017; pp. 5–8. [CrossRef]

43. Riente, F.; Ziemys, G.; Mattersdorfer, C.; Boche, S.; Turvani, G.; Raberg, W.; Luber, S.; Breitkreutz-v Gamm, S. Controlled data storage for non-volatile memory cells embedded in nano magnetic logic. *AIP Adv.* **2017**, *7*, 055910. [CrossRef]

44. Becherer, M.; Gamm, S.B.V.; Eichwald, I.; Žiemys, G.; Kiermaier, J.; Csaba, G.; Schmitt-Landsiedel, D. A monolithic 3D integrated nanomagnetic co-processing unit. *Solid-State Electron.* **2016**, *115*, 74–80. [CrossRef]

45. Eichwald, I.; Kiermaier, J.; Breitkreutz, S.; Wu, J.; Csaba, G.; Schmitt-Landsiedel, D.; Becherer, M. Towards a Signal Crossing in Double-Layer Nanomagnetic Logic. *IEEE Trans. Magn.* **2013**, *49*, 4468–4471. [CrossRef]

46. Eichwald, I.; Breitkreutz, S.; Kiermaier, J.; Csaba, G.; Schmitt-Landsiedel, D.; Becherer, M. Signal crossing in perpendicular nanomagnetic logic. *J. Appl. Phys.* **2014**, *115*, 17E510. [CrossRef]

47. Eichwald, I.; Breitkreutz, S.; Ziemys, G.; Csaba, G.; Porod, W.; Becherer, M. Majority logic gate for 3D magnetic computing. *Nanotechnology* **2014**, *25*, 335202. [CrossRef] [PubMed]

48. Cofano, M.; Santoro, G.; Vacca, M.; Pala, D.; Causapruno, G.; Cairo, F.; Riente, F.; Turvani, G.; Roch, M.R.; Graziano, M.; et al. Logic-in-Memory: A Nano Magnet Logic Implementation. In Proceedings of the 2015 IEEE Computer Society Annual Symposium on VLSI, Montpellier, France, 8–10 July 2015; pp. 286–291. [CrossRef]

49. Riente, F.; Ziemys, G.; Turvani, G.; Schmitt-Landsiedel, D.; Gamm, S.B.; Graziano, M. Towards Logic-In-Memory circuits using 3D-integrated Nanomagnetic logic. In Proceedings of the 2016 IEEE International Conference on Rebooting Computing (ICRC), San Diego, CA, USA, 17–19 October 2016; pp. 1–8. [CrossRef]

50. Garlando, U.; Riente, F.; Turvani, G.; Ferrara, A.; Santoro, G.; Vacca, M.; Graziano, M. Architectural exploration of perpendicular Nano Magnetic Logic based circuits. *Integration* **2018**, *63*, 275–282. [CrossRef]

51. Santoro, G.; Vacca, M.; Bollo, M.; Riente, F.; Graziano, M.; Zamboni, M. Exploration of multilayer field-coupled nanomagnetic circuits. *Microelectron. J.* **2018**, *79*, 46–56. [CrossRef]

52. Vacca, M.; Graziano, M.; Wang, J.; Cairo, F.; Causapruno, G.; Urgese, G.; Biroli, A.; Zamboni, M. *NanoMagnet Logic: An Architectural Level Overview*; LNCS, Lecture Notes in Computer Science (including Subseries Lecture Notes in Artificial Intelligence and Lecture Notes in Bioinformatics); Springer: Berlin/Heidelberg, Germany, 2014; Volume 8280, pp. 223–256.

53. Riente, F.; Garlando, U.; Turvani, G.; Vacca, M.; Roch, M.R.; Graziano, M. MagCAD: A Tool for the Design of 3D Magnetic Circuits. *IEEE J. Explor. Solid-State Comput. Devices Circuits* **2017**, *3*, 65–73. [CrossRef]

54. Turvani, G.; Riente, F.; Graziano, M.; Zamboni, M. A quantitative approach to testing in Quantum dot Cellular Automata: NanoMagnet Logic case. In Proceedings of the 2014 10th Conference on Ph.D. Research in Microelectronics and Electronics (PRIME), Grenoble, France, 30 June–3 July 2014; pp. 1–4. [CrossRef]

55. Turvani, G.; Tohti, A.; Bollo, M.; Riente, F.; Vacca, M.; Graziano, M.; Zamboni, M. Physical design and testing of Nano Magnetic architectures. In Proceedings of the 2014 9th IEEE International Conference on Design & Technology of Integrated Systems in Nanoscale Era (DTIS), Santorini, Greece, 6–8 May 2014; pp. 1–6. [CrossRef]

56. Turvani, G.; Riente, F.; Cairo, F.; Vacca, M.; Garlando, U.; Zamboni, M.; Graziano, M. Efficient and reliable fault analysis methodology for nanomagnetic circuits. *Int. J. Circuit Theory Appl.* **2016**, *45*, 660–680. [CrossRef]

 *micromachines* 

Article

A Flexible Hybrid BCH Decoder for Modern NAND Flash Memories Using General Purpose Graphical Processing Units (GPGPUs)

Arul Subbiah * and Tokunbo Ogunfunmi

Department of Electrical Engineering, Santa Clara University, 500 El Camino Real, Santa Clara, CA 95053, USA; togunfunmi@scu.edu
* Correspondence: asubbiah@scu.edu; Tel.: +1-408-368-5099

Received: 20 April 2019; Accepted: 23 May 2019; Published: 31 May 2019

Abstract: Bose–Chaudhuri–Hocquenghem (BCH) codes are broadly used to correct errors in flash memory systems and digital communications. These codes are cyclic block codes and have their arithmetic fixed over the splitting field of their generator polynomial. There are many solutions proposed using CPUs, hardware, and Graphical Processing Units (GPUs) for the BCH decoders. The performance of these BCH decoders is of ultimate importance for systems involving flash memory. However, it is essential to have a flexible solution to correct multiple bit errors over the different finite fields ($GF(2^m)$). In this paper, we propose a pragmatic approach to decode BCH codes over the different finite fields using hardware circuits and GPUs in tandem. We propose to employ hardware design for a modified syndrome generator and GPUs for a key-equation solver and an error corrector. Using the above partition, we have shown the ability to support multiple bit errors across different BCH block codes without compromising on the performance. Furthermore, the proposed method to generate modified syndrome has zero latency for scenarios where there are no errors. When there is an error detected, the GPUs are deployed to correct the errors using the iBM and Chien search algorithm. The results have shown that using the modified syndrome approach, we can support different multiple finite fields with high throughput.

Keywords: BCH; decoder; iBM; GPU; hybrid; flash memory; Galois field; CUDA

---

1. Introduction

NAND flash memories are widely used in many electronic devices. These devices face reliability issues because of the densely-populated memory cells [1]. In fact, the 3D method used to manufacture flash memories, discussed in detail by Spinelli et al. [2], enforces the necessity to have high throughput error correction techniques. Bose–Chaudhuri–Hocquenghem (BCH) codes [3] are the most common error correction mechanisms for flash memory devices and other digital communications like optical networks. The increasing efficiency and throughput of the flash memory systems have drawn researchers to provide highly-efficient BCH decoders. The three major categories of the BCH decoders proposed are Central Processing Units (CPUs), hardware circuits, and Graphical Processing Units (GPUs). Cho proposed an efficient CPU-based implementation in [4], and Poolakkaprambil discussed multi-bit error using Hamming, BCH, and Low-Density Parity Check (LDPC) codes in [5]. Later, Lee et al. proposed a high throughput hardware architecture in [6]. Moreover, Zhang discussed different hardware implementation techniques in [7]. Qi et al. [8] proposed a GPU-based BCH decoder; later, we proposed an efficient algorithm for BCH decoders using GPUs in [9]. In addition to the requirement of high throughput, modern BCH decoders are required to support multiple bit error correction across various block sizes, which is the focus of this paper. Technology scaling has rendered the ability to integrate multiple GPUs within a System On Chip (SOC), which has enabled researchers to use GPU for non-graphical applications. In fact, the term General Purpose Graphical Processing Unit (GPGPU) refers to the application of GPU for nongraphical applications. We use the term GPU instead of GPGPU since these terms are interchangeable in practice. Streaming Multiprocessors (SMs) are the building blocks of these GPUs, which has multiple CPUs within them. Each of the instantiated SMs is capable of handling multiple threads, which are scheduled by a warp scheduler. Therefore, we need an exclusive compiler like the Computer Unified Device Architecture (CUDA) C [10] software to program these SMs. The CUDA software creates the necessary grid of kernel routines, which in turn create the same instruction that operates on a different data path; this technique is referred to as the single instruction multiple data (SIMD) stream. The kernel subroutines are executed across multiple cores and in a multiple thread fashion. The GPU-based BCH decoders [9] are flexible, and they can support multiple BCH block sizes.

We have organized this paper as follows: Section 2 discusses the background and previous works. Section 3 describes our proposed hybrid method using GPUs and hardware design. Section 4 presents the results observed, and we conclude in Section 5.

2. Background

BCH codes are cyclic block codes encoded by the generator polynomial $g(x)$ over the $GF(2)$. The roots of this polynomial equation reside in the extended field, also known as the splitting field, $GF(2^m)$. Let $\phi_i(x)$ be the minimal polynomial of an arbitrary element β^i, then the generator polynomial for BCH code with t error correction capability is given by the following equation:

$$g(x) = LCM(\phi_1(x), \phi_2(x), ..., \phi_{2t}(x)) \tag{1}$$

Narrow sense BCH codes use primitive element α^i for the minimal polynomial with i starting from one. For simplicity, the narrow sense BCH code decoder is discussed and implemented in this paper, and it could be easily extended for other general BCH codes [3]. The parity bits are then generated using the equation $p(x) = m(x)\, mod\, g(x)$, and these parity bits are concatenated to form the message polynomial $m(x)$. This concatenation is given as:

$$c(x) = m(x) \cdot x^{deg(g(x))} + m(x)\, mod\, g(x) \tag{2}$$

The generated parity bits are then stored in the spare area allocated in the page within the flash memory device. In general, the hard decision BCH decoder has three steps in the decoding process: syndrome generation, key-equation solver, and an error locator.

2.1. Encoder

The main issue when using large BCH codes, i.e., t greater than 30, is the fan-out issue created by implementing the Linear Feedback Shift Register (LFSR) method of the generator polynomial. Parhi has addressed this fan-out issue by breaking down the LFSR register into multiple cascaded LFSRs by realizing the circuit in the Z-domain [11]. Hao addressed the same issue by using the Chinese Remainder Theorem (CRT) method [12], but this method requires more computation on the encoder and is applicable for encoders that have t higher than 100. Later, Tang et al. proposed a hybrid approach for long BCH encoders that is area efficient [13]. The authors of this paper had proposed an area efficient method by sharing the hardware between the encoder and syndrome generator [14].

2.2. Decoder

The BCH decoders can be categorized as hard decision and soft decision decoders [15,16]. These decoders' realization can be broadly classified into three categories: Central Processing Unit (CPU) [4], Very Large Scale implementation (VLSI) [17], and GPU implementation [8]. Various hardware implementations of BCH decoders were discussed by Zhang [7]. BCH decoders can be categorized by the place of the decoders. The decoders can be either located on-chip within memory device [18] or outside the memory device [19]. The focus of this paper is on the decoder being outside the memory device.

The syndrome generator is the first step of the BCH decoding process [6,20]. The syndromes S_i of the received vector $r(x)$ are given as:

$$S_i = r(\alpha^i) \tag{3}$$

In other words, the syndrome generator checks if the received code vector $r(x) = r_{n-1}x^{n-1} + \ldots + r_1 + r_0$ has the roots as α^1, α^2, ..., α^{2t}. If so, then there are no errors in the received code vector. In the case of an error, the key-equation solver and the error locator steps are executed. For t bit error correction on a narrow sense BCH code, it is sufficient to find t syndromes, because the elements of a conjugacy class have the same minimal polynomial $\phi_i(x)$. We have discussed an alternate approach to share the syndrome generator and encoders in [14]; Figure 1 depicts the area sharing between the encoder and the syndrome generator presented in [14]. This method requires separate error protection to the parity bits, and one proposal is to use a Single-Level Cell (SLC) for the parity bits to reduce error probability.

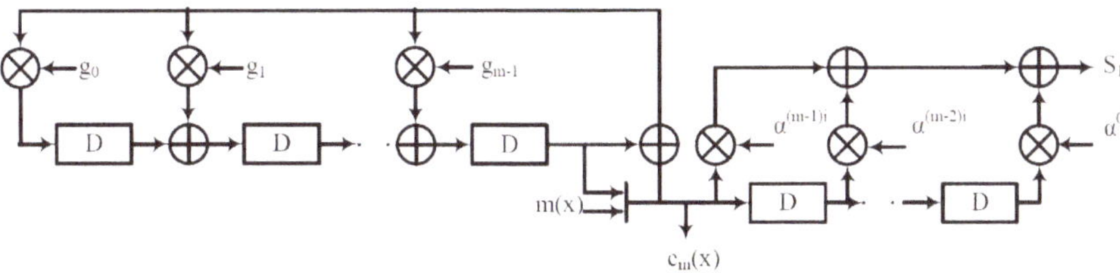

Figure 1. Area-efficient syndrome generator.

An error locator polynomial $\Lambda(x)$, which has dependency on the error location, gives a hint about the error location, and it is given by the equation:

$$\Lambda(x) = \sum_{i=0}^{t} \Lambda_i x^i = (1 - X_1 x)(1 - X_2 x)...(1 - X_t x) \tag{4}$$

where X_i represents the error location of the vector $r(x)$. The key equation:

$$S(x).\Lambda(x) = \Omega(x)\, mod\, 2t \tag{5}$$

shows the relationship between the error locator polynomial and the error evaluator polynomial; moreover, Newton identities [3] show the relation between the error locator polynomial $\Lambda(x)$ and the syndromes S_i. There have been many algorithms like Berlekamp–Massey (BM), Peterson, and others proposed to solve the key equation [3,7], but the inversion-less BM (iBM) algorithm is predominantly used in high throughput architectures [6,21]. Park et al. [22] proposed a novel folded method to reduce the area in the hardware architecture, but the proposed method takes more clock cycles and is proportional to the folding factor. For the final step, the Chien Search (CS) algorithm is used to locate the error position from the error locator polynomial equation. Yoo et al. proposed a low power and high throughput parallel CS algorithm in [23].

2.3. Motivation

This paper intends to propose a solution that can address two configurable parameters of a BCH decoder. First, the solution should be scalable across different GF fields, i.e., it should be able to support different GF field extensions ($GF(2^m)$). Second, the solution should be able to scale across different bit errors t. Different configurable BCH decoder solutions have been proposed [20,24], but they lack support for both configurable parameters of the BCH decoders. Inspired by the attempt to solve BCH decoders for multiple GF dimensions in [20], we propose an alternate hybrid approach to have a flexible solution. In [20], a hardware solution was proposed to support multiple BCH codes; however, the circuit area increases in order to support multiple GF dimensions because of the dual-mBCH decoders' requirement. We have proposed a method to share the hardware logic between the BCH encoder and BCH syndrome generator by modifying the encoding method in [14]. In this paper, we extend our previous work by using a programmable modified syndrome generator algorithm and GPU to have a decoder that works with multiple GF dimensions.

3. Hybrid Method

We propose a high throughput system that can correct t bit errors over different BCH codes (n, k, t), i.e., error correction over different finite field dimensions, using hardware design and GPU kernel routines. Figure 2 depicts the architectural block diagram for our proposed hybrid method. The flash memory interface is a physical interface to a flash memory device, and the host interface is a standard bus interface, which communicates with the GPU. The GPU could either reside inside the host interface (system on chip) or external to the host system. It is important to note that the GPU system in the system is used for dual purposes, i.e., for the graphical display and error correction. Furthermore, in our proposed method, the GPUs are only deployed when there is an error detected in the page, and the GPUs are not used for pages without error. It is assumed that the host system exercises a memory copy routine to transfer data whenever there is an interaction between the host system and the GPU system. The syndrome generation, proposed in [14], is split into two modules: the Syndrome Residual Unit (SRU) and the syndrome kernel. Then, we propose to implement the modules SRU and the FIFO (shaded area) in hardware and to use GPU kernel routines for the modules' syndrome calculator, key-equation solver, and an error corrector.

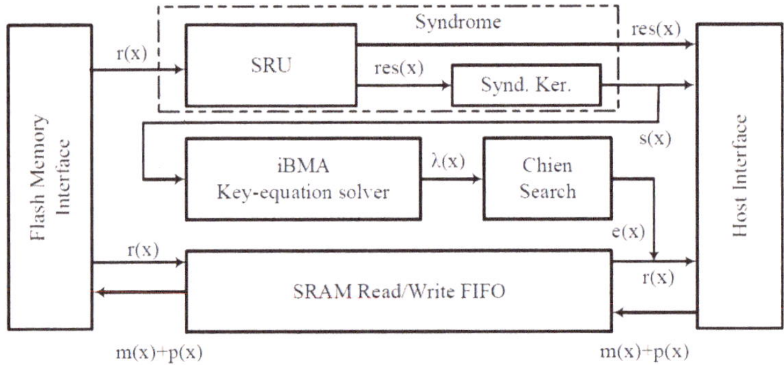

Figure 2. Hybrid BCH decoder block diagram.

3.1. Flowchart

Figure 3 depicts the flow of our proposed hybrid method. Initially, the GPUs create a LUT memory for faster GF multiplication; this method has been proven to be faster on GPUs than threads spawning sub-kernel routines [9]. Next, a page read command is initiated to the flash memory interface. The SRU calculates the t residuals of the minimal polynomial, while the data are written into the FIFO. If all the residuals are zero, then we conclude that there are no errors detected in the received vector, and the host shall transfer the data to the application layer. If there are non-zero residuals, then the host calls the Syndrome Calculation Kernel (SK) routine to calculate the syndrome and then calls the Key Equation Kernel (KEK) to form the error locator polynomial $\Lambda(x)$. Once the $\Lambda(x)$ is formed, the Chien Search Kernel (CSK) is executed for each bit location. The final error vector is then added to the data in the FIFO to correct the bit errors and then copied to the host memory. Until all the intended data from the flash memory are read, we repeat the previously mentioned steps (Node 1 in Figure 3).

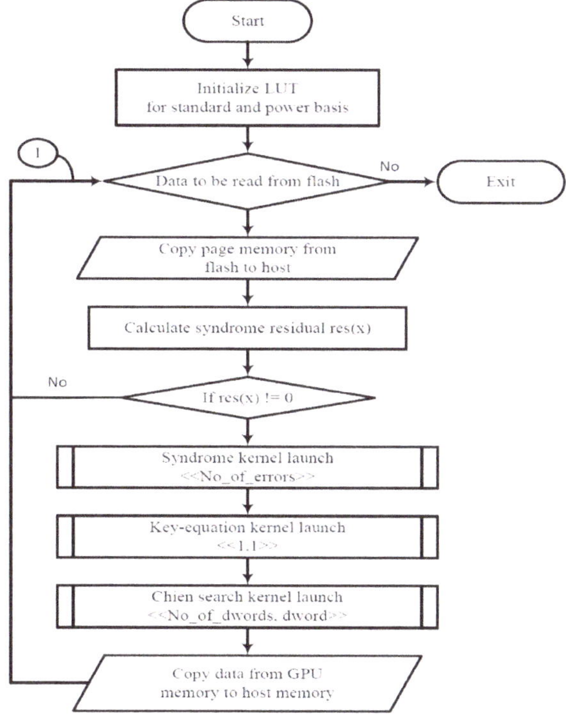

Figure 3. Flow chart for the hybrid system.

3.2. Modified Syndrome Generator

The conventional syndrome generator as discussed in Section 2.2 can be split into two steps: First, the residual polynomial $res_i(x)$, of the received code word $r(x)$, is generated by the equation $res_i(x) = r(x) \bmod \phi_i(x)$, where the minimal polynomial $\phi_i(x) = g_{i,m}x^m + \ldots + g_{i,0}$ and residual polynomial $res_i(x) = res_{i,m-1}x^{m-1} + \ldots + res_{i,0}$. Second, the syndrome can be calculated by substituting the primitive element α^i in the residual polynomial and is expressed as:

$$S_i = \sum_{k=0}^{deg(\phi_i(x))-1} res_{i,k} \cdot (\alpha^i)^k \tag{6}$$

It is clear that by splitting the syndrome generation, $res_i(x)$ does not have any dependency on the field extensions. In fact, the polynomial division used in $res_i(x)$ is identical to a Linear Feedback Shift Register (LFSR) with its coefficient from $\phi_i(x)$. We introduce the idea to have the coefficients of the LFSR as programmable. Figure 4 represents a hardware realization of the SRU array in a serial fashion with programmable feedback coefficients. In most cases, depending on the data interface width of the flash memory interface, we can unfold the serial interpretation of the SRU to process more bits in parallel. Because of the relationship between the conjugacy class and $\phi_i(x)$ [7], it is sufficient to generate t SRU units. These SRUs can compute $res_i(x)$ of different minimal polynomials in tandem. Once the residuals are computed, the values of the $res_i(x)$ are compared for non-zero values. An error is triggered if any of the $res_i(x)$ has non-zero coefficients in them, and the GPU kernel routines for the other stages of the BCH decoder are executed.

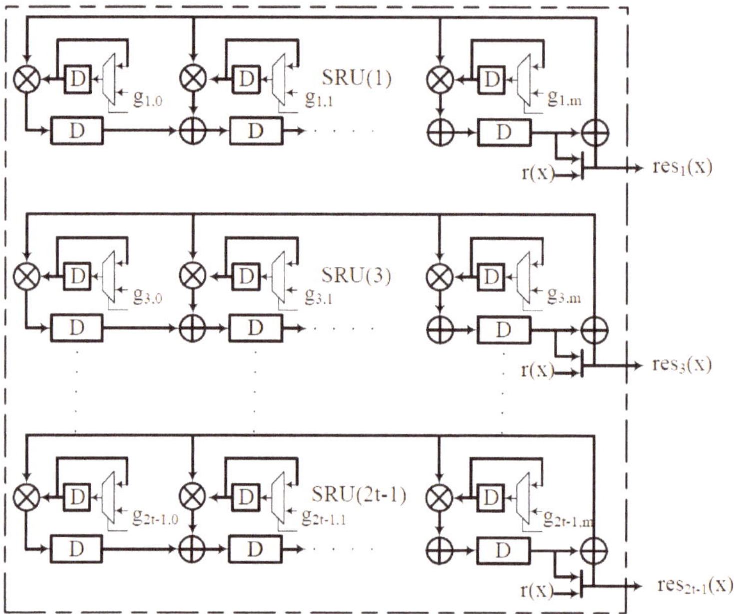

Figure 4. Array of the syndrome residual unit.

3.3. GPU Kernel Routines

Kernel routines are the fundamental sub routines, representing the SIMDtype of parallelism, executed by the GPU for our proposed decoder. Figure 5 illustrates a systematic execution of the kernel routines where PG2 and PG4 represent pages with errors. PG0, PG1, and PG3 represent pages without error. When there are no errors, the latency incurred is the computation time consumed by the SRU systolic array, as shown in Figure 5. Furthermore, to achieve better throughput, the SRU units can compute $res_i(x)$ for the next page, while the GPU kernel routines are triggered during an error scenario. For *GF* multiplication in the algorithm, the multiplicand and multiplier are converted to the power basis by referring to the LUT in the global memory. Thus, the multiplication is transformed into a simple XOR operation in the power basis domain. After the multiplication is computed, a reverse transformation is performed by referring to another basis converter LUT in the global memory. The three basic GPU kernel routines used in our approach are explained in detail below.

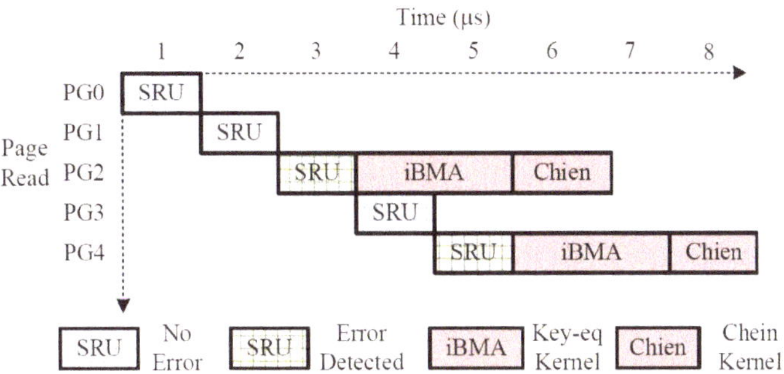

Figure 5. Decoder execution sequence. PG, Page.

3.3.1. Syndrome Kernel

In this routine, the S_i is calculated by substituting the α_i in the equation $res_i(x)$. Since there are no dependencies on the syndromes, t parallel SK routines are launched within the GPU. Algorithm 1 represents the pseudocode for the syndrome routine. The *atomicXor* operation is required to synchronize the value updated by the SK routines across multiple threads.

Algorithm 1 Syndrome kernel.

1: **procedure** SYND KERNEL(res_i, S_i)
2:     $sum \leftarrow 0$
3:     **for** $j \leftarrow 0, deg(\phi_i(x)) - 1$ **do**
4:         $sum \leftarrow sum + res_{i,j}.\alpha^{i \cdot j}$
5:     **end for**
6:     $atomicXor(S_i, sum)$ ▷ synchronize between threads
7: **end procedure**

3.3.2. Key-Equation Kernel

The KEK is the only single thread routine, in our proposal, because of the iterative nature of the iBM algorithm. Algorithm 2 represents the pseudocode for iBM implementation in the GPU routine. There are other methods like the simplified iBM (siBM) algorithm [7] proposed for the key-equation solver module, but experimental results have proven that siBM does not have significant improvement on the performance of the GPU kernel routines.

Algorithm 2 Key-equation kernel.

1: **procedure** KEQ EQ KERNEL(Λ, S)

2: $\quad \Lambda^{(0)} \leftarrow 1 + S_1 x$

3: $\quad$ **if** $S_1 = 0$ **then**

4: $\quad\quad d_p \leftarrow 1; \beta^{(1)} \leftarrow x^3; l_1 \leftarrow 0$

5: $\quad$ **else**

6: $\quad\quad d_p \leftarrow S_1; \beta^{(1)} \leftarrow x^2; l_1 \leftarrow 1$

7: $\quad$ **end if**

8: $\quad$ **for** $r \leftarrow 1, t - 1$ **do**

9: $\quad\quad d_r \leftarrow \sum\limits_{i=1}^{t} \Lambda_i^{(r)} S_{2r-i+1}$

10: $\quad\quad \Lambda^{(r)} \leftarrow d_p \Lambda^{(r-1)} + d_r \beta^{(r)}$

11: $\quad\quad$ **if** $d_r = 0 \, or \, r < l_r$ **then**

12: $\quad\quad\quad \beta^{(r+1)} \leftarrow x^2 \beta^{(r)}; l_{r+1} \leftarrow l_r; d_p \leftarrow d_p$

13: $\quad\quad$ **else**

14: $\quad\quad\quad \beta^{(r+1)} \leftarrow x^2 \Lambda^{(r)}; l_{r+1} \leftarrow l_r + 1; d_p \leftarrow d_r$

15: $\quad\quad$ **end if**

16: $\quad$ **end for**

17: **end procedure**

3.3.3. Chien Search Kernel

The CS algorithm is the final step within the decoder. The primitive element $\alpha^{pos^{-1}}$ is checked if it is a root for the error locator polynomial $\Lambda(x)$ as specified in [9]. This kernel routine is an ideal candidate for the GPU because of the parallelism it offers. Each element of the finite field is evaluated in the equation $\Lambda(x)$ as shown in Algorithm 3. This evaluation kernel routine is independent for each GF element; hence these routines can be launched in parallel threads. Similar to the SK routine, the memory within the GPU device is shared between threads, so the *atomicXOR* operation is used to avoid writing overlap by different CSK routines. Once the error vector is formed, the error is masked with the data in the memory to yield corrected data.

Algorithm 3 Chien search. Kernel

1: **procedure** CHIEN KERNEL(Λ, pos, err)

2: $\quad sum \leftarrow 1$ $\qquad\qquad\qquad\qquad\qquad\qquad\qquad\qquad\qquad\qquad$ ▷ Always KEQ is minimal

3: $\quad$ **for** $j \leftarrow 0, deg(\phi_i(x)) - 1$ **do**

4: $\quad\quad sum \leftarrow sum + \Lambda_j (\alpha^{pos^{-1}})^j$

5: $\quad$ **end for**

6: $\quad$ **if** $sum = 0$ **then** $\qquad\qquad\qquad\qquad\qquad\qquad\qquad\qquad$ ▷ $\alpha^{pos^{-1}}$ is a root

7: $\quad\quad atomicXor(err[pos], sum)$ $\qquad\qquad\qquad\qquad\qquad$ ▷ Prevent overlap write

8: $\quad$ **end if**

9: **end procedure**

---

4. Experimental Results and Analysis

The proposed hybrid approach was compared against conventional GPU [8,9] and hardware [20] architectures. The hardware implementation of the syndrome generator was synthesized for 28-nm technology, and it achieved an operational frequency of 1 GHz. The setup used for the GPU implementation is given in Table 1. In our experiments, we analyzed the performance and the area consumed for different BCH code sizes. We have used the finite field dimension of $m = 12, 13, 14, 15$ in our comparison, which corresponds to block sizes of $256, 512, 1024, 2048$ bytes. Furthermore, we have analyzed the results for different bit errors ($t = 2, ..., 40$) in our experiments.

Table 1. Experimental setup.

	GPGPU	CPU
Platform	Geforce GTX 760. 1152 cores	Intel Xeon i7
Clock Freq.	1.033 GHz	3.7 GHz
Memory	GDDR5(2 GB), 6 Gbps	DDR2 (32 GB), 102.4 Gbps

4.1. Error Analysis

The error correction capability increased with n, but the larger the n, the higher the probability of random bit error. Based on the raw bit error probability p, parity bits, and message code $cpar = 2 \cdot t \cdot m + |m(x)|$, the sector with correctable error (P_{secErr}) might increase and is given as:

$$P_{secErr} = 1 - \sum_{i=t+1}^{cpar} \binom{cpar}{i} \cdot p^i \cdot (1-p)^{(cpar)-i} \tag{7}$$

Figure 6 plots the bit error vs. sector error for different BCH codes. We can also see that the P_{secErr} decreased as p decreased. Furthermore, there was a slight increase in P_{secErr} when compared against different m. This was due to the increase in the probability of error within bigger sector sizes.

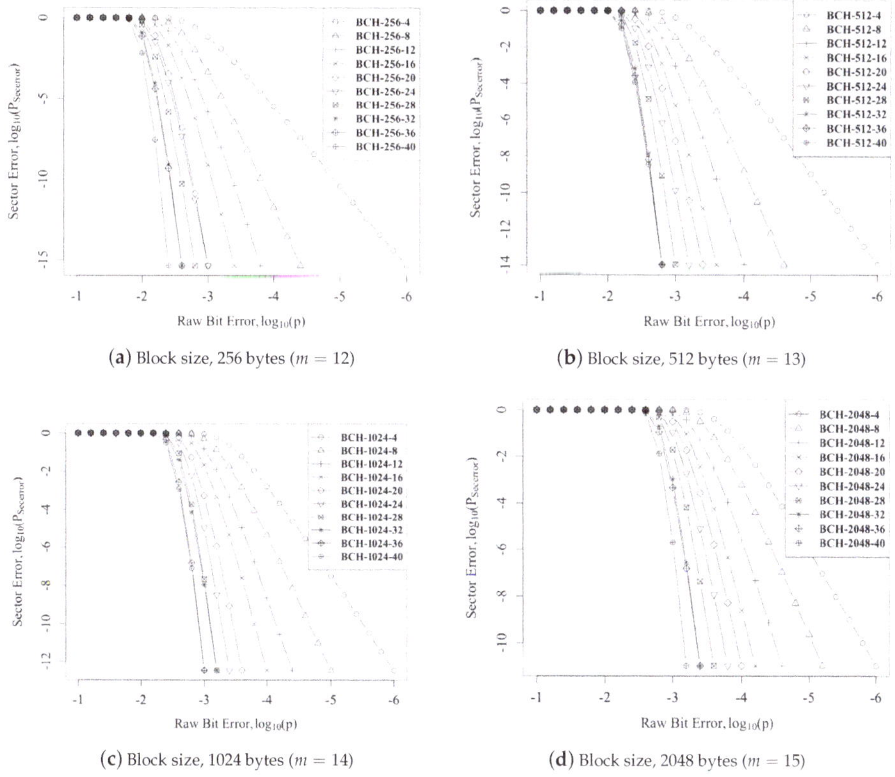

(**a**) Block size, 256 bytes ($m = 12$)

(**b**) Block size, 512 bytes ($m = 13$)

(**c**) Block size, 1024 bytes ($m = 14$)

(**d**) Block size, 2048 bytes ($m = 15$)

Figure 6. Raw bit error vs. sector error.

4.2. Syndrome Generation Analysis

Syndrome generation is the critical area where the proposed hybrid method provides an advantage over the GPU methods [8,9]. Figure 7 shows the plot of the syndrome computation time vs. different bit errors ($t = 2, ..., 40$) across different finite fields ($m = 12, 13, 14, 15$) for different architectures: GPU [9], hardware [21], and hybrid (proposed). The computation of the SRU engine depended on the number of clock cycles required for a page read. Since all the $res_i(x)$ that were necessary for the key-equation solver were calculated in tandem, the latency only depended on the read cycles for flash memory. The hardware architecture for syndrome generation consumed the same clock cycle as the hybrid approach since the approach to syndrome generation was similar. It should be noted that the GPU unit was used as a display unit, so the results of kernel profiling depended on the load of the GPU during the execution of the kernel routine. The execution time for the syndrome on the GPU architecture depended on the number of threads getting executed, and typically, it was from 30–100 μs.

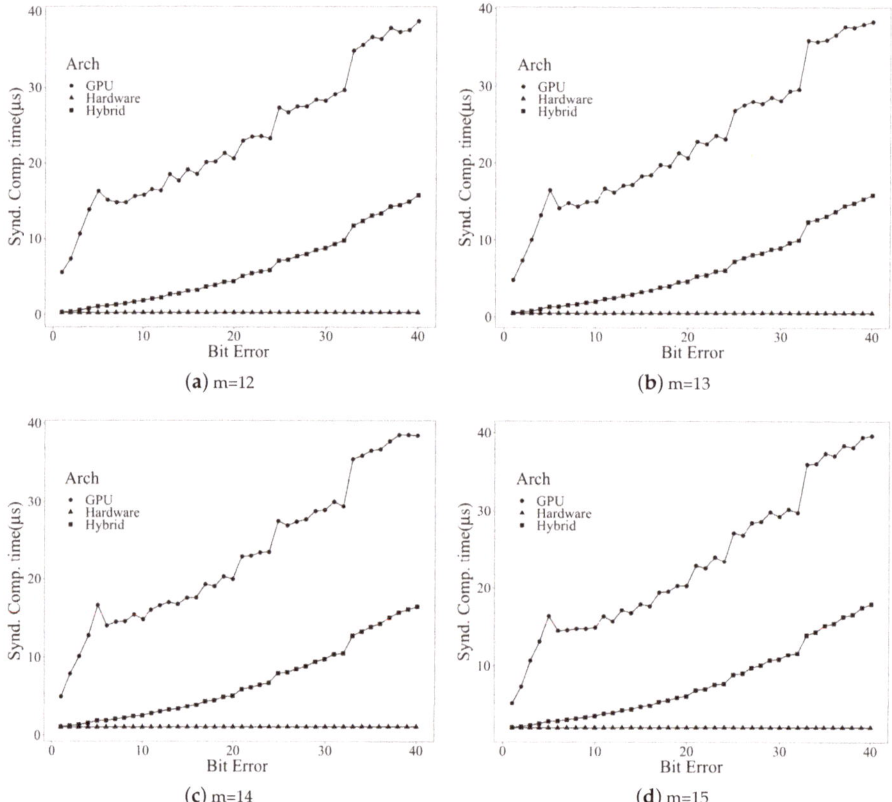

Figure 7. Syndrome computation time for different arches.

4.3. VLSI Analysis

Table 2 compares the hardware area required for different methods of the syndrome generator ([20] and the proposed). Since the GPUs were employed for key-equation and Chien search, the hardware was only compared for the syndrome generator unit for multi-bit error correction for different GF dimensions. This is a fair comparison since the area of the GPUs was already accounted for systems with graphical display. The hardware implementations were targeted for 28-nm and met a frequency of 1 GHz. In order to support different finite field ($m = 12, 13, 14, 15$) and 40-bit error correction, the hardware method [20] consumed 30,247 μm^2, whereas our proposed hybrid architecture consumed 10,633 μm^2, thus saving two-fold of the area. This area savings was due to the splitting of the syndrome generation into two units (SRU and syndrome kernel). When there were no errors in the page, the total time taken by the proposed decoder was less than 5 μs (Figure 7), which was less than the average read latency of 100 μs. Table 3 compares the power consumed by the conventional method [20] and our proposed method. The last entry in the table provides the power consumption required to support error correction over different fields ($m = 12, 13, 14, 15$) and until 40-bit error correction. There was a savings of 4 mW in our proposed method.

Table 2. Area comparison for different syndrome generators vs. the proposed SRU.

Setup		Area for t (μm^2)									
		4	*8*	*12*	*16*	*20*	*24*	*28*	*32*	*36*	*40*
m = 12	[20]	606	1287	2079	2871	3256	3661	3959	4252	4611	4917
	Prop.	853	1704	2550	3399	4209	5061	5903	6747	7592	8436
m = 13	[20]	858	1863	2962	4043	4648	5246	5814	6387	6988	7573
	Prop.	924	1846	2767	3682	4607	5532	6390	7305	8308	9134
m = 14	[20]	916	2002	3167	4330	5016	5723	6375	7048	7714	8390
	Prop.	994	1985	2976	3966	4966	5959	6948	7942	8935	9927
m = 15	[20]	988	2172	3435	4707	5485	6292	7043	7818	8587	9367
	Prop.	1064	2125	3186	4249	5318	6383	7448	8504	9570	10,633
m = 12, ..., 15	[20]	3368	7324	116,43	15,951	18,405	20,922	23,191	25,505	27,900	*30,247*
	Prop.	1064	2125	3186	4249	5318	6383	7448	8504	9570	*10,633*

Table 3. Power comparison for different syndrome generators vs. the proposed SRU.

Setup		Power for t (mW)									
		4	*8*	*12*	*16*	*20*	*24*	*28*	*32*	*36*	*40*
m = 12	[20]	0.179	0.374	0.599	0.819	0.897	0.978	1.037	1.097	1.170	1.232
	Proposed	0.167	0.336	0.499	0.673	0.841	1.014	1.185	1.354	1.524	1.695
m = 13	[20]	0.226	0.491	0.773	1.054	1.178	1.304	1.426	1.546	1.674	1.797
	Proposed	0.178	0.355	0.534	0.714	0.889	1.061	1.248	1.424	1.615	1.778
m = 14	[20]	0.231	0.503	0.704	1.068	1.213	1.363	1.499	1.640	1.780	1.923
	Proposed	0.187	0.373	0.561	0.747	0.938	1.122	1.309	1.493	1.678	1.863
m = 15	[20]	0.235	0.512	0.812	1.110	1.270	1.438	1.593	1.755	1.915	2.078
	Proposed	0.194	0.388	0.588	0.776	0.975	1.168	1.370	1.564	1.755	1.953
m = 12, ..., 15	[20]	0.692	1.506	2.369	3.232	3.661	4.105	4.518	4.941	5.369	*5.798*
	Proposed	0.194	0.388	0.588	0.776	0.975	1.168	1.370	1.564	1.755	*1.953*

4.4. Performance Analysis

The comparison of the total time taken, in case of an error, is compared for the hardware [20], GPU [8,9], and hybrid architecture (proposed) in Figure 8 (the variable *be* represents bit error). We can see that the hybrid approach was better than the GPU method because of the SRU implementation in hardware. However, the hardware implementation took less than 1 μs because of the high performance (which is indistinguishable in Figure 8). We can observe a gain of more than 25% when the system has errors. We can find the probability of sector error from a given bit error rate using Equation (7). For a given $P_{secError}$, the throughput was calculated for a second's worth of data transfer.

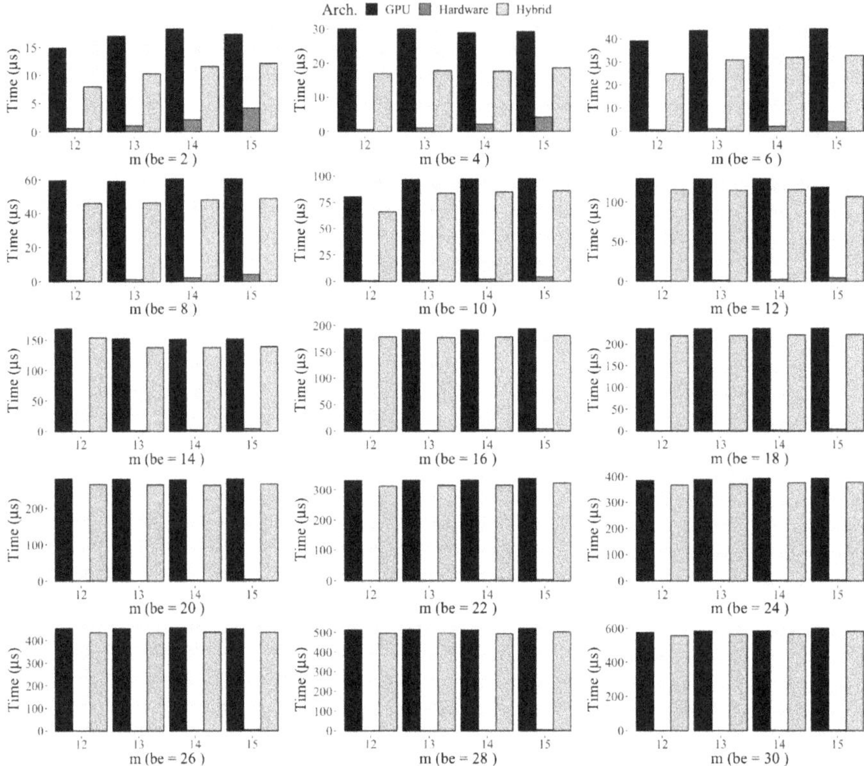

Figure 8. Total computation time for different architectures and different finite fields.

Figure 9 represents the plot for throughput vs. bit error rate for different finite fields ($m = 12, 13, 14, 15$) and different $t = 4, ..., 40$. We can see that for $m = 12, 13$, the throughput was sustained till 10^{-3}, and for $m = 14, 15$, the throughput was sustained till $10^{-3.5}$. This throughput is sustainable for flash memories that have an Uncorrectable Bit Error Rate (UBER) of 10^{-15}, an it is also sustainable for the end of life for flash memories which is greater than 10^{-5}.

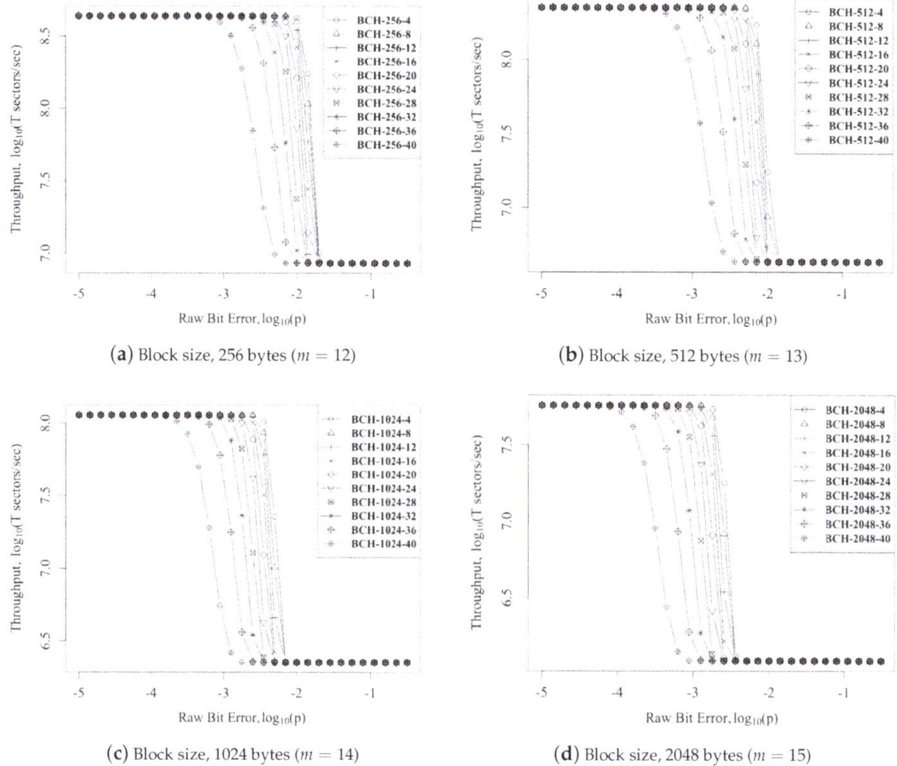

(**a**) Block size, 256 bytes ($m = 12$)

(**b**) Block size, 512 bytes ($m = 13$)

(**c**) Block size, 1024 bytes ($m = 14$)

(**d**) Block size, 2048 bytes ($m = 15$)

Figure 9. Raw bit error vs. throughput (sectors/s).

5. Conclusions

In this paper, we have proposed a novel hybrid method to implement an efficient BCH decoder for different finite field extensions by having the SRU module in hardware and the rest implemented in GPU kernel routines. By using this method, we have given flexibility on two parameters: first, the flexibility over different finite fields $GF(2^m)$; second, the flexibility over different bit error support. The flexibility over $GF(2^m)$ was achieved by splitting the syndrome into the SRU unit and the syndrome kernel. The SRU module resided on the Euclidean domain of $GF(2)$ polynomials, thus making it programmable across multiple finite fields. The syndrome kernel was executed only when an error was encountered. The latency taken by our method, without error, was superior to the CPU and GPU implementations and was equal to the performance observed in [20]. Besides, we had two-fold area savings in the SRU unit to achieve flexibility over $GF(2^m)$ and bit errors. Therefore, this hybrid approach is a pragmatic solution to have a flexible error correction for modern NAND flash devices.

Author Contributions: Conceptualization, A.S. and T.O.; methodology, A.S.; software, A.S.; validation, A.S.; formal analysis, A.S. and T.O.; investigation, A.S. and T.O.; resources, T.O.; data curation, T.O.; writing, original draft preparation, A.S.; writing, review and editing, A.S. and T.O.; visualization, A.S.; supervision, T.O.; project administration, T.O.

Funding: This research received no external funding.

Conflicts of Interest: The authors declare no conflict of interest.

Abbreviations

The following abbreviations are used in this manuscript:

BCH	Bose–Chaudhuri–Hocquenghem
iBMA	inversion-less Berlekamp-Massey algorithm
CPU	Central Processing Unit
CS	Chien Search
CSK	Chien Search Kernel
KEK	Key Equation Kernel
GPU	Graphical Processing Unit
GPGPU	General Purpose Graphical Processing Unit
LDPC	Low Density Parity Check
LFSR	Linear Feedback Shift Register
MLC	Multi-Level Cell
RS	Reed–Solomon
SK	Syndrome calculation Kernel
SLC	Single-Level Cell
SRU	Syndrome Residual Unit
UBER	Uncorrectable Bit Error Rate

References

1. Micheloni, R.; Marelli, A.; Crippa, L. *Inside NAND Flash Memories*; Springer: New York, NY, USA, 2010; doi:10.1007/978-90-481-9431-5.
2. Spinelli, A.; Compagnoni, C.; Lacaita, A. Reliability of NAND Flash Memories: Planar Cells and Emerging Issues in 3D Devices. *Computers* **2017**, *6*, 16, doi:10.3390/computers6020016. [CrossRef]
3. Costell, S.L.; Costello, D. *Error Control Coding—Fundamentals and Applications*, 2nd ed.; Prentice-Hall: Englewood Cliffs, NJ, USA, 2004; pp. 192–233.
4. Cho, J.; Sung, W. Efficient software-based encoding and decoding of BCH codes. *IEEE Trans. Comput.* **2009**, *58*, 878–889, doi:10.1109/TC.2009.27. [CrossRef]
5. Poolakkaparambil, M.; Mathew, J.; Jabir, A. Multiple Bit Error Tolerant Galois Field Architectures over GF (2m). *Electronics* **2012**, *1*, 3–22, doi:10.3390/electronics1010003. [CrossRef]
6. Lee, Y.; Yoo, H.; Yoo, I.; Park, I.C. High-throughput and low-complexity BCH decoding architecture for solid-state drives. *IEEE Trans. Very Large Scale Integr. Syst.* **2014**, *22*, 1183–1187, doi:10.1109/TVLSI.2013.2264687. [CrossRef]
7. Zhang, X. *VLSI Architectures for Modern Error-Correcting Codes*, 2nd ed.; CRC Press: Boca Raton, FL, USA, 2016; pp. 189–225.
8. Qi, X.; Ma, X.; Li, D.; Zhao, Y. Implementation of accelerated BCH decoders on GPU. In Proceedings of the 2013 International Conference on Wireless Communications and Signal Processing (WCSP), Hangzhou, China, 24–26 October 2013; pp. 1–6, doi:10.1109/WCSP.2013.6677084. [CrossRef]
9. Subbiah, A.K.; Ogunfunmi, T. Efficient implementation of BCH decoders on GPU for flash memory devices using iBMA. In Proceedings of the 2016 IEEE International Conference on Consumer Electronics (ICCE), Las Vegas, NV, USA, 7–11 January 2016; pp. 275–278, doi:10.1109/ICCE.2016.7430612. [CrossRef]
10. NVIDIA. *Cuda C Programming Guide*; NVIDIA: Santa Clara, CA, USA, 2015; PMCID:PMC3074485, NIHMSID:Nihms253063, doi:10.1016/j.pedhc.2005.10.011.
11. Parhi, K.K. Eliminating the fan-out bottleneck in parallel long BCH encoders. *IEEE Trans. Circuits Syst. I Regul. Pap.* **2004**, *51*, 512–516, doi:10.1109/TCSI.2004.823655. [CrossRef]
12. Chen, H. CRT-based high-speed parallel architecture for long BCH encoding. *IEEE Trans. Circuits Syst. II Express Briefs* **2009**, *56*, 684–686, doi:10.1109/TCSII.2009.2024247. [CrossRef]
13. Tang, H.; Jung, G.; Park, J. A hybrid multimode BCH encoder architecture for area efficient re-encoding approach. In Proceedings of the IEEE International Symposium on Circuits and Systems, Lisbon, Portugal, 24–27 May 2015; Volume 2015, pp. 1997–2000, doi:10.1109/ISCAS.2015.7169067. [CrossRef]

14. Subbiah, A.K.; Ogunfunmi, T. Area-effcient re-encoding scheme for NAND Flash Memory with multimode BCH Error correction. In Proceedings of the 2018 IEEE International Symposium on Circuits and Systems (ISCAS), Florence, Italy, 27–30 May 2018; pp. 1–5, doi:10.1109/ISCAS.2018.8351503. [CrossRef]
15. Zhang, X. An efficient interpolation-based chase BCH decoder. *IEEE Trans. Circuits Syst. II: Express Briefs* **2013**, *60*, 212–216, doi:10.1109/TCSII.2013.2251941. [CrossRef]
16. Yang, C.H.; Huang, T.Y.; Li, M.R.; Ueng, Y.L. A 5.4 uw soft-decision bch decoder for wireless body area networks. *IEEE Trans. Circuits Syst. I: Regul. Pap.* **2014**, *61*, 2721–2729, doi:10.1109/TCSI.2014.2312478. [CrossRef]
17. Jamro, E. The Design of a Vhdl Based Synthesis Tool for Bch Codecs. Ph.D. Thesis, University of Huddersfield, Huddersfield, UK, 1997.
18. Sun, F.; Devarajan, S.; Rose, K.; Zhang, T. Design of on-chip error correction systems for multilevel NOR and NAND flash memories. *IET Circuits Devices Syst.* **2007**, *1*, 241–249, doi:10.1049/iet-cds. [CrossRef]
19. Sun, F.; Rose, K.; Zhang, T. On the Use of Strong BCH Codes for Improving Multilevel NAND Flash Memory Storage Capacity. In Proceedings of the IEEE Workshop on Signal Processing, Banff, AB, Canada, 2–4 October 2006; pp. 1–5.
20. Park, B.; Park, J.; Lee, Y. Area-Optimized Fully-Flexible BCH Decoder for Multiple GF Dimensions. *IEEE Access* **2018**, *6*, 14498–14509, doi:10.1109/ACCESS.2018.2815640. [CrossRef]
21. Wei, L.; Junrye, R.; Wonyong, S. Low-power high-throughput BCH error correction VLSI design for multi-level cell NAND flash memories. In Proceedings of the 2006 IEEE Workshop on Signal Processing Systems Design and Implementation (SIPS), Banff, AB, Canada, 2–4 October 2006; pp. 303–308, doi:10.1109/SIPS.2006.352599. [CrossRef]
22. Park, B.; An, S.; Park, J.; Lee, Y. Novel folded-KES architecture for high-speed and area-efficient BCH decoders. *IEEE Trans. Circuits Syst. II Express Briefs* **2017**, *64*, 535–539, doi:10.1109/TCSII.2016.2596777. [CrossRef]
23. Yoo, H.; Lee, Y.; Park, I.C. Low-Power Parallel Chien Search Architecture Using a Two-Step Approach. *IEEE Trans. Circuits Syst. II: Express Briefs* **2016**, *63*, 269–273, doi:10.1109/TCSII.2015.2482958. [CrossRef]
24. Freudenberger, J.; Spinner, J. A Configurable Bose–Chaudhuri–Hocquenghem Codec Architecture for Flash Controller Applications. *J. Circuits Syst. Comput.* **2013**, *23*, 1450019, doi:10.1142/s0218126614500194. [CrossRef]

 *micromachines*

Article

Investigation of Intra-Nitride Charge Migration Suppression in SONOS Flash Memory

Seung-Dong Yang [1], Jun-Kyo Jung [1], Jae-Gab Lim [1], Seong-gye Park [2], Hi-Deok Lee [1] and Ga-Won Lee [1,*]

[1] Department of Electronics Engineering, Chungnam National University, Daejeon 305-764, Korea; sdyang83@gmail.com (S.-D.Y.); jjk1006@cnu.ac.kr (J.-K.J.); jaegabi@cnu.ac.kr (J.-G.L.); hdlee@cnu.ac.kr (H.-D.L.)

[2] SK Hynix Inc., Gyeongchung-daero, Bubal-eub, Icheon-si 17336, Korea; pskg@sk.com

* Correspondence: gawon@cnu.ac.kr; Tel.: +82-42-821-7702; Fax: +82-42-823-9544

Received: 24 April 2019; Accepted: 27 May 2019; Published: 29 May 2019

Abstract: In order to suppress the intra-nitride charge spreading in 3D Silicon-Oxide-Nitride-Oxide-Silicon (SONOS) flash memory where the charge trapping layer silicon nitride is shared along the cell string, N_2 plasma treated on the silicon nitride is proposed. Experimental results show that the charge loss decreased in the plasma treated device after baking at 300 °C for 2 h. To extract trap density according to the location in the trapping layer, capacitance-voltage analysis was used and N_2 plasma treatment was shown to be effective to restrain the interface trap formation between blocking oxide and silicon nitride. Moreover, from X-ray Photoelectron Spectroscopy, the reduction of Si-O-N bonding was observed.

Keywords: SONOS; flash memory; charge spreading; plasma treatment; Oxygen-related trap; data retention

1. Introduction

The NAND flash memory market is continuously growing by the successive introduction of mass data storage applications in portable electronic devices, such as USB memory and solid-state drives for tablet PCs and laptops [1]. The cell price as well as bit density are key factors in this application. Until now, it has been possible to reduce the bit cost and increase the bit density through the linear scaling down of cell size, which has been achieved by advanced lithography [2]. Recently, however, the NAND Flash memory industry has faced a scaling limitation of the conventional floating gate (FG) NAND cell. In order to find an alternative technology, Silicon-Oxide-Nitride-Oxide-Silicon (SONOS) device has received attention from researchers, as it provides simpler process steps, lower cell to cell coupling, and virtual immunity to stress-induced leakage current (SILC), when compared to FG [3–5]. However, the down-scaling process is still challenging in SONOS when attempted beyond the 30nm generation. To overcome the problem, SONOS has been fabricated with 3-dimesional (3D) structures such as BiCS [6], P-BiCS [7], TCAT [8], VG-NAND [9] and SMArT [10]. However, in the 3D SONOS structure, the charge trapping layer is not isolated but shared in a cell string, as shown in Figure 1. Due to this continuous trapping layer structure in the 3D scheme, the intra-nitride charge spreading can be a serious problem for data retention properties [11,12]. Charge spreading in silicon nitride has previously been studied in NROM devices, where a trapped charge is locally distributed, and recent research has reported that charge spreading is driven by the spatial concentration difference [13,14]. Figure 2a shows the probable charge spreading mechanism in silicon nitride. For trapped charges in deep-level sites, hopping can happen, yet the possibility is very low because of the long distance between deep-level sites. In the case of shallow-level sites, however, the hopping possibility increases due to relatively high concentration of trapping sites. Charge spreading via the shallow trap sites can be accelerated by conduction band diffusion of thermionic emitted carriers from the trap sites. Figure 2b shows trap energy levels in silicon nitride, and we can see that substitutional oxygen atoms at nitrogen vacancy causes a shallow-level trap site. Considering that the oxygen incorporation is active near the oxide/nitride interface, it is reasonable to estimate that the oxygen and nitrogen vacancy related defects will be formed near the nitride/oxide interface and that they are mainly located in shallow energy level, as reported in [15–17]. Figure 3 shows comparison results on the total number of bulk (N_{Bulk}) and interface traps (N_{int}) according to the channel radius of a cylinder type 3D SONOS device. Assuming $N_{Bulk} = 1.0 \times 10^{18}$ cm^{-3}, relative importance of N_{int} increases as the channel radius decreases. Therefore, when the energy level of N_{int} is shallow, like as the oxygen related traps, the charge spreading via the interface trap sites becomes more critical with shrinkage of device dimension occurring.

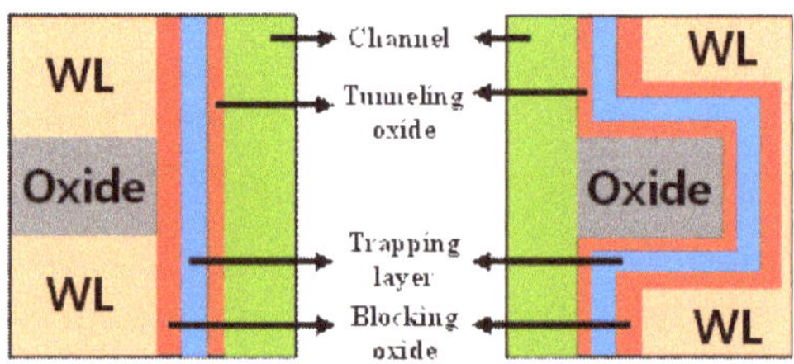

Figure 1. The charge trapping layer structure of (**a**) BiCS 3D NAND and (**b**) TCAT 3D NAND.

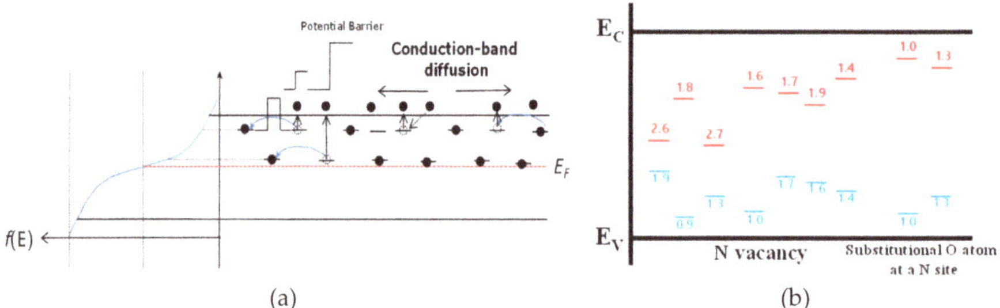

Figure 2. (**a**) Conduction mechanism of programmed Silicon-Oxide-Nitride-Oxide-Silicon (SONOS) memories, (**b**) energy level of silicon nitride.

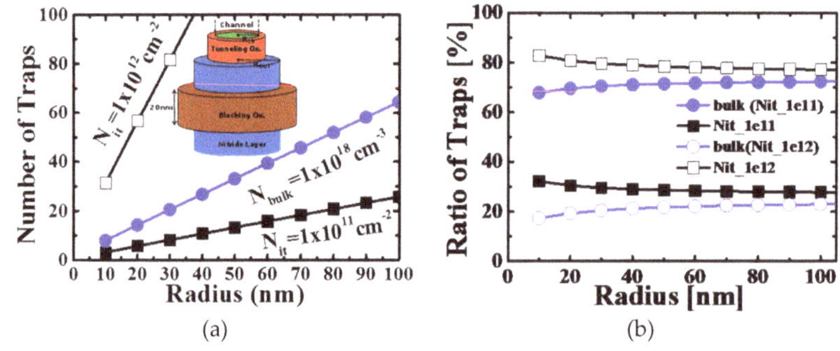

Figure 3. (**a**) Total real number of interfaces and bulk traps and (**b**) the percentage of traps depending on the channel radius of the cylindrical 3D SONOS device. Here, the radius (R_{in} in inset figure) was in the range of 10 to 100 nm, trapping layer thickness was 5 nm and gate length was 20 nm.

In this study, N$_2$ plasma treatment on silicon nitride is proposed to suppress the intra-nitride charge spreading by controlling the interface trap formation. To extract the trap density quantitatively, the capacitance-voltage (C–V) analysis was made based on the measurement results by a LCR meter

(HP 4284A, Agilent, Santa Clara, CA, USA) at a small signal frequency of 1 MHz. To find the bonding state changes induced by plasma treatment, X-ray Photoelectron Spectroscopy (XPS) was also measured with a K-Alpha+ spectrometer (ThermoFisher Scientific, East Grinstead, UK).

2. Experiments

To fabricate SONOS structure, 6 nm SiO_2 for tunneling oxide was thermally grown on a prime grade p-type Si substrate with high-purity oxygen gas via dry oxidation furnace. After the oxidation of Si, N_2 plasma treatment was carried out for 30 sec. The flow rate of nitrogen gas was 45 sccm at a pressure of 10 mTorr, and a plasma power of 200 W. Silicon nitride as a charge storage layer was deposited by low-pressure chemical vapor deposition (LPCVD) at 825 °C with a gas flow rate of SiH_2Cl_2:NH_3 = 170:70 sccm on the tunneling oxide. In this experiment, the nitride thickness varied between 7 nm, 15 nm, and 20 nm to extract the trap density by C–V analysis. Following this, N_2 plasma treatment was performed once again on the top of nitride. Then, blocking oxide of 10nm was deposited by LPCVD at 680 °C, and 100 nm titanium (Ti) was deposited by RF-sputter for gate electrode. The test devices have a gate width by length of 100/100 µm. In order to investigate the impact of lateral charge migration on data retention, different gate stack structures were fabricated using a lithography mask, as shown in Figure 4. In extended structure (Ext. 10), the charge-trapping layer was extended to 10 µm in every direction of the gate electrode. In Ext. 10 structure, the gate etch was stopped on the blocking oxide layer, while the charge trapping layer was etched self-aligned with the gate in the reference devices (Ref.).

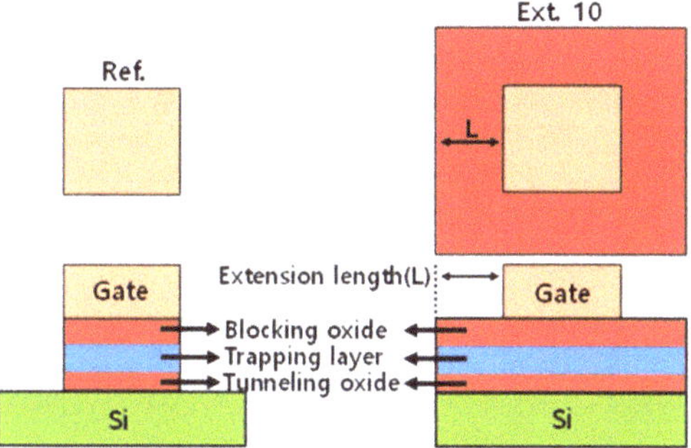

Figure 4. Lithography mask layout to fabricate the test device with a cross-sectional view of the device. Here, Ext. 10 means the extension length of the charge trapping layer was 10 µm. In the case of Ref., the charge trapping layer was etched and self-aligned with the gate and the extension length is 0 µm.

3. Results and Discussion

The program and retention behavior of the fabricated devices with and without N_2 plasma treatment were measured as shown in Figure 5 and the charge loss during retention mode were calculated and are summarized in Table 1. The devices with extended trapping layer showed a larger memory window than the reference device, regardless of N_2 plasma treatment. The reason for this is thought to be due to the fringe field effect of the extended devices. Furthermore, the over-etching issue has been shown to occur during the wet etching process in the reference devices, which in turn lowers program efficiency. However, the charge loss was larger after baking at 300 °C for 2 h. implying the intra-nitride charge spreading effect. The lateral charge loss of the extended devices was estimated to be about 28% in total charge loss. After N_2 plasma treatment, the amount of charge loss deceased in the extended devices and the portion of lateral charge loss was 16%. For the quantitative comparison, nitride/oxide interface trap density was extracted using C–V method. When the positive bias was forced to the gate during C–V measurement, the charge was injected from the substrate and the flatband voltage (V_{FB}) shifted due to charges captured at the traps. The V_{FB} shift (ΔV_{FB}) enlarged with the increase in the ratio of occupied traps, and was finally saturated when all the traps were occupied. From the saturated ΔV_{FB}, according to the trapping layer thickness as shown in Figure 6, a respective trap density of the silicon nitride can be calculated based on the formula below [18].

$$
\begin{aligned}
\Delta V_{FB} &= \frac{q N_{BO/TL}}{\varepsilon_{SiO_2}\varepsilon_0} T_{BO} + \frac{q}{\varepsilon_{SiN}\varepsilon_0} \int_0^{T_{TL}} x N_{Bulk}(x)dx \\
&+ \frac{q T_{BO}}{\varepsilon_{SiO_2}\varepsilon_0} \int_0^{T_{TL}} N_{Bulk}(x)dx + \left(\frac{T_{TL}}{\varepsilon_{SiN}\varepsilon_0} + \frac{T_{BO}}{\varepsilon_{SiO_2}\varepsilon_0}\right) q N_{BO/TL} \\
&= \frac{q N_{Bulk}}{2\varepsilon_{SiN}\varepsilon_0} T_{TL}^2 + \left(\frac{q T_{BO} N_{Bulk}}{\varepsilon_{SiO_2}\varepsilon_0} + \frac{q N_{BO/TL}}{\varepsilon_{SiN}\varepsilon_0}\right) T_{TL} + \frac{q T_{BO} N_{BO/TL}}{\varepsilon_{SiO_2}\varepsilon_0} + \frac{q T_{BO} N_{TO/TL}}{\varepsilon_{SiO_2}\varepsilon_0}
\end{aligned}
\tag{1}
$$

where T_{BO}, T_{TL}, and T_{TO} are the thickness of blocking oxide, trapping layer and tunneling oxide. N_{Bulk} (cm^{-3}) is the trap density of trapping layer and $N_{BO/TL}$ (cm^{-2}) and $N_{TO/TL}$ (cm^{-2}) are the interface trap density of blocking oxide/trapping layer and tunneling oxide/trapping layer, respectively, as shown in inset of Figure 6. From the dependency of ΔV_{FB} on the trapping layer thickness, N_{Bulk} can be assumed to be negligible and then, Equation (1) is expressed as follows.

$$
\Delta V_{FB} = \frac{q T_{TL} N_{BO/TL}}{\varepsilon_{SiN}\varepsilon_0} + \frac{q T_{BO} N_{BO/TL}}{\varepsilon_{SiO_2}\varepsilon_0} + \frac{q T_{BO} N_{TO/TL}}{\varepsilon_{SiO_2}\varepsilon_0}
\tag{2}
$$

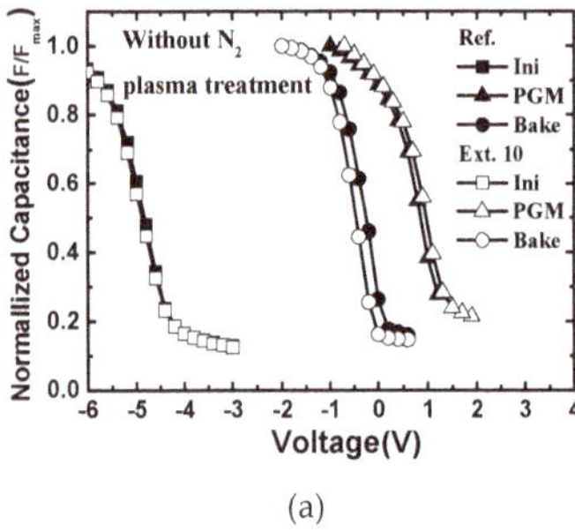

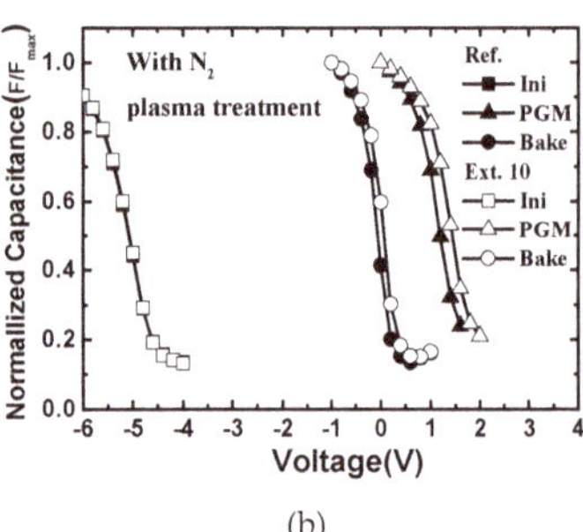

(a) (b)

Figure 5. Measurement results of program and data retention characteristics of the fabricated devices (**a**) without N$_2$ plasma treatment and (**b**) with treatment. Here, the retention properties were measured after baking at 300 °C for 2 h.

Table 1. Extracted trap density based on C–V analysis. Here, $N_{BO/TL}$ and $N_{TO/TL}$ are the interface trap density of blocking oxide/trapping layer and tunneling oxide/trapping layer, respectively.

Sample	$N_{BO/TL}$ (cm^{-2})	$N_{TO/TL}$ (cm^{-2})	Charge Loss [%]
Ref.	2.53×10^{12}	8.91×10^{11}	18.6
Ext.10	4.36×10^{12}	7.32×10^{11}	25.7
N$_2$ plasma treated Ref.	4.35×10^{11}	1.11×10^{12}	17.3
N$_2$ plasma treated Ext.10	5.21×10^{11}	1.18×10^{12}	20.5

Based on Equation (2), the extracted trap densities are summarized in Table 1.

We can see that there was a distinct interface trap reduction in N$_2$ plasma treatment, especially at blocking oxide and trapping layer (BO/TL) interface. Thus, charge loss decreased by 5.2% in extended N$_2$ plasma devices. In the tunneling oxide and trapping layer (TO/TL) interface, the additional nitrogen supply effect by N$_2$ plasma was ambiguous, but this may be because the nitrogen contributed to Si-O-N bonding formation on tunneling oxide, rather than curing the N vacancy in the nitride as the nitride was deposited after oxide formation. More consideration is needed to evaluate the accurate nitrogen behavior according to the underlying layer, but the results show that N$_2$ plasma treatment was effective in reducing the interface trap between blocking oxide and silicon nitride while maintaining the nitride bulk trap.

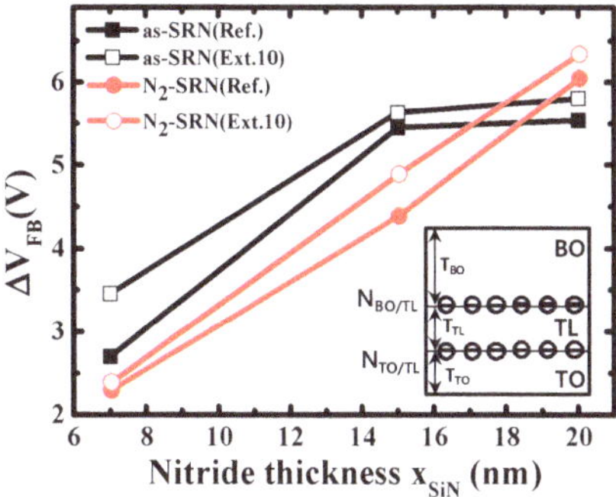

Figure 6. Extracted results of V_{FB} shift in capacitance-voltage curve according to the trapping layer thickness. Inset shows the oxide/trapping layer interface trap sites in the SONOS device structure.

For the physical analysis on N_2 plasma effect, XPS was also measured on the oxide/nitride interface to find the bonding state changes caused by plasma treatment. Figure 7 shows the XPS multi-peak fitting results. After N_2 plasma treatment, the reduction of Si-O-N bonding was observed. The results show that when the additional nitrogen was incorporated into the nitride layer by the plasma treatment, N vacancies in nitride decreased, suppressing subsequent O interactions. This shows that N_2 plasma treatment can be effective method to reduce the aforementioned O-related traps that are located at oxide/nitride interface.

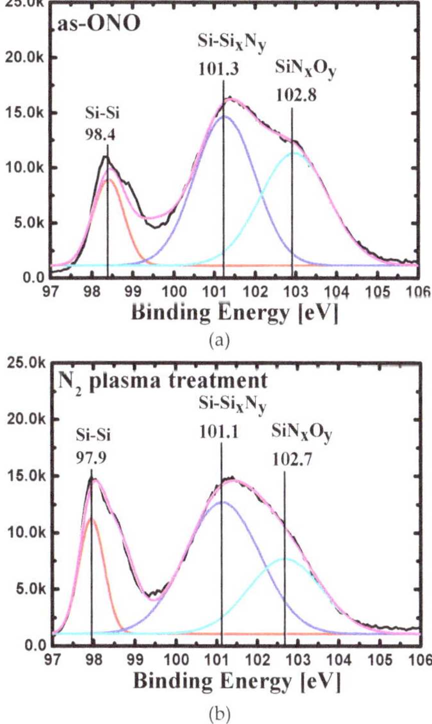

Figure 7. X-ray Photoelectron Spectroscopy (XPS) results of Si2p multi peak fitting of nitride/oxide interface (**a**) as-nitride (without N_2 plasma treatment) and (**b**) N_2 plasma treated nitride.

4. Conclusions

In this paper, N_2 plasma treatment on silicon nitride is proposed as a solution to suppress the interface trap formation and charge spreading in a SONOS device. In order to investigate the impact of intra-nitride charge spreading on data retention in a 3D SONOS device where the charge trapping layer is shared in a cell string, different gate structures were fabricated using a lithography mask, and the charge loss appeared to be much more severe after baking at 300 °C for 2 h. After N_2 plasma treatment, both before and after a silicon nitride formation, charge loss was found to decrease. To extract the trap density quantitatively, C–V analysis method was used, which showed an apparent trap decrease, especially in blocking oxide and the trapping layer interface. XPS also showed the reduction of Si-O-N bonding after plasma treatment. The results indicate that N_2 plasma treatment on silicon nitride is effective to control the shallow O-related interface trap and improve the data retention characteristics of SONOS memory devices.

Author Contributions: Methodology, Formal Analysis, Investigation, Writing—original draft preparation, S.-D.Y.; Data Curation, Visualization, J.-K.J. and J.-G.L.; Conceptualization, S.-g.P.; Methodology, H.-D.L.; Conceptualization, Methodology, Writing—review and editing and Funding acquisition, Supervision G.-W.L.

Funding: This research was financially supported by Hynix semiconductor and the National Research Foundation of Korea (NRF) grant, funded by the Korea government (MSIP) (2017R1D1A1B03033601) and by Nano·Material Technology Development Program through the National Research Foundation of Korea (NRF) funded by the Ministry of Science, ICT and Future Planning. (2009-0082580).

Acknowledgments: The authors would like to thank Kyu-Suk Cho and Mun-sik Seo for technical comments and advice on the research.

Conflicts of Interest: The authors declare no conflict of interest.

References

1. Choi, J.; Seol, K.S. 3D approaches for non-volatile memory. In *2011 Symposium on VLSI Technology-Digest of Technical Papers*; IEEE: Piscataway, NJ, USA, 2011; pp. 178–179.
2. Jung, S.-M.; Jang, J.; Cho, W.; Cho, H.; Jeong, J.; Chang, Y.; Kim, J.; Rah, Y.; Son, Y.; Park, J.; et al. Three dimensionally stacked NAND flash memory technology using stacking single crystal Si layers on ILD and TANOS structure for beyond 30nm node. In *2006 International Electron Devices Meeting*; IEEE: Piscataway, NJ, USA, 2006; pp. 1–4.
3. Cho, M.K.; Kim, D.M. High performance SONOS memory cells free of drain turn-on and over-erase: Compatibility issue with current flash technology. *IEEE Electron Device Lett.* **2000**, *21*, 399–401.
4. Park, Y.-W.; Choi, J.; Kang, C.; Lee, C.; Shin, Y.; Choi, B.; Kim, J.; Jeon, S.; Sel, J.; Park, J.; et al. Highly Manufacturable 32Gb Multi – Level NAND Flash Memory with 0.0098 μm^2 Cell Size using TANOS (Si - Oxide - Al2O3 - TaN) Cell Technology. In *2006 International Electron Devices Meeting*; IEEE: Piscataway, NJ, USA, 2006; pp. 1–4.
5. De Salvo, B.; Geradi, C.; van Schaijk, R.; Lombardo, S.A.; Corso, D.; Plantamura, C.; Serafino, S.; Ammendola, G.; van Duuren, M.; Goarin, P.; et al. Performance and reliability features of advanced nonvolatile memories based on discrete traps (silicon nanocrystals, SONOS). *IEEE Trans. Device Mater. Reliab.* **2004**, *4*, 377–389. [CrossRef]
6. Tanaka, H.; Kido, M.; Yahashi, K.; Oomura, M.; Katsumata, R.; Kito, M.; Fukuzumi, Y.; Sato, M.; Nagata, Y.; Matsuoka, Y.; et al. Bit Cost Scalable Technology with Punch and Plug Process for Ultra High Density Flash Memory. In *2007 IEEE Symposium on VLSI Technology*; IEEE: Piscataway, NJ, USA, 2007; pp. 14–15.
7. Katsumata, R.; Kito, M.; Fukuzumi, Y.; Kido, M.; Tanaka, H.; Koromi, Y.; Ishiduki, M.; Matsunami, J.; Fujiwara, T.; Nagata, Y.; et al. Pipe-shaped BiCS flash memory with 16 stacked layers and multi-level-cell operation for ultra-high density storage devices. In *2009 Symposium on VLSI Technology*; IEEE: Piscataway, NJ, USA, 2009; pp. 136–137.
8. Jang, J.-H.; Kim, H.-S.; Cho, W.; Cho, H.; Kim, J.; Shim, S.I.; Jeong, J.H.; Son, B.-K.; Kim, D.W.; Shim, J.-J.; et al. Vertical Cell Array using TCAT (Terabit Cell Array Transistor) Technology for Ultra High Density NAND Flash Memory. In *2009 Symposium on VLSI Technology*; IEEE: Piscataway, NJ, USA, 2009; pp. 192–193.

9. Kim, W.-J.; Choi, S.; Sung, J.; Lee, T.; Park, C.; Ko, H.; Jung, J.; Yoo, I.; Park, Y. "Multi-layered Vertical Gate NAND Flash overcoming stacking limit for terabit density storage. In *2009 Symposium on VLSI Technology*; IEEE: Piscataway, NJ, USA, 2009; pp. 188–189.

10. Choi, E.S.; Park, S.K. Device considerations for high density and highly reliable 3D NAND flash cell in near future. In *2012 International Electron Devices Meeting*; IEEE: Piscataway, NJ, USA, 2012; pp. 9.4.1–9.4.4.

11. Maconi, A.; Arreghini, A.; Compagnoni, C.M.; Spinelli, A.S.; van Houdt, J.; Lacaita, A.L. Impact of lateral charge migration on the retention performance of planar and 3D SONOS devices. In *2011 Proceedings of the European Solid-State Device Research Conference (ESSDERC)*; IEEE: Piscataway, NJ, USA, 2011; pp. 195–198.

12. Kang, C.-S.; Choi, J.; Sim, J.; Lee, C.; Shin, Y.; Park, J.; Sel, J.; Jeon, S.; Park, Y.; Kim, K.-N. Effects of lateral charge spreading on the reliability of TANOS (TaN/AlO/SiN/Oxide/Si) NAND flash memory. In *2007 IEEE International Reliability Physics Symposium Proceedings. 45th Annual*; IEEE: Piscataway, NJ, USA, 2007; pp. 167–170.

13. Maconi, A.; Arreghini, A.; Compagnoni, C.M.; Spinelli, A.S.; van Houdt, J.; Lacaita, A.L. Comprehensive investigation of the impact of lateral charge migration on retention performance of planar and 3D SONOS devices. *Solid-State Electron.* **2012**, *74*, 64–70. [CrossRef]

14. Liu, L.; Arreghini, A.; Pan, L.; van Houdt, J. Comprehensive understanding of charge lateral migration in 3D SONOS memories. *Solid-State Electron.* **2016**, *116*, 95–99. [CrossRef]

15. Morokov, Y.N.; Novikov, Y.N.; Gritsenko, V.A.; Wong, H. Two-fold coordinated nitrogen atom: an electron trap in MOS devices with silicon oxynitride as the gate dielectric. *Microelectron. Eng.* **1999**, *48*, 175–178. [CrossRef]

16. Wong, H.; Gritsenko, V.A. Defects in silicon oxynitride gate dielectric films. *Microelectron. Reliab.* **2002**, *42*, 597–605. [CrossRef]

17. Perera, R.; Ikeda, A.; Hattori, R.; Kuroki, Y. Effects of post annealing on removal of defect states in silicon oxynitride films grown by oxidation of silicon substrates nitrided in inductively coupled nitrogen plasma. *Thin Solid Films* **2003**, *423*, 212–217. [CrossRef]

18. Ishida, T.; Okuyama, Y.; Yamada, R. Characterization of charge traps in metal-oxide-nitride-oxide-semiconductor (MONOS) structures for embedded flash memories. In *2006 IEEE International Reliability Physics Symposium Proceedings*; IEEE: Piscataway, NJ, USA, 2006; pp. 516–522.

Article

Matrix Mapping on Crossbar Memory Arrays with Resistive Interconnects and Its Use in In-Memory Compression of Biosignals

Yoon Kyeung Lee, Jeong Woo Jeon, Eui-Sang Park, Chanyoung Yoo, Woohyun Kim, Manick Ha and Cheol Seong Hwang *

Department of Materials Science and Engineering at Seoul National University, Seoul 08826, Korea; greense9@snu.ac.kr (Y.K.L.); wjd2153@snu.ac.kr (J.W.J.); euispark@snu.ac.kr (E.-S.P.); cyyoo0117@snu.ac.kr (C.Y.); kimkwh@snu.ac.kr (W.K.); manick.ha@snu.ac.kr (M.H.)
* Correspondence: cheolsh@snu.ac.kr; Tel.: +82-2-880-7535

Received: 4 April 2019; Accepted: 5 May 2019; Published: 7 May 2019

Abstract: Recent advances in nanoscale resistive memory devices offer promising opportunities for in-memory computing with their capability of simultaneous information storage and processing. The relationship between current and memory conductance can be utilized to perform matrix-vector multiplication for data-intensive tasks, such as training and inference in machine learning and analysis of continuous data stream. This work implements a mapping algorithm of memory conductance for matrix-vector multiplication using a realistic crossbar model with finite cell-to-cell resistance. An iterative simulation calculates the matrix-specific local junction voltages at each crosspoint, and systematically compensates the voltage drop by multiplying the memory conductance with the ratio between the applied and real junction potential. The calibration factors depend both on the location of the crosspoints and the matrix structure. This modification enabled the compression of Electrocardiographic signals, which was not possible with uncalibrated conductance. The results suggest potential utilities of the calibration scheme in the processing of data generated from mobile sensing or communication devices that requires energy/areal efficiencies.

Keywords: resistive memory; crossbar; in-memory computing; analogue computing; matrix-vector multiplication; ECG

1. Introduction

Emerging classes of mobile electronic devices offer attractive capabilities for real-time analytics of the physical world through the connection to central computing systems. One of the critical challenges in this emerging Internet of Things (IoT) is the instantaneous extraction of relevant information from the abundant data with the limited power and communication bandwidth for data transmission. This challenge demands smart components on the edge of the mobile devices that can filter, compress, or classify the data outputs onsite [1–4]. This pre-processing needs to be extremely power efficient and quick to handle the large volume of data continuously generated from the surrounding world.

A subset of the processing operations can be categorized as a linear transformation which can be expressed as a matrix-vector multiplication (MVM). The MVM can be performed in an analogue domain using a resistive memory crossbar array by storing the matrix values as the conductance of the memory cell. The operation can take a constant time complexity ($O(1)$), and be energy efficient owing to the functional integration of the processing and memory units [5–7]. The scalability of the crossbar structure down to $4F^2$ (F: feature size of a technology node) is also beneficial for the device miniaturization. Envisioned applications include linear equation solver and training of or inference on neural networks as demonstrated recently [1,7–11].

Prior studies have shown that the throughputs per area and the energy efficiency can exceed today's von Neumann computing scheme, but computational accuracy remained as a non-trivial challenge for high-precision analogue-based MVM. In device levels, output errors can be originated from the variations of the electrical characteristics between the cells, non-linear current-voltage relationship, and stochasticity in resistance switching process. Separate from the efforts in development of the reliable devices, it is also important to optimize the conductance mapping scheme using realistic crossbar arrays. Finite conductivity of interconnecting wire has been suggested as one of the important factors causing errors in the crossbar-based MVM [9,12]. Empirical calibration methods that are based on the comparison between the desired output and real measurements have shown to improve the accuracy level although the origin of the discrepancy of the measurement values was not clearly identified [1]. To overcome the limitation of such hardware-based methods, model-based theoretical analysis attempted more systematic approach to understand the computational error [9,12]. Hu et al. first introduced a comprehensive crossbar array model for MVM, and applied it to the training of neural network for pattern recognition [9]. This simulation-based optimization of the conductance minimizes the time and power consumption to post-process the outputs and provides explanation for the computational outputs with given circuits.

This work implemented a mapping algorithm of memory conductance for MVM using a crossbar model with finite wire resistance, and analyzed the calibration performance for the compression of electrocardiographic (ECG) signals. An iterative software simulation calculates the matrix-specific local junction voltages at each cross-point, and calculate the ratio between the junction voltages and input voltage applied from the source. The ratio becomes a calibration factor to update the memory conductance to systematically compensates the voltage drop. The results indicate that the calibration factors both depend on the location of the junctions and matrix structure. This correction enabled the in-memory compression of ECG signals whose reconstruction error is comparable to the double precision calculation. The findings suggest a possible route to overcome difficulties in analogue computing in realizing diverse edge computing devices for onsite data processing.

2. Methods

2.1. Calibration Factor for Matrix Mapping on Proposed Crossbar Model

Figure 1a shows a schematic representation of the crossbar model that includes interconnection line resistance to calculate the local potential at each cross-point. The model incorporates both the cell-to-cell resistance and the access resistance from a voltage source to the first column/row metal lines. The analogue-based MVM using a crossbar array assuming an ideal behavior has the current output from the column (or bit) line (BL) as follows.

$$I_j^{ideal} = G_{1,j}V_{1,app} + \cdots + G_{m,j}V_{m,app} \tag{1}$$

Here, I_j^{ideal} is the current output from j^{th} BL. $G_{i,j}$ is the conductance of memory cell located at a crosspoint of the ith word and the jth bit lines. The conductance ($G_{i,j}$) represents a linear-transformed matrix element to map the matrix values within the range of the achievable conductance of the device. $V_{i,app}$ is input voltage to the ith word lines (WL). (BLs are assumed to be grounded.) Equation (1) holds true only if the series resistance of the interconnection wires is negligible. Considering the resistivity of conventional metal wires ($\rho = 10^{-8}$ to 10^{-7} $\Omega \cdot$m), the resistance between the nearest cells ($R = \rho \cdot F/(F \cdot d)$, F: feature size, d: metal thickness) ranges from 10^0 to 10^1 Ω when d is assumed ~10 nm. The wire resistance may further increase due to lower density caused by vapor deposition. For a $4F^2$ crossbar structure, the interconnect resistance between two adjacent cells can be estimated to be ~4.53, 2.97, and 1.55 Ω under 16 nm, 22 nm, and 32 nm technology node, respectively, according to the International Technology Roadmap for Semiconductors 2013 [12]. Simple calculation estimates the voltage drop can be a significant source of errors considering the realistic conductivity of the resistive memories. For example, if we assume ~100 by 100 bits of crossbar arrays and 0.1 to 1 mA total current along the word line, iR drop at the end of the word line can be 0.01 to 0.1V. (e.g., 0.1–1 mA × R(cell-cell) × 100 → 0.01–0.1 V). In this realistic case, the current output needs to be modified as

$$I_j^{real} = G_{1,j}V_{1,j} + G_{2,j}V_{2,j} + \cdots + G_{m,j}V_{m,j} \tag{2}$$

instead of Equation (1) with $V_{i,app}$ terms to conform with the Ohm's law. Here, $V_{i,j}$ is the local junction potentials across the memory cell at (i, j) crosspoint. Since $V_{i,j}$ is not guaranteed to be equal to the applied voltage to the ith WL due to voltage drop, I_j becomes small compared to the ideal case as observed in previous studies [1,9].

One way to compensate the smaller current output can be the increase of the conductance level of the memory according to the local voltage drop. If the voltage drop for arbitrary WL and BL input voltages can be estimated, the conductance of the memory can be set as

$$G'_{i,j} = G_{i,j}\frac{V_{i,app}}{V_{i,j}} \tag{3}$$

instead of $G_{i,j}$. With the calibrated conductance ($G'_{i,j}$), the current outputs become the ideal current as follows.

$$I_j^{real} = G_{1,j}\frac{V_{1,app}}{V_{1,j}} \cdot V_{1,j} + \cdots + G_{m,j}\frac{V_{m,app}}{V_{m,j}} \cdot V_{m,j} = I_j^{ideal} \tag{4}$$

Thus, the ratio ($V_{i,app}/V_{i,j}$) can be considered as a calibration factor for the memory conductance for in-memory MVM when the junction potential deviates from the applied voltage. There can be other approaches that use equilvalent conductance terms multiplied by the applied voltage to describe the measured current. This approach may be useful if measurement data are available and the calibration algorithm to drive the real current to the ideal one is developed. Yet, the current work is more focused on the calibration based only on theoretical model circuits without requirement for any real measurements.

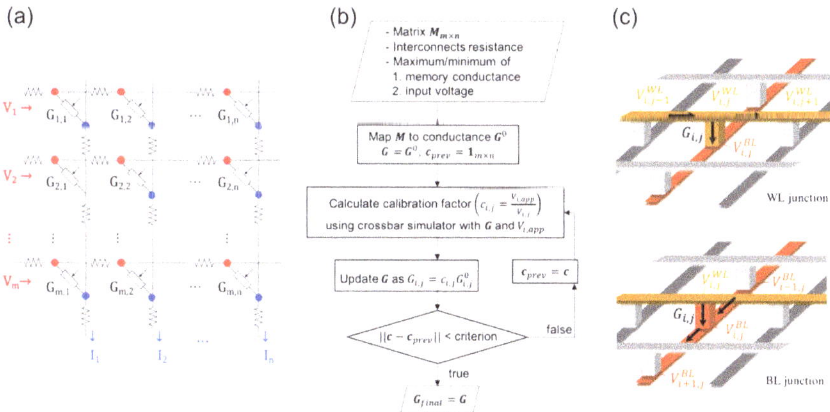

Figure 1. (**a**) Simulation model for resistive memory crossbar array with finite conductance of interconnects. (**b**) Conductance calibration algorithm for mapping of an m × n matrix using a crossbar simulator. (**c**) Local currents at word lines (WL) and bit line (BL) junctions in accordance with Kirchhoff's law.

2.2. Iterative Calibration Based on Crossbar Simulation

An iterative algorithm was developed to progressively increase conductance values based on the simulated $V_{i,j}$ at individual junctions. Figure 1b summarizes the procedure of the calibration process.

Through the iterations, $V_{i,j}$'s are updated by solving the $2mn$ Kirchhoff's relations (mn WL junctions + mn BL junctions) that need to be simultaneously satisfied with given memory conductance and the voltage inputs [13]. Figure 1c, for example, illustrates the local currents on the WL junction that follow the equation below.

$$G_w\left(V_{i,j}^{WL} - V_{i,j-1}^{WL}\right) = G_{i,j}\left(V_{i,j}^{BL} - V_{i,j}^{WL}\right) + G_w\left(V_{i,j+1}^{WL} - V_{i,j}^{WL}\right) \tag{5}$$

Here, G_w is a cell-to-cell conductance, and $V_{i,j}^{WL}$ and $V_{i,j}^{BL}$ are voltages at (i, j) crosspoint on WL and BL, respectively. $2mn$ Kirchhoff's equations can be arranged in a simple matrix form whose details are described in the Appendix A. Since the calibrated conductance ($G'_{i,j}$) is higher than the previous conductance ($G_{i,j}$), the overall current increases, and the voltage drops need to be recalculated with this new $G'_{i,j}$ by the next iteration of the simulation. The iteration is repeated until the conductance (or $V_{i,app}/V_{i,j}$ ratios) converge, and the final ratios determine the conductance level of the memory to represent the arithmetic matrix elements. The simulation code is implemented in MATLAB and each iteration takes ~1 sec with single 3.5 GHz Intel Core i7 for 64×64 crossbar arrays. The calibration factors were converged after 10 to 20 iterations depending on the cell-to-cell resistance and termination criteria. The runtime and error depend on the termination criteria, and assumed to be a similar level to the previous report [9].

3. Results and Discussion

The in-memory MVM can be used for low-power data processing, such as compression or high- or low-pass filtering. Here, as an example, the discrete wavelet transform (DWT) matrix is mapped to the final memory conductance ranging from 0.01 to 70 µS [14,15]. The cell-to-cell resistance (R) and the access resistance from the voltage source to the crossbar are assumed to be 1 Ω and 100 Ω, respectively. Larger R (10 Ω) is also studied for comparison. Voltages are supplied from the left for WLs and the bottom for BLs. For the calculation of the voltage drops at each junction, the supply voltage of 0.1 V was assumed for all WLs. (The calibration factors were insensitive to the voltage (0.1 to 0.5 V) since $V_{i,j}$ ~ $V_{i,app} - iR$ where iR varies approximately with the same factor as $V_{i,app}$). The operation parameters were set to be consistent with the practical values reported in the previous PRAM-based studies [7].

Figure 2 presents the simulation results of the conductance mapping of 64×64 DWT matrix using biorthogonal filters with 4-level of decomposition. Figure 2a describes the change in the calibration factors through the iteration represented by the 2-norm of the difference matrix. The conductance is quickly converged, and the norm values less than 10^{-4} were achieved after 10 cycles (R = 1 Ω) and 16 cycles (10 Ω). Figure 2b compares the initial conductance ($G_{i,j}^0$) and final conductance for R = 10 Ω case. Figure 2c plots the final calibration factors to visualize the voltage drop across the crossbar. (R = 1 Ω (left), 10 Ω (right)) Calibration factors range from 1.1 to 1.4 for 1 Ω case, and 1.1 to 2.2 for 10 Ω case. 10 Ω resistance shows larger dependency of the calibration factor on the distance from the voltage source. The location dependency of the calibration factors implies that the effect of possible fluctuation in the resistance of nanoscale wires can be averaged over the long distance from the voltage source for the junctions with large calibration factors. The colormaps also reveal the large values for the first four columns and small values for every four rows. As depicted in Figure 2d, the calibration factors reflect the matrix structure. The conductance sum ($\sum_i G_{i,j}^0$) is large for the first four columns, which results in a large current gathered along the four BLs. For the same reason, the small conductance sum ($\sum_j G_{i,j}^0$) for every four rows result in small overall current along the WLs: thus, smaller calibration factors. This variation in the overall current along the metal line causes different level of iR drop, resulting in matrix-dependent calibration factors.

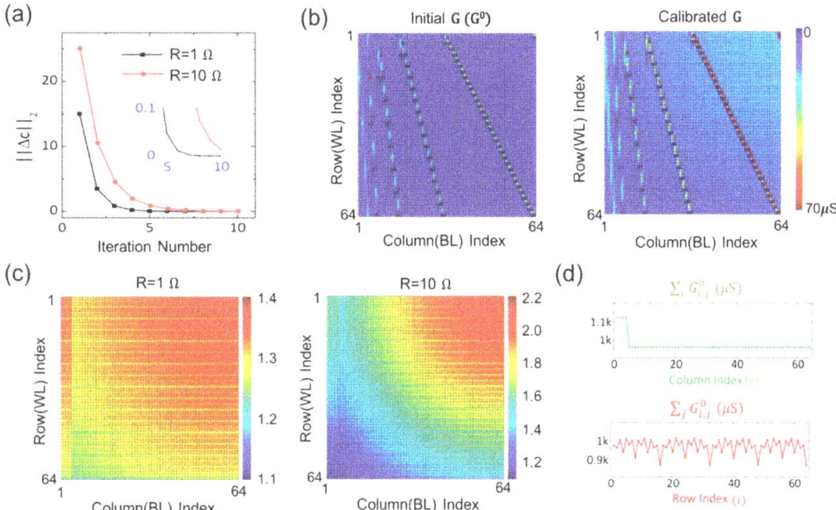

Figure 2. Conductance mapping of 64 × 64 matrix for discrete wavelet transform (DWT). (**a**) Convergence of calibration factors though the iterations for 1 Ω and 10 Ω cell-cell resistance. (**b**) Colored map of cell conductance of a crossbar before/after calibration. (R = 10 Ω). (**c**) Matrix-specific calibration factors at individual cross-points for R = 1 Ω (left) and R = 10 Ω (right). (**d**) Conductance sum of each column (top) or row (bottom) of the initial conductance.

Figure 3 summarizes the effect of the conductance calibration on the data compression and reconstruction performance. Rescaled ECG signals from the MIT-BIH database were applied as the input voltage (0–0.3 V) for DWT [16]. Figure 3a,b show the coefficients of the DWT converted from the simulated currents from the BLs for R = 1 Ω and 10 Ω, respectively. The black squares present the exact coefficients calculated in double-precision (64 bits), and the green diamond lines present the simulated coefficients with the initial memory conductance before calibration. The negatively shifted values of the simulated coefficients result from the small currents due to the voltage drop along the resistive metal interconnects. This shift fails the threshold-based compression of data where the small coefficients are cut off based on their absolute quantity (distance from zero). The larger negative slope in Figure 3b compared to Figure 3a reflects a severe reduction in current outputs for the columns located far from the voltage source due to the larger R (10 Ω). The other lines in the figures show the coefficients calculated with the calibrated memory conductance at different stages of iteration. The red lines in Figure 3a,b show that the fully calibrated coefficients well match to the exact values for both R values. The 2-norms of the difference between the exact and the experimental coefficient vectors were 4.2 (1 Ω) and 8.6 (10 Ω), and the maximum difference were 3.5 (1Ω) and 7.2 (10 Ω) at the peak of the coefficient (exact coefficient value: 224.8, index: 29). Figure 3c shows the reconstructed ECG signals using the calibrated coefficients. (ECG signals were vertically shifted for visibility of individual lines.) The magenta line shows the reconstructed signals from the 15 largest exact coefficients out of 64. By filtering of the small coefficients, the noise in the original signal was removed as the case with exact coefficients. Figure 3d plots the error of the reconstructed signal. The reconstructed signal-to-noise ratios, defined as $20log_{10}(\|x\|_2 / \|x - \hat{x}\|_2)$ (x: original ECG, $\hat{x}$: reconstructed ECG), were 28.2/43.4 (1 Ω) and 27.8/37.1 (10 Ω) with/without cut-off, respectively, compared to 28.3 for the reconstruction using 15 largest exact coefficients.

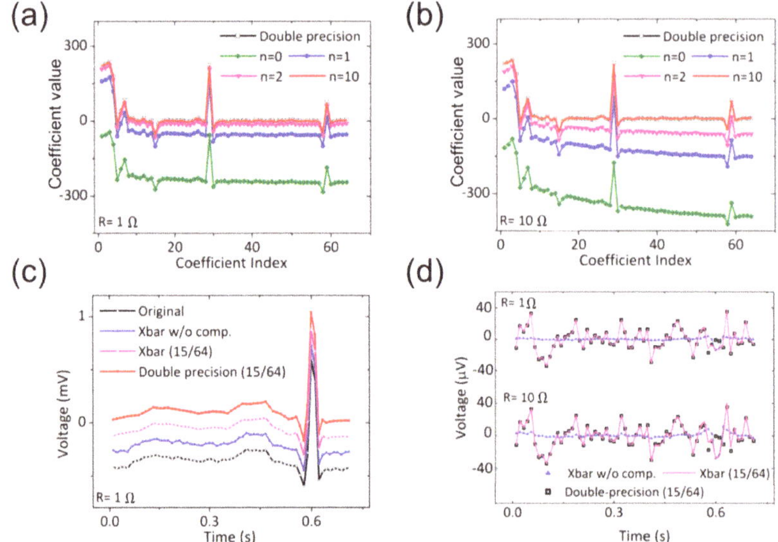

Figure 3. Electrocardiographic (ECG) signal compression using in-memory computing. (**a,b**) Coefficients of ECG signal after DWT using crossbar (Xbar) conductance determined by simulation. n: iteration number of simulation for conductance calibration. (**a**) R = 1 Ω. (**b**) 10 Ω. (**c**) Reconstruction of ECG from the coefficients. Compression ratio = 15/64. (**d**) Reconstruction error.

4. Conclusions

A conversion algorithm of a matrix to conductance was proposed in a crossbar memory array when the metal interconnects have finite conductance. The iterative simulation systematically compensates for the voltage drop along the interconnects by increasing the memory conductance. The calibration enables in-memory data compression. Considering the power limit in healthcare-related mobile devices, the proposed real-time compression using a memory crossbar can have potential as pre-processing units in such devices for diagnosis/therapeutic purposes.

Author Contributions: Conceptualization: Y.K.L., J.W.J.; formal analysis: Y.K.L.; software: Y.K.L., J.W.J.; validation: J.W.J., E.-S.P., W.K., M.H., C.S.H.; writing: Y.K.L., C.S.H.

Funding: This research was funded by the Basic Science Research Program through the National Research Foundation of Korea (NRF) funded by the Ministry of Education, grant number 2018R1A6A3A01012588.

Conflicts of Interest: The authors declare no conflict of interest.

Appendix A

The crossbar model aims to calculate junction potentials at each cross-point. Since we can build one Kirchhoff's equation for each junction, $2mn$ relations (Figure A1, mn junctions on WL+ mn junctions on BL) need to be simultaneously satisfied with given memory resistances and the applied WL and BL applied potentials. Here, G_w and $G_{i,j}$ are the wire and memory conductance, and $V_{i,j}^{WL}$ and $V_{i,j}^{BL}$ are the local voltages at the junctions in a real system with finite conductance of the interconnects.

$$(WL, (i,j)) \qquad G_w\left(V_{i,j}^{WL} - V_{i,j-1}^{WL}\right) - G_{i,j}\left(V_{i,j}^{BL} - V_{i,j}^{WL}\right) - G_w\left(V_{i,j+1}^{WL} - V_{i,j}^{WL}\right) = 0 \qquad (A1)$$

$$(WL, j = 1) \quad G_{i,access}^{WL}\left(V_{i,1}^{WL} - V_{i,applied}^{WL}\right) - G_{i,1}\left(V_{i,1}^{BL} - V_{i,1}^{WL}\right) - G_w\left(V_{i,2}^{WL} - V_{i,1}^{WL}\right) = 0 \qquad (A2)$$

$$(WL, j = n) \qquad G_w\left(V_{i,n}^{WL} - V_{i,n-1}^{WL}\right) - G_{i,n}\left(V_{i,n}^{BL} - V_{i,n}^{WL}\right) = 0 \qquad (A3)$$

$$(BL, (i,j)) \qquad G_w\left(V_{i+1,j}^{BL} - V_{i,j}^{BL}\right) - G_{i,j}\left(V_{i,j}^{BL} - V_{i,j}^{WL}\right) - G_w\left(V_{i,j}^{BL} - V_{i-1,j}^{BL}\right) = 0 \tag{A4}$$

$$(BL,\ i = m) \quad G_{j,access}^{BL}\left(V_{j,applied}^{BL} - V_{m,j}^{BL}\right) - G_{m,j}\left(V_{m,j}^{BL} - V_{m,j}^{WL}\right) - G_w\left(V_{m,j}^{BL} - V_{m-1,j}^{BL}\right) = 0 \tag{A5}$$

$$(BL,\ i = 1) \qquad G_w\left(V_{2,j}^{BL} - V_{1,j}^{BL}\right) - G_{i,j}\left(V_{1,j}^{BL} - V_{1,j}^{WL}\right) = 0 \tag{A6}$$

When the equations are arranged in the order as described in Figure A1, the equations can be simplified as the following matrix formulation:

$$A_{mn\times mn}v_{WL} + B_{mn\times mn}v_{BL} = E_{WL} \text{ (for WL junctions)} \tag{A7}$$

$$C_{mn\times mn}v_{WL} + D_{mn\times mn}v_{BL} = E_{BL} \text{ (for BL junctions)} \tag{A8}$$

$$\begin{bmatrix} A & B \\ C & D \end{bmatrix}\begin{bmatrix} v_{WL} \\ v_{BL} \end{bmatrix} = \begin{bmatrix} E_{WL} \\ E_{BL} \end{bmatrix} \tag{A9}$$

where

$$v_{WL,mn\times 1} = \left[V_{1,1}^{WL}, V_{1,2}^{WL}, \cdots, V_{1,n}^{WL}, V_{2,1}^{WL}, \cdots, V_{m,n}^{WL}\right]^T = \left[v_{WL,i\,=\,1}, v_{WL,i\,=\,2}, \cdots, v_{WL,i\,=\,m}\right]^T \tag{A10}$$

$$v_{BL,mn\times 1} = \left[V_{1,1}^{BL}, V_{1,2}^{BL}, \cdots, V_{1,n}^{BL}, V_{2,1}^{BL}, \cdots, V_{m,n}^{BL}\right]^T = \left[v_{BL,i\,=\,1}, v_{BL,i\,=\,2}, \cdots, v_{BL,i\,=\,m}\right]^T. \tag{A11}$$

$$E_{WL,mn\times 1} = \left[G_{1,access}^{WL}V_{1,app}^{WL}, 0, \cdots, G_{2,access}^{WL}V_{2,app}^{WL}, 0, \cdots, G_{m,access}^{WL}V_{m,app}^{WL}, 0, \cdots\right]^T \tag{A12}$$

$$E_{BL,mn\times 1} = -\left[G_{1,access}^{BL}V_{1,app}^{BL}, G_{2,access}^{BL}V_{2,app}^{BL}, \cdots, G_{n,access}^{BL}V_{n,app}^{BL}\, 0, \cdots\right]^T \tag{A13}$$

Here, A and D are sparse matrices whose nonzero elements are the ones that are multiplied by the local potentials adjacent to the junction under consideration along the WL (for A) or BL (for D). For example, the Kirchhoff's law on the (i,j) WL junction is described by

$$A_{(i-1)\times j+j^{th}\ row}\ v_{WL} + B_{(i-1)\times j+j^{th}\ row}\ v_{BL} = E_{WL,(i-1)\times j+j^{th}\ row} \tag{A14}$$

The only nonzero elements of $(i-1) \times j + j^{th}$ row of A are $j-1$, j, $j+1^{th}$ elements of the row. B and C are $mn \times mn$ diagonal matrices related to the conductance of the resistive memory to describe the currents flow through the memory layer. More details are available in [13] although the structure of the matrices A, B, C, D, E_{WL} and E_{BL} depends on the order of the Kirchhoff's equations that correspond to the individual junctions.

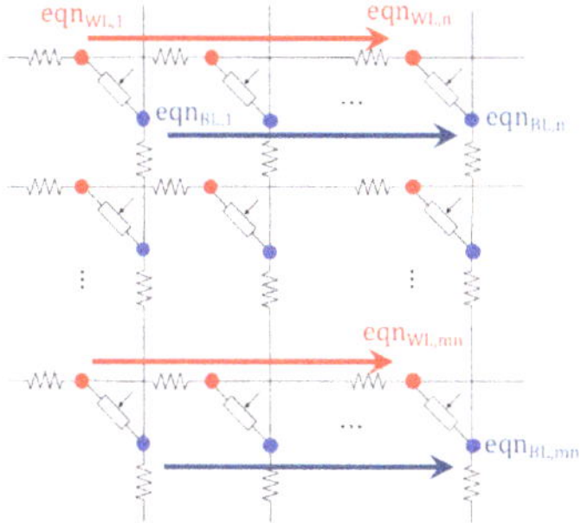

Figure A1. Kirchhoff's law produces $2mn$ equations.

For the simulation where all the applied potentials to the WL and BL are set, local potentials at the crossbar junctions can be obtained in two steps by solving the following two equations:

$$\left(B - AC^{-1}D\right)v_{BL} \;=\; E_{WL} - AC^{-1}E_{BL} \tag{A15}$$

$$v_{WL} \;=\; C^{-1}(E_{BL} - Dv_{BL}) \tag{A16}$$

References

1. Li, C.; Hu, M.; Li, Y.; Jiang, H.; Ge, N.; Montgomery, E.; Zhang, J.; Song, W.; Dávila, N.; Graves, C.E.; et al. Analogue Signal and Image Processing with Large Memristor Crossbars. *Nat. Electron.* **2017**, *1*, 52. [CrossRef]
2. Wright, C.D. Precise Computing with Imprecise Devices. *Nat. Electron.* **2018**, *1*, 212–213. [CrossRef]
3. Hussein, A.F.; Hashim, S.J.; Aziz, A.F.A.; Rokhani, F.Z.; Adnan, W.A.W. A Real Time ECG Data Compression Scheme for Enhanced Bluetooth Low Energy ECG System Power Consumption. *J. Ambient Intell. Humaniz. Comput.* **2017**. [CrossRef]
4. Yu, B.; Yang, L.; Chong, C.C. ECG Monitoring over Bluetooth: Data Compression and Transmission. In Proceedings of the IEEE Wireless Communication and Networking Conference, Sydney, NSW, Australia, 18–21 April 2010; pp. 1–5.
5. Gallo, M.; Sebastian, A.; Cherubini, G.; Giefers, H.; Eleftheriou, E. Compressed Sensing With Approximate Message Passing Using In-Memory Computing. *IEEE Trans. Electron. Devices* **2018**, *99*, 1–9. [CrossRef]
6. Wang, Y.; Li, X.; Xu, K.; Ren, F.; Yu, H. Data-Driven Sampling Matrix Boolean Optimization for Energy-Efficient Biomedical Signal Acquisition by Compressive Sensing. *IEEE Trans. Biomed. Circuits Syst.* **2017**, *11*, 255–266. [CrossRef] [PubMed]
7. Le Gallo, M.; Sebastian, A.; Mathis, R.; Manica, M.; Giefers, H.; Tuma, T.; Bekas, C.; Curioni, A.; Eleftheriou, E. Mixed-Precision In-Memory Computing. *Nat. Electron.* **2017**, *1*, 246. [CrossRef]
8. Burr, G.W.; Shelby, R.M.; Sidler, S.; Di Nolfo, C.; Jang, J.; Boybat, I.; Shenoy, R.S.; Narayanan, P.; Virwani, K.; Giacometti, E.U.; et al. Experimental Demonstration and Tolerancing of a Large-Scale Neural Network (165,000 Synapses) Using Phase-Change Memory as the Synaptic Weight Element. *IEEE Trans. Electron. Devices* **2015**, *62*, 3498–3507. [CrossRef]
9. Hu, M.; Strachan, J.P.; Li, Z.; Grafals, E.M.; Davila, N.; Graves, C.; Lam, S.; Ge, N.; Williams, R.S.; Yang, J.; et al. Dot-Product Engine for Neuromorphic Computing: Programming 1T1M Crossbar to Accelerate Matrix-Vector Multiplication. In Proceedings of the 53rd Annual Design Automation Conference, Austin, TX, USA, 5–9 June 2016.
10. Zidan, M.A.; Jeong, Y.; Lee, J.; Chen, B.; Huang, S.; Kushner, M.J.; Lu, W.D. A General Memristor-Based Partial Differential Equation Solver. *Nat. Electron.* **2018**, *1*, 411–420. [CrossRef]
11. Ambrogio, S.; Narayanan, P.; Tsai, H.; Shelby, R.M.; Boybat, I.; Di Nolfo, C.; Sidler, S.; Giordano, M.; Bodini, M.; Farinha, N.C.P.; et al. Equivalent-Accuracy Accelerated Neural-Network Training Using Analogue Memory. *Nature* **2018**, *558*, 60–67. [CrossRef] [PubMed]
12. Gu, P.; Li, B.; Tang, T.; Yu, S.; Cao, Y.; Wang, Y.; Yang, H. Technological Exploration of RRAM Crossbar Array for Matrix-Vector Multiplication. In Proceedings of the 20th Asia and South Pacific Design Automation Conference, Chiba, Japan, 19–22 January 2015.
13. Chen, A.; Member, S. Solutions for Line Resistance and Nonlinear Device Characteristics. *IEEE Trans. Electron. Devices* **2013**, *60*, 1–9. [CrossRef]
14. Sabarimalai Sur, M.; Dandapat, S. Wavelet-Based Electrocardiogram Signal Compression Methods and Their Performances: A Prospective Review. *Biomed. Signal Process. Control* **2014**, *14*, 73–107.
15. Uvi_wave Toolbox. Available online: https://Github.Com/Uviwave/Uvi_wave (accessed on 10 March 2015).
16. Moody, G.B.; Mark, R.G. The Impact of the MIT-BIH Arrhythmia Database. *IEEE Eng. Med. Biol. Mag.* **2001**, *20*, 45–50. [CrossRef] [PubMed]

 *micromachines*

Article

In-DRAM Cache Management for Low Latency and Low Power 3D-Stacked DRAMs

Ho Hyun Shin [1,2] and Eui-Young Chung [2,*]

1   Samsung Electronics Company, Ltd., Hwasung 18448, Korea; hhshin@yonsei.ac.kr
2   School of Electrical and Electronic Engineering, Yonsei University, Seoul 03722, Korea
*   Correspondence: eychung@yonsei.ac.kr; Tel.: +82-2-2123-5866

Received: 24 December 2018; Accepted: 5 February 2019; Published: 14 February 2019

Abstract: Recently, 3D-stacked dynamic random access memory (DRAM) has become a promising solution for ultra-high capacity and high-bandwidth memory implementations. However, it also suffers from memory wall problems due to long latency, such as with typical 2D-DRAMs. Although there are various cache management techniques and latency hiding schemes to reduce DRAM access time, in a high-performance system using high-capacity 3D-stacked DRAM, it is ultimately essential to reduce the latency of the DRAM itself. To solve this problem, various asymmetric in-DRAM cache structures have recently been proposed, which are more attractive for high-capacity DRAMs because they can be implemented at a lower cost in 3D-stacked DRAMs. However, most research mainly focuses on the architecture of the in-DRAM cache itself and does not pay much attention to proper management methods. In this paper, we propose two new management algorithms for the in-DRAM caches to achieve a low-latency and low-power 3D-stacked DRAM device. Through the computing system simulation, we demonstrate the improvement of energy delay product up to 67%.

Keywords: 3D-stacked; DRAM; in-DRAM cache; low-latency; low-power

---

1. Introduction

The latency of dynamic random access memory (DRAM) has been a critical issue for two primary reasons [1]. Firstly, while the processing speed of central processing unit (CPU) has been continuously improved, DRAM latency has remained relatively unchanged for decades. This speed gap, called the memory wall, causes significant bottlenecks in the overall computing performance [2,3]. As shown in Figure 1a, while the capacity and bandwidth have increased 16 and 6 times over time, respectively, the timing constraints representing the DRAM latency, row address to column address delay (*tRCD*) and row cycle time (*tRC*), have only been improved by 11.2% and 20.0%, respectively [4–7].

Secondly, the processing speed of big data workloads is affected by the memory latency, as well as bandwidth. Russell et al. proved that the instructions per cycle of the applications dealing with big data could be significantly improved by reducing the DRAM latency [8]. This is because the data stream of big data is likely to have large dependency between its elements. In particular, on-line transaction processing (OLTP), which supports high transaction-oriented applications, is a representative example of latency-sensitive applications [9]. In addition, recent AI applications require large amounts of memory to handle large amounts of data, and require low latency to provide real-time data processing. In other words, we expect to see an increasing number of applications that simultaneously demand high capacity and low latency.

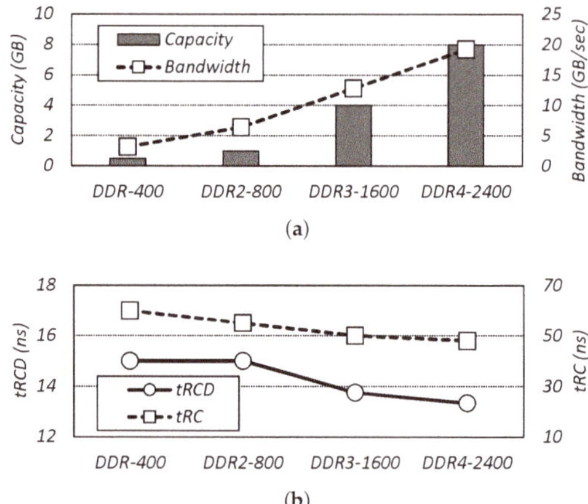

Figure 1. Comparison of dynamic random access memory (DRAM) capacity, bandwidth, and latency improvement by DRAM generation [4–7]. (**a**) Capacity and bandwidth of DRAM. (**b**) DRAM access latency: row address to column address delay (*tRCD*) and row cycle time (*tRC*).

DRAM devices are being transformed into various structures as a result of recent developments in die stacking through silicon via (TSV) [10]. For example, the die stacking of homogeneous DRAM chips extends their capacity without power and performance losses [11,12]. Moreover, a heterogeneous combination of logic and DRAM dies, such as for a high-bandwidth memory (HBM) or hybrid memory cube (HMC), increases the data bandwidth without a significant power overhead [13,14]. The meaning of the power implied above is precisely the power relative to the performance value, such as capacity and bandwidth. For example, when comparing Graphic Double Data Rate 5 (GDDR5) and HBM with the same capacity and bandwidth performance, HBM's power consumption is significantly smaller. However, though they have enhanced the memory sub-system in terms of capacity and bandwidth, the latency improvements have been neglected.

In order to overcome the long latency problem of DRAM, many computers embed numerous caches in the CPU. The cache not only overcomes the long latency of DRAM, but it also provides data locality for the pre-fetched pages. Thus, it offers large bandwidth locally in a CPU. However, since a typical cache is implemented using static random access memory (SRAM), it incurs large costs and consumes a high amount of leakage power. As a result, it is essential to reduce the DRAM latency itself to improve memory access latency (In this paper, DRAM latency refers to the time required for a DRAM controller to read or write data to a DRAM device, and memory access latency represents the latency required to access the data of the cache or DRAM by the processor instructions.).

The in-DRAM cache, which is embedded in a DRAM device, has several unique characteristics that differ from the processor cache [15]. First, the cache itself is placed in the DRAM, but its operation is managed by the DRAM controller. This is because the interface between the controller and the DRAM follows the DRAM timing constraints specified in joint electron device engineering council (JEDEC), which maintains high compatibility with the current computing system. Of course, there are various ways to implement the in-DRAM cache and its manager, such as operating systems (OS) or processor modifications. However, such methods require many modifications to the current computing system, and eventually degrade compatibility. We designed the manager to the DRAM controller so that the proposed method could follow the JEDEC specification, and implemented the in-DRAM cache in the DRAM device.

Secondly, the capacity of the in-DRAM cache increases proportionally to the DRAM capacity and is much larger than the processor cache. For example, when hundreds of gigabytes of DRAM are mounted in a system, while the memory capacity of the processor cache remains constant at several hundred megabytes, the capacity of the in-DRAM cache can be up to tens of gigabytes. However, this large-capacity in-DRAM cache requires a larger tag size. This results in long tag access latency, which in turn increases the overall memory access latency. To overcome this problem, the data transfer granularity between the DRAM and in-DRAM cache, which is called cache block size, must be increased. However, this causes significant power consumption.

Power issues in DRAMs are very important in terms of minimizing the energy consumed by the DRAM chip itself, and are also critical parameters for 3D-stacked DRAMs from a thermal point of view. Since a 3D-stacked DRAM chip consists of several dies, it is very difficult to emit the heat generated inside the chip to the outside. This heat degrades the retention characteristics of the DRAM cells, and thus DRAM requires a shorter refresh cycle. However, reducing the refresh cycle of the high-capacity 3D-stacked DRAM results in more heat, which causes the retention time of the DRAM cell to decrease again. Therefore, thermal problems in 3D-stacked DRAMs are very sensitive design parameters and must be overcome.

Considering various properties of the in-DRAM cache, this paper proposes two new in-DRAM cache management algorithms for the data replacement, particularly to maximize its efficiency and minimize its energy consumption. In addition, the proposed management algorithms are not tied to a specific in-DRAM cache architecture, and can be appropriately adapted to general architectures.

2. Background and In-Dynamic Random Access Memory (DRAM) Cache Architecture

A DRAM chip consists of the DRAM cell array area and peripheral circuits, including several in-out ports (Figure 2). Here, the DRAM cell region is composed of a plurality of sub-arrays, including DRAM cells and bit-line sense amplifiers. As mentioned in Section 1, DRAM latency improvements are very slow, and there are many reasons for this. The reason for the slow latency improvement is directly related to cost and power consumption [16,17]. In order to reduce the sensing and pre-charge time, for example, the number of cells connected per bit-line should be reduced [18]. However, this leads to an increase in the number of bit-line sense amplifiers, and thus increases the chip size. Moreover, timing constraints, such as CAS latency (*tCL*) are mainly influenced by the speed of the data path. In order to improve this speed, the capacitive metal loading of the data path signal should be decreased, or its driver strength should be increased. However, these approaches may increase the cost or power consumption. Consequently, the latency of a DRAM device must be optimized with the simultaneous consideration of multiple side-effects. In this paper, we focus on the in-DRAM cache among various skills to reduce the latency of DRAM, and discuss its management method.

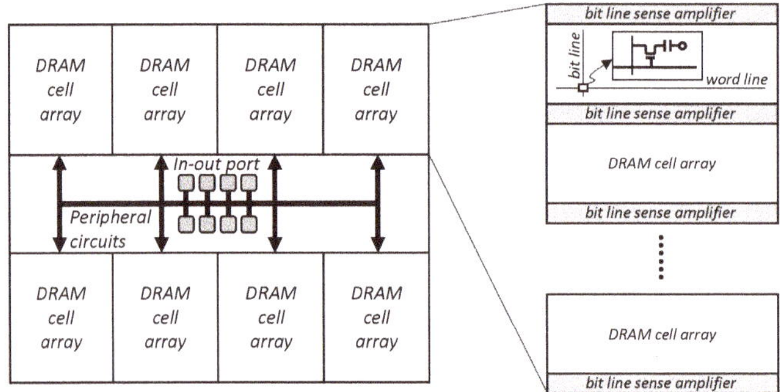

Figure 2. Conventional DRAM structure.

We deal with three types of in-DRAM cache structures based on recently published tiered-latency DRAM (TL-DRAM) and center high-aspect-ratio mats (CHARM) [19,20].

- TL-DRAM: This divides the bit line of the DRAM array into two segments and uses the long one as the DRAM memory, and the short one as the in-DRAM cache [19,21]. Here, the TL-DRAM exploits the characteristic that the short bit line improves the sensing and the pre-charge speed, and uses it as a cache memory. Figure 3a shows the TL-DRAM architecture, which is the same in terms of the overall DRAM structure. However, the DRAM array belonging to one bank is different from the conventional one.
- Cache-die: This utilizes a single die among the 3D-stacked dies as the cache (Figure 3b). The in-DRAM cache can be implemented as SRAM or DRAM, but only the DRAM is covered in this paper. This architecture has the advantage of being able to implement a significant amount of cache capacity, but it has the disadvantage of requiring a large area overhead.
- Cache-bank: This is similar to the CHARM structure [20]. Some DRAM banks are used as low-latency DRAM caches, and this paper calls them cache banks (Figure 3c). It has a smaller cache capacity than the cache die, but it can significantly reduce the latency because the cache banks are close to the input/output interfaces of the DRAM.

In this work, we consider the three types of in-DRAM architecture described above at the same time. This is because the purpose of this paper is not to propose a new in-DRAM architecture, but to describe its efficient management algorithms. The cache replacement policy is also important. The most representative cache replacement algorithms are fist-in-first-out (FIFO) and least-recently-used (LRU). The FIFO policy removes the first block accessed the first time, regardless of how often or how many times the cache is accessed. Conversely, LRU discards the least recently used items first, and is a commonly used policy because it generally exhibits better hit-ratio characteristics. However, since it takes a long time to find the appropriate replacement items, it is not appropriate for in-DRAM caches that are very sensitive to latency. Therefore, we chose to adopt the FIFO policy as the default replacement policy for the in-DRAM cache due to its fast operating time. We tackle these issues in Sections 3 and 4 in more detail.

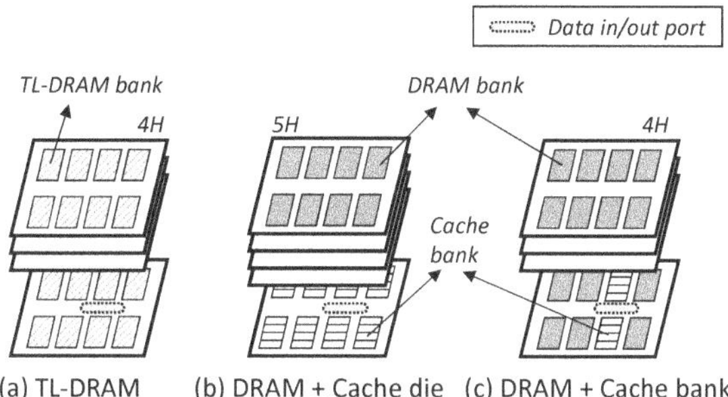

Figure 3. In-DRAM cache architectures.

3. Exploration of in-DRAM Cache Management

To design the in-DRAM cache and its management scheme, it is important to distinguish between the properties of the typical caches and the in-DRAM caches [1]. This is because in-DRAM cache management techniques are fundamentally based on the processor cache. This section describes the key design parameters of the in-DRAM cache that are distinct from the processor cache.

3.1. Trade-Off between Capacity and Latency

The capacity of the in-DRAM cache is generally much larger than the processor cache. While the processor cache, which is implemented by SRAM, has limited capacity growth due to the power consumption and area overhead, since the in-DRAM cache is configured by DRAM cells, the capacity can be expanded at a low cost. However, the capacity of such an in-DRAM cache is in a trade-off relation with latency depending on how many cells are connected to a bit line. This is because as more cells are connected to one bit line, the capacity of the DRAM increases, while the sensing speed decreases. Figure 4 shows the simulation program with integrated circuit emphasis (SPICE) simulation results of the *tRCD* and *tRP* representing the sensing and pre-charge speed, respectively. The figure shows that when 64 cells are connected to a bit line, *tRCD* and *tRP* are set to saturation. In addition, Figure 5 shows the waveform of the bit line and cell node for the 512 and 64 cells per bit line. Based on these results, we assumed the 64 cells per bit line as the basic configuration of the in-DRAM cache.

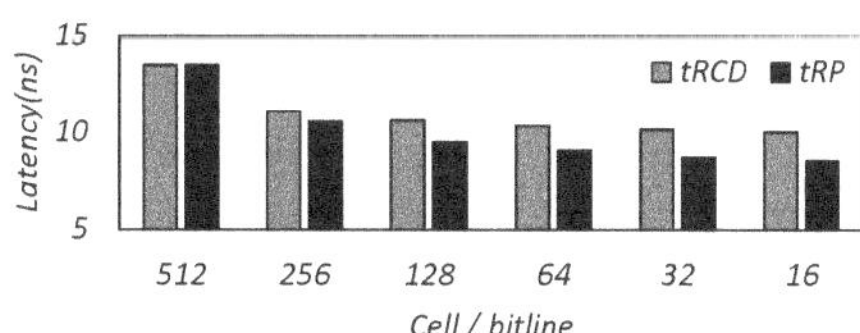

Figure 4. Changes in *tRCD* and *tRP* according to the various cells per bit line.

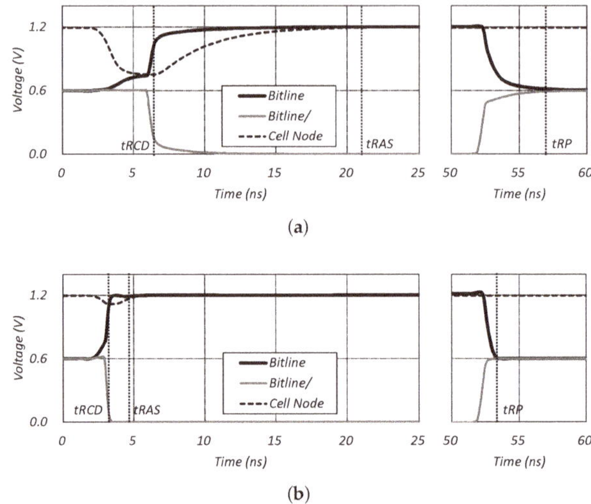

Figure 5. (**a**) SPICE simulation waveform with 512 cells per bit line, (**b**) Spice simulation waveform with 64 cells per a bit-line and pre-charge time.

3.2. Trade-Off between Tag Size and Power Consumption

The processor cache consists of data and tags in the CPU, which greatly reduces hundreds of nanoseconds of memory-access latency to tens of nanoseconds. Therefore, even though the size of the tag is large and its read-speed is somewhat slow, it is not a big deal on the overall memory access time. On the other hand, although the capacity of the in-DRAM cache is large and its hit ratio is thus quite high, the latency that can be reduced by the in-DRAM cache is only several nanoseconds. Therefore, it is very important to minimize the tag access time.

The access time of the tag is influenced by the block size and the capacity of the in-DRAM cache. The larger the cache capacity or the smaller the cache block size, the larger the tag size. Figure 6 shows that the tag size grows from several KB to tens of MB depending on the block size and the capacity.

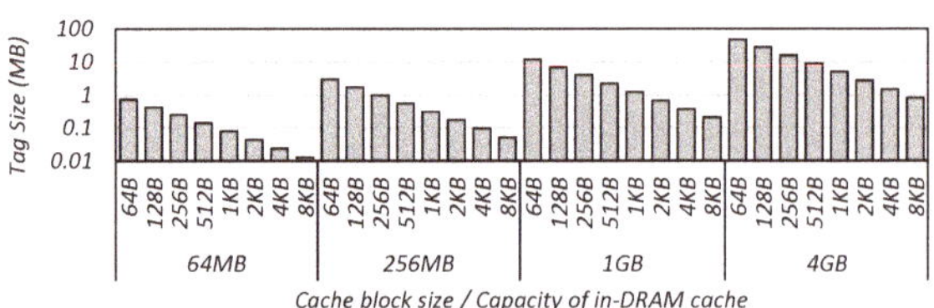

Figure 6. Tag-size variation with the block size and the capacity of in-DRAM cache.

There are two ways to reduce the tag size. One is to reduce the capacity of the cache, and the other is to increase the cache block size. However, the former is not the ultimate goal of an in-DRAM cache. Therefore, we should increase the cache block size, which has other side-effects. Firstly, a cache block size which is too large can cause significant time overhead and power consumption for the data transfer. Secondly, for applications with low locality, it lowers the hit ratio of the in-DRAM cache. Therefore, it needs to design very sophisticated cache management techniques that considers these aspects.

4. Proposed In-DRAM Cache Management Algorithms

Typical DRAMs use a rank and bank interleaving policy to maximize data bandwidth. It maximizes the reuse rate of any pre-activated row address. This property motivated us to define the block size of the in-DRAM cache as the total data contained in a specific row address of all ranks and banks in the 3D-stacked DRAM. This method is disadvantageous in terms of time and power consumption, because a single data transfer operation moves hundreds of KB of data at the same time. On the other hand, it has the advantage that the tag access time can be reduced by minimizing the tag size. Therefore, it is important to maximize the hit ratio of the in-DRAM cache and to minimize the performance and power damage caused by the transfer. We discuss how to effectively utilize the in-DRAM cache by proposing two new in-DRAM cache management algorithms in the sections below.

4.1. Critical Data Detection and Evaluation Scheme

The Critical Data Detection and Evaluation (CDDE) scheme is designed to maximize the hit ratio of an in-DRAM cache. This is a technique that evaluates and replaces the criticality of new data, rather than replacing it with new data unconditionally when a cache miss occurs. Therefore, the proposed technique is divided into the critical data detection stage and evaluation stage. Figure 7 shows the brief description of the proposed algorithm. A unit cycle to determine a data transfer is defined by multiple activation counts, called T_1. T_1 is divided into four steps, as shown below.

- Step 1: The algorithm finds the most frequently accessed row address (*First_Row*).
- Step 2: The in-DRAM cache manager selects a candidate entry (*Replace_Row*) to be replaced in the tag, where the replacement policy can be the least recently used (LRU) or first-in first-out (FIFO) that are similar to the legacy replacement policy [1]. In this paper, we use the FIFO, which can minimize the time delay for the candidate selection.
- Step 3: It measures the reuse counts for the *First_Row* and *Replace_Row*, called *RC_FR* and *RC_RR*, respectively, to define the more valuable one in terms of reuse.
- Step 4: The manager compares *RC_FR* and *RC_RR* and starts the transfer if *RC_FR* is larger than *RC_RR*.

The CDDE scheme is an algorithm that allows in-DRAM caches to operate very carefully to maximize hit ratios, but does not consider power consumption due to mass transfer. Therefore, we propose a new in-DRAM cache management scheme that considers power consumption.

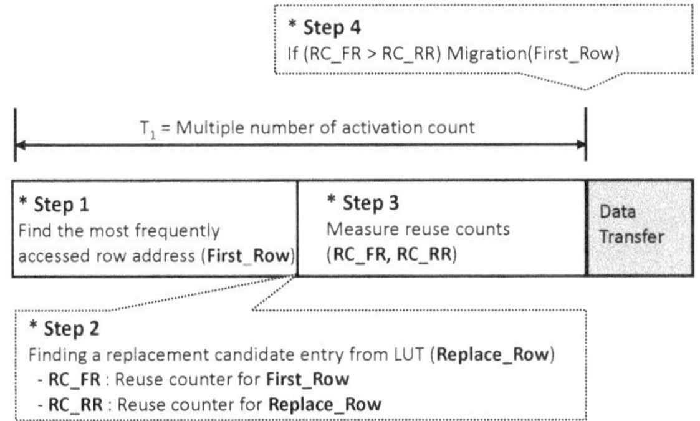

Figure 7. Descriptions of the proposed Critical Data Detection and Evaluation (CDDE) algorithm.

4.2. Power-Aware in-DRAM Cache Management Algorithm

Although the operation of the in-DRAM cache increases the power consumption as a result of massive data transfer, it also decreases the operating power owing to the reduced capacitance of the bit-line or shortened signal line between the core and I/O pads. These facts provide us an opportunity to compensate for the increase in transfer power. In other words, if the hit rate of the in-DRAM cache is sufficiently high enough to compensate for the increased transfer power, the overall power of the DRAM device can be maintained constant. In this work, we define several parameters. P_N and αP_N represent the amounts of power consumed to access the normal DRAM and in-DRAM cache, respectively. Furthermore, we define the transfer power as P_M and the hit ratio of the in-DRAM cache as HR. Along with the defined parameters, the total DRAM access energy over time of any activation count (C_A) is calculated as Equation (1).

$$E_{acc} = \{P_N \times tRC(1 - HR) + \alpha P_N \times tRC(HR)\} \times C_A \tag{1}$$

Equation (1) indicates that, as the hit rate of the in-DRAM cache increases, the overall access energy decreases. We will fill the reduced energy with transfer energy.

$$E_{tran} = P_T \times T_T \tag{2}$$

The transfer energy is calculated as shown in Equation (2), where P_T represents the transfer power consumed when the rows of all the ranks and banks are migrated. In addition, the T_T indicates the time needed for a data transfer.

In this paper, we limit the total energy of the proposed scheme to be less than that of normal DRAM devices. Finally, Equation (3) shows the limiting condition.

$$E_{acc} + E_{tran} < P_N \times tRC \times C_A \tag{3}$$

From Equations (1)–(3), we conclude that the transfer counts are limited, as shown in Equation (4).

$$T_T < \frac{P_N \times tRC \times C_A(1 - \alpha)HR}{P_T} \tag{4}$$

In Equation (4), all the parameters except HR of the right terms are predefined design parameters. Therefore, if the proposed scheme can monitor HR in real time, the available T_T can be calculated periodically. The in-DRAM cache manager in the DRAM controller controls the T_1 according to Equation (4).

Figure 8 shows the hardware implementation of the proposed scheme. The shaded part—the in-DRAM cache manager—must be added to the normal DRAM controller. The manager controls the timing constraints, such as $tRCD$, tRP, tAA, tWR, and $tRAS$ when the addresses of the issued commands are included in the tag. The active counter identifies the four stages of CDDE, and the first row detector determines the most frequently accessed row address. Finding the *First_Row* is done in real time whenever an active row address is entered. The first row detector has as many counters as the number of bits in a row address. For example, if a row address is configured from 0 to 15, there will be a total of 16 counters. Therefore, when a row address is input, only the counters of bits corresponding to 1 out of the 16 bits are incremented by 1. At the end of Step 1 of the CDDE algorithm, the first row detector compares the total number of active inputs and the number of 1's in each bit during step 1, and sets only the row address bits that are more than half of the active counts to 1. Finally, it returns *First_Row* consisting only of bits defined as 1 out of 16 bits. The reuse counter has two registers, one for storing the address of *First_Row* and the other for storing the row address to be replaced. In addition, it has a counter for each register, which increments each counter whenever a row address equal to the value of each register is input. Finally, it defines the more valuable row address in terms of hit rate with the counter output. Our proposed approach is applicable regardless of whether it is an open- or closed-page policy. In other words, the DRAM controllers using an open-page policy do not send multiple active commands continuously for a single row address. However, due to the specification of DRAM which requires only one row address to be activated in one bank, even if the open-page policy is used, there is a high possibility of accessing the same row address discontinuously. The data transfer controller contains a hit history queue (*HitQ*) and a transfer history queue (*TransQ*). It finally determines whether or not to execute a transfer according to the power-aware management algorithm.

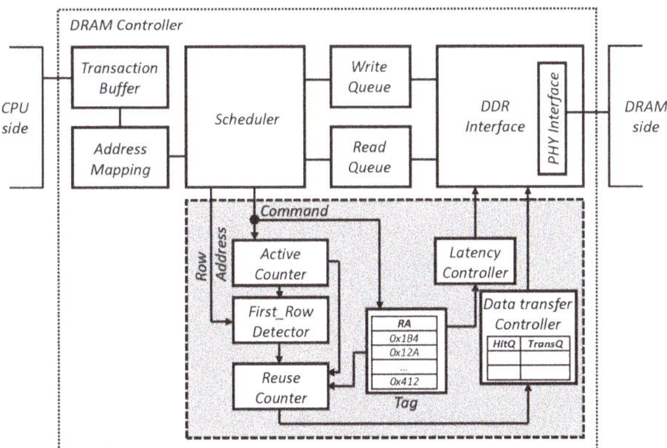

Figure 8. Implementations of in-DRAM cache manager on the DRAM controller.

Our proposed in-DRAM cache structure consists only of tags and data. This is because it can minimize tag access time, which is one of the most important factors of in-DRAM cache. Secondly, because it is not a multilevel structure like a typical cache, the tags do not need bits to store various information. In addition, our proposed in-DRAM cache operates in a write-through manner, minimizing the complexity of the cache itself and eliminating the latency penalty.

The biggest overhead in the in-DRAM cache manager is a tag that occupies from 1.125 KB to 4.5 KB. We used the CATTI tool to calculate its area and leakage power [22]. According to the CACTI tool, for a 32 nm technology, the tag requires 0.05 mm^2 and consumes 1.2 mW standby leakage power. In addition, the time overhead of the tag is expected to be 2 ns, which can be minimized because the tag does not have any special information other than the row address and operates in a direct-mapped manner. Since the *HitQ* and *TransQ* each consist of 64 entries, we assumed that the area or time overhead could be ignored. The size of the tag may vary depending on the size of the in-DRAM cache. In contrast, the size of the *HitQ* and *TransQ* does not depend on the capacity of the in-DRAM cache, which is one of the design parameters.

5. Experimental Results and Discussion

In this paper, we have proposed two new in-DRAM cache management techniques. The ultimate goal of the both is to reduce DRAM latency by achieving maximum in-DRAM cache efficiency within a given energy budget. To evaluate the performance of the proposed techniques, we modeled a computing system including various 3D-stacked DRAM architectures using *gem5* and *DRAMSim2*, a modular platform for computer system architectures [23,24]. Table 1 shows the system and DRAM configurations used in the system simulation of this paper. The cache block size of 256 KB is equal to the total data size contained in a row address of all ranks and banks in the 3D-stacked DRAM. The tag for the in-DRAM cache is implemented in the DRAM controller with a direct-mapped manner by SRAM. We verify the effectiveness of the proposed schemes for various workloads of the *PARSEC* benchmark suite consisting of multi-threaded programs [25]. Table 2 summarizes the timing constraints for the normal DRAM and in-DRAM cache, where the *tAA* and *tWR* of the in-DRAM cache are only applied to the cache-bank architecture.

Table 1. System and dynamic random access memory (DRAM) configurations.

CPU Frequency	2 GHz
DRAM Types	DDR3 1600 (800 MHz)
DRAM Capacity	2 GB
in-DRAM Cache Capacity	TL-DRAM: 256 MB Cache-die: 512 MB Cache-bank: 128 MB
Cache Block Size	256 KB
Tag Size (DRAM controller)	TL-DRAM: 2.25 KB Cache-die: 4.5 KB Cache-bank: 1.125 KB
Row Buffer Policy	Adaptive Open Page
DRAM cells per a bit line	512 (DRAM) 64 (in-DRAM cache)
DRAM cells per a word line	1024
Refresh Rate	64 ms
Bit line array structure	Open bit-line
Transfer time per a row	128 * tCCD (5 ns) = 640 ns

Table 2. Timing constraints of the normal DRAM and the in-DRAM cache. ACT–activate, PRE–pre-charge, RD–read, WR–write.

Paramter	Symbol	Normal DRAM	in-DRAM Cache
Clock cycle	*tCK*	1.25 ns	1.25 ns
ACT to internal RD or WR delay	*tRCD*	13.75 ns	8.75 ns
PRE command period	*tRP*	13.75 ns	8.75 ns
ACT-to-PRE command period	*tRAS*	35.0 ns	15.0 ns
ACT-to-ACT command period	*tRC*	48.75 ns	23.75 ns
Internal RD command to data	*tAA*	13.75 ns	8.75 ns
Write recovery time	*tWR*	15.0 ns	10.0 ns

Figure 9 shows the energy delay product (EDP) results for the TL-DRAM, cache die, and cache bank architectures, which are managed by the conventional FIFO cache management (In this paper, all experimental results are normalized for a typical 3D-stacked DRAM without an in-DRAM cache.). As shown in Figure 9, TL-DRAM, which requires low transfer latency and power, has an average of 54% improvement in EDP across all workloads, even when using a conventional cache management scheme. However, for the cache die and cache bank, EDP increases by 2 and 1239 times, respectively, when the most memory-intensive workload *canneal* is running. That is, if the data locality of the workload is low, data transfer between the cache and the DRAM is more frequent and energy consumption due to the transfer becomes more serious. In particular, such a phenomenon is exacerbated in a cache bank-like structure having a small cache capacity. These results show that typical cache management schemes are not suitable for cache die and cache bank structures, although they may be appropriate for TL-DRAM, and require new algorithms for them.

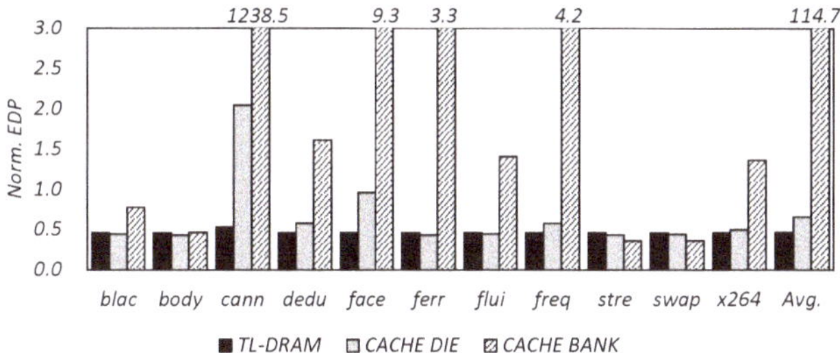

Figure 9. Normalized energy delay product (EDP) results for the TL-DRAM, cache die, and cache bank architecture which are managed by conventional FIFO cache management.

To evaluate the effectiveness of the CDDE scheme, we experimented with the latency, energy, and EDP performance of 3D-stacked DRAMs with the TL-DRAM, cache die, and cache bank structures for various transfer cycles (T_1), and Figure 10 shows the results. As shown in Figure 10, TL-DRAM exhibits better latency and EDP performance as the T_1 is smaller, but the cache die and cache bank structure have an optimal T_1 in terms of EDP depending on the properties of the workloads. Since the CDDE scheme helps prevent unnecessary data transfer between the in-DRAM cache and the DRAM, it can achieve better EDP performance over conventional cache management techniques. In addition, CDDE minimizes the EDP performance variation across the workloads compared to conventional management. When applying the conventional management, the difference of normalized EDP is shown to be 0.5 to 1238, according to the data locality (Figure 9). However, when CDDE is applied, it is shown to be 0.5 to 0.9. Despite the benefits of CDDE, it suffers from low EDP efficiency because it has to use a fixed T_1, even though different T_1s have to be applied to each application.

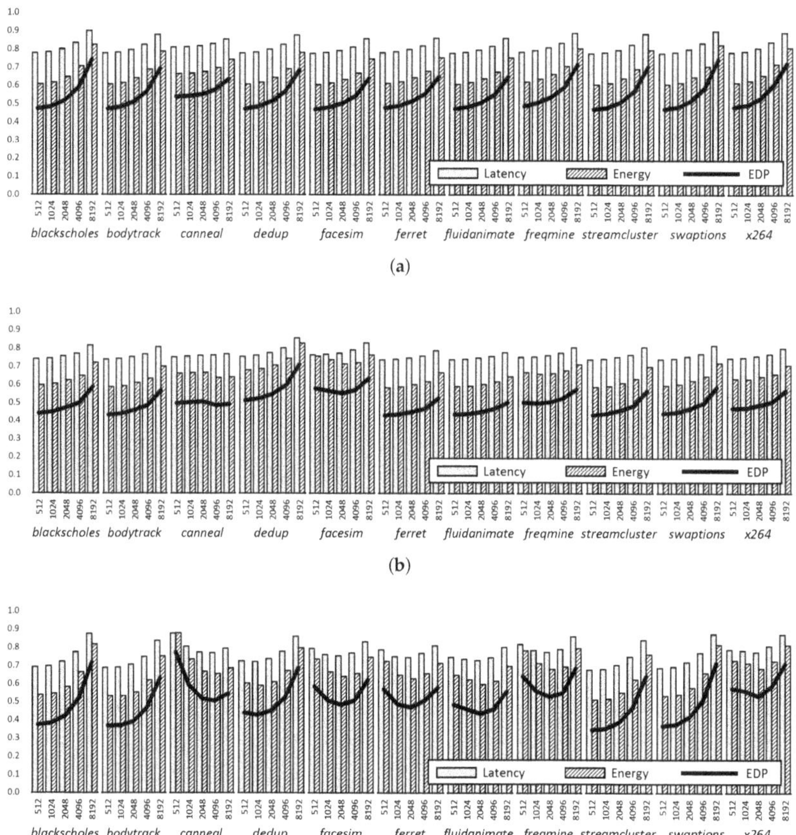

Figure 10. Normalized latency, energy, and EDP of TL-DRAM (**a**), cache die (**b**), and cache bank (**c**) structures for various unit cycles (T_1) with CDDE.

In order to the overcome the drawbacks of CDDE, we implemented the power-aware in-DRAM cache management algorithm and evaluated its performance. Figure 11 shows that the average latency of 3D-stacked DRAMs improved by 22%, 25%, and 28% for the TL-DRAM, cache die, and cache bank, respectively, and EDP by 53%, 53%, and 67%, respectively. Applying the conventional cache management techniques to the in-DRAM cache, TL-DRAM had the best performance with 23% and 54% improvements in latency and EDP, respectively. However, when the proposed CDDE and power-aware management schemes were applied, the EDP of cache bank architecture showed 28% and 67% improvements in latency and EDP, respectively. This implies that although the TL-DRAM has low time and energy consumption for the data transfer, it is not sufficient to improve DRAM latency. In addition, adaptive management techniques, such as CDDE and power-aware which were proposed in this paper, can more effectively reduce DRAM latency in a structure that can basically maximize latency improvement, like cache die and cache bank.

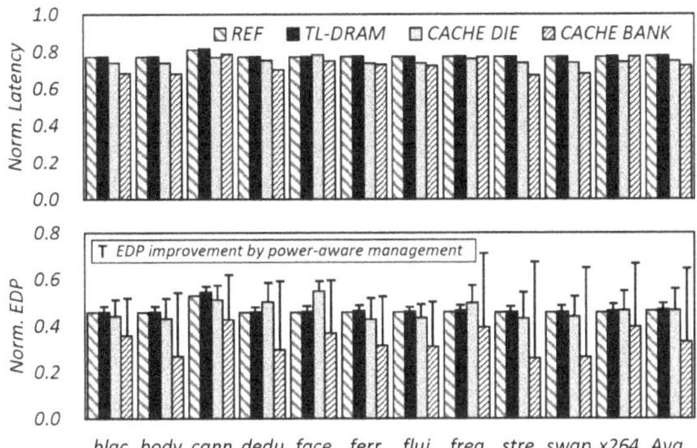

Figure 11. Normalized latency and EDP for the TL-DRAM, cache die, and cache bank architecture with the proposed algorithm. REFs are the latency and EDP results of TL-DRAM with conventional management.

6. Conclusions

Despite the recent introduction of various in-DRAM cache architectures, there was a lack of interest in how to manage them. In this paper, we studied how to derive optimal EDP by maximizing the hit ratio of In-DRAM cache and reducing power consumption due to data transfer. As a result, we achieved an improved EDP of 3D-stacked DRAM up to 67% compared to the conventional cache management scheme. Typical cache management techniques have several limitations when applied to the in-DRAM cache, and the effect depends on the architecture. However, the approach proposed in this paper demonstrates consistent improvements across all architectures.

Author Contributions: H.H.S. designed the architecture and algorithm, and performed the experimental testing. E.-Y.C. supervised the work and provided expertise.

Funding: This work was funded by the National Research Foundation of Korea (NRF), by the Korea government (MSIP) (grant number 2016R1A2B4011799), by the Ministry of Trade, Industry & Energy (MOTIE) (grant number 10080722) and Korea Semiconductor Research Consortium (KSRC) support program for the development of the future semiconductor device and by Samsung Electronics Company, Ltd., Hwasung, Korea.

Conflicts of Interest: The authors declare no conflict of interest.

Abbreviations

The following abbreviations are used in this manuscript:

DRAM	dynamic random access memory
OLTP	on-line transaction processing
TSV	through silicon via
HBM	high-bandwidth memory
HMC	hybrid memory cube
SRAM	static random access memory
CDDE	critical data detection and evaluation
LRU	least recently used
FIFO	first-in first-out
EDP	energy delay product

References

1. Jacob, B.; Ng, S.; Wang, D. *Memory Systems: Cache, DRAM, Disk*; Morgan Kaufmann Publishers: Burlington, MA, USA, 2010.
2. Wulf, W.A.; McKee, S.A. Hitting the Memory Wall: Implications of the Obvious. *SIGARCH Comput. Archit. News* **1995**, *23*, 20–24. [CrossRef]
3. Wilkes, M.V. The Memory Gap and the Future of High Performance Memories. *SIGARCH Comput. Archit. News* **2001**, *29*, 2–7. [CrossRef]
4. JEDEC. *DDR SDRAM STANDARD*; JEDEC: Arlington, VA, USA, 2008.
5. JEDEC. *DDR2 SDRAM STANDARD*; JEDEC: Arlington, VA, USA, 2009.
6. JEDEC. *DDR3 SDRAM STANDARD*; JEDEC: Arlington, VA, USA, 2012.
7. JEDEC. *DDR4 SDRAM STANDARD*; JEDEC: Arlington, VA, USA, 2017.
8. Clapp, R.; Dimitrov, M.; Kumar, K.; Viswanathan, V.; Willhalm, T. A Simple Model to Quantify the Impact of Memory Latency and Bandwidth on Performance. In Proceedings of the 2015 ACM SIGMETRICS International Conference on Measurement and Modeling of Computer Systems, Portland, OR, USA, 15–19 June 2015; ACM: New York, NY, USA, 2015; pp. 471–472. [CrossRef]
9. Zhang, H.; Chen, G.; Ooi, B.C.; Tan, K.L.; Zhang, M. In-Memory Big Data Management and Processing: A Survey. *IEEE Trans. Knowl. Data Eng.* **2015**, *27*, 1920–1948. [CrossRef]
10. Xie, Y.; Loh, G.H.; Black, B.; Bernstein, K. Design space exploration for 3D architectures. *ACM J. Emerg. Technol. Comput. Syst. (JETC)* **2006**, *2*, 65–103. [CrossRef]
11. Kang, U.; Chung, H.J.; Heo, S.; Ahn, S.H.; Lee, H.; Cha, S.H.; Ahn, J.; Kwon, D.; Kim, J.H.; Lee, J.W.; et al. 8Gb 3D DDR3 DRAM using through-silicon-via technology. In Proceedings of the 2009 IEEE International Solid-State Circuits Conference—Digest of Technical Papers, San Francisco, CA, USA, 8–12 February 2009; pp. 130–131. [CrossRef]
12. Oh, R.; Lee, B.; Shin, S.W.; Bae, W.; Choi, H.; Song, I.; Lee, Y.S.; Choi, J.H.; Kim, C.W.; Jang, S.J.; et al. Design technologies for a 1.2V 2.4Gb/s/pin high capacity DDR4 SDRAM with TSVs. In Proceedings of the 2014 Symposium on VLSI Circuits Digest of Technical Papers, Honolulu, HI, USA, 10–13 June 2014; pp. 1–2. [CrossRef]
13. JEDEC. *HIGH BANDWIDTH MEMORY (HBM) DRAM*; JEDEC: Arlington, VA, USA, 2012.
14. Pawlowski, J.T. Hybrid Memory Cube (HMC). In Proceedings of 2011 IEEE Hot Chips 23 Symposium (HCS), Stanford, CA, USA, 17–19 August 2011.
15. Zhang, Z.; Zhu, Z.; Zhang, X. Cached DRAM for ILP processor memory access latency reduction. *IEEE Micro* **2001**, *21*, 22–32. [CrossRef]
16. Kimuta, T.; Takeda, K.; Aimoto, Y.; Nakamura, N.; Iwasaki, T.; Nakazawa, Y.; Toyoshima, H.; Hamada, M.; Togo, M.; Nobusawa, H.; et al. 64 Mb 6.8 ns random ROW access DRAM macro for ASICs. In Proceedings of the 1999 IEEE International Solid-State Circuits Conference, San Francisco, CA, USA, 17 February 1999; pp. 416–417. [CrossRef]
17. Micron Technology. *RLDRAM 2 and 3 Specifications*; Micron Technology: Boise, ID, USA, 2004.
18. Sharroush, S.M.; Abdalla, Y.S.; Dessouki, A.A.; El-Badawy, E.S.A. Dynamic random-access memories without sense amplifiers. *e i Elektrotechnik und Informationstechnik* **2012**, *129*, 88–101. [CrossRef]
19. Lee, D.; Kim, Y.; Seshadri, V.; Liu, J.; Subramanian, L.; Mutlu, O. Tiered-latency DRAM: A low latency and low cost DRAM architecture. In Proceedings of the 2013 IEEE 19th International Symposium on High Performance Computer Architecture (HPCA), Shenzhen, China, 23–27 February 2013; pp. 615–626. [CrossRef]
20. Son, Y.H.; Seongil, O.; Ro, Y.; Lee, J.W.; Ahn, J.H. Reducing Memory Access Latency with Asymmetric DRAM Bank Organizations. In Proceedings of the 40th Annual International Symposium on Computer Architecture, Tel-Aviv, Israel, 23–27 June 2013; ACM: New York, NY, USA, 2013; pp. 380–391. [CrossRef]
21. Kim, Y.; Seshadri, V.; Lee, D.; Liu, J.; Mutlu, O. A case for exploiting subarray-level parallelism (SALP) in DRAM. In Proceedings of the 2012 39th Annual International Symposium on Computer Architecture (ISCA), Portland, OR, USA, 9–13 June 2012; pp. 368–379. [CrossRef]

22. Muralimanohar, N.; Balasubramonian, R.; Jouppi, N.P. CACTI 6.0: A Tool to Model Large Caches. In Proceedings of the 40th Annual IEEE/ACM International Symposium on Microarchitecture, Chicago, IL, USA, 1–5 December 2007.

23. Binkert, N.; Beckmann, B.; Black, G.; Reinhardt, S.K.; Saidi, A.; Basu, A.; Hestness, J.; Hower, D.R.; Krishna, T.; Sardashti, S.; et al. The Gem5 Simulator. *SIGARCH Comput. Archit. News* **2011**, *39*, 1–7. [CrossRef]

24. Rosenfeld, P.; Cooper-Balis, E.; Jacob, B. DRAMSim2: A Cycle Accurate Memory System Simulator. *IEEE Comput. Archit. Lett.* **2011**, *10*, 16–19. [CrossRef]

25. Bienia, C.; Kumar, S.; Singh, J.P.; Li, K. The PARSEC benchmark suite: Characterization and architectural implications. In Proceedings of the 2008 International Conference on Parallel Architectures and Compilation Techniques (PACT), Toronto, ON, Canada, 25–29 October 2008; pp. 72–81.

MDPI

St. Alban-Anlage 66

4052 Basel

Switzerland

Tel. +41 61 683 77 34

Fax +41 61 302 89 18

www.mdpi.com

Micromachines Editorial Office

E-mail: micromachines@mdpi.com

www.mdpi.com/journal/micromachines